Dictyostelium

The Dictyostelia are soil amoebae capable of extraordinary feats of survival, motility, chemotaxis, and development. Known as the "social amoebae" or "cellular slime molds," these organisms have been the subjects of serious study since the 1930s. Research in this area has been instrumental in shaping general views of differentiation, morphogenesis, and communication.

Beginning with the history of Dictyostelids, this book considers the problems of the evolution of this multicellular organism, which is characterized by its ability to transform from a single-celled organism into an elaborate assemblage of thousands of synchronously moving cells. Each stage of this development is treated in a separate chapter. The special properties of the Dictyostelid genome are rigorously analyzed, and the methods available to manipulate genes are presented in detail. Research techniques that enable many cell biology problems to be approached are also presented. Throughout, the emphasis is on combining classical experiments with modern molecular findings, and this book represents the only modern synthesis of such material.

Richard H. Kessin is Professor of Anatomy and Cell Biology at Columbia University.

Developmental and Cell Biology Series
SERIES EDITORS
Jonathan B. L. Bard, *Department of Anatomy, Edinburgh University*
Peter W. Barlow, *Long Ashton Research Station, University of Bristol*
David L. Kirk, *Department of Biology, Washington University*

The aim of the series is to present relatively short critical accounts of areas of developmental and cell biology where sufficient information has accumulated to allow a considered distillation of the subject. The fine structure of cells, embryology, morphology, physiology, genetics, biochemistry and biophysics are subjects within the scope of the series. The books are intended to interest and instruct advanced undergraduates and graduate students and to make an important contribution to teaching cell and developmental biology. At the same time, they should be of value to biologists who, while not working directly in the area of a particular volume's subject matter, wish to keep abreast of developments relevant to their particular interests.

Dedicated to:

The Kessins—Galene, Nat, Ruth, Lois, Zach, and Jessica

and

The Frankes—Gely, Ettaly, and Arva

Dictyostelium

Evolution, Cell Biology, and the Development of Multicellularity

RICHARD H. KESSIN
Columbia University

Bibliography by Jakob Franke
Columbia University

CAMBRIDGE
UNIVERSITY PRESS

PUBLISHED BY THE PRESS SYNDICATE OF THE UNIVERSITY OF CAMBRIDGE
The Pitt Building, Trumpington Street, Cambridge, United Kingdom

CAMBRIDGE UNIVERSITY PRESS
The Edinburgh Building, Cambridge CB2 2RU, UK   http://www.cup.cam.ac.uk
40 West 20th Street, New York, NY 10011-4211, USA   http://www.cup.org
10 Stamford Road, Oakleigh, Melbourne 3166, Australia
Ruiz de Alarcón 13, 28014 Madrid, Spain

First published 2001

Printed in the United Kingdom at the University Press, Cambridge

Typeface Times Roman 10/12   *System* 3B2   [KW]

*A catalog record for this book is available from
the British Library*

Library of Congress Cataloging-in-Publication Data

Kessin, Richard H., 1944–
    Dictyostelium: evolution, cell biology, and the development of multicellularity /
    Richard H. Kessin; bibliography by Jakob Franke.
        p.   cm – (Developmental and cell biology series)
    Includes bibliographical references.
    1. Dictyostelium.   I. Title.   II. Series.
    QK635.D5 K47   2001
    579.4′32 – dc21   00-037820

ISBN 0 521 58364 0 hardback

Contents

Preface

I was an undergraduate when I first saw a film of *Dictyostelium* aggregating, and decided that it was something I had to study. It would be great to say, 30-odd years later, that I had a clear vision of great issues in biology that could be addressed with such an organism, but actually I just thought it was neat. I suppose it had the correct combination of interest and mystery. Like hundreds, if not thousands of people before and since, I was fascinated by its regularity, its rhythms, and for someone schooled in bacteriophage, it seemed simple. If you are blessed with youth and think *Dictyostelium* development is simple, I hope this book will help you to realize your error without discouraging you.

The thousands who have been taken by this strange little organism include better minds than ours. John Tyler Bonner tells of being a young assistant professor at Princeton who one day received a call to show his slime mold films to Albert Einstein. John showed up with his 16 mm projector and he thinks Einstein and his colleagues were suitably impressed, but he does not actually know. John speaks wonderful French, but Einstein conversed in German.

When I first started to work on this organism, with Maurice and Raquel Sussman at Brandeis University, it was thought remarkable that a eukaryotic organism induced genes and enzymes, just like *E. coli* induced the gene for β-galactosidase. The natural synchrony of development let the Sussmans follow the enzymes responsible for polysaccharide synthesis. Before the advent of cloning, we were limited in our studies of gene induction to adding actinomycin D and cycloheximide, and for many years this was a great frustration. There was a whole counter-school that explained changes in enzyme activity on the basis of substrate fluxes. I remember lots of differential equations. Now, after cloning, Northern blots and RT–PCR, it is assuming to realize that there was ever such a conflict, but at the time it was vicious. The book that follows presents quite a lot of history of the field. I hope that readers will find that interesting. I was a history major and perhaps I overemphasize it. Skip it if you like, but Andre Lwoff once said that it is dangerous to parachute young scientists into a new field without some idea of what came before, and I believe him. People in our little field have remarkable longevity, so if you are giving a talk, say, at a conference in 2007, and you have missed the fact that what you are discussing was done by John Bonner in 1957 or Ikuo Takeuchi in 1973, someone is likely to remind you.

Research with *Dictyostelium* has always focused on development, but now that our amoebae have become so amenable to molecular and genetic techniques, they are being used to study a variety of other problems. We can investigate cell motility, transporters, pathogenesis of certain bacteria, or osmoregulation, without ever letting the organisms develop. With the sequencing of the genome, the pace of research will increase. It is already possible to obtain important sequence information from the database. As more interesting sequences appear, my prediction for the future is a greater use of *Dictyostelium* to investigate problems in cell biology. But I am not sure my predictive skills are any better now than they were thirty years ago.

I hope this book will be useful, no matter what the problem under investigation. I have organized it into modules that separate subjects that are normally lumped together. Pattern formation in the mound is separate from the patterning in slugs. These subjects are considered in strict developmental chronology because I think it is less confusing that way. Many of the overlapping roles of the cytoskeleton have been considered separately. Where controversy exists – and there are many – I have tried to point these out. I do not necessarily take a side. In all cases the reader can find the literature, and I hope the book will present the context – if not the solution.

Naturally, I have many people to thank. In addition to the Sussmans, there are my old teachers at Yale – Chris Mathews and Gerry Wyatt. Peter Newell survived my arrogance as a post-doc and has contributed greatly to the current volume. Frank Rothman welcomed me to his laboratory at Brown, and Bill Gelbart was a great friend at Harvard.

For many years it has been my good fortune to work with Jakob Franke, who is good at everything that I cannot do. He has provided the bibliography on which this book is based. Nearly all of the researchers in the field have used his bibliography and realize its value. Jakob has also been the copy-editor of the book, but any mistakes that remain are my fault. There are probably more than a few and I am sure that my friends and colleagues will point out failures due to omission. There are more than 1000 references in the bibliography, but I am sure that I have left out a favorite paper of nearly everyone. I hope there are relatively few factual errors; Jakob has hunted most of them down. I have also to thank Tristan Smith, formerly of the Bronx High School of Science, now an undergraduate at the University of Chicago, for helping me assemble the figures. Ellen Cohen, an undergraduate at Cornell, was also a great help.

Many people have read chapters. There is enough that is controversial so that no amount of reading can smooth out all errors of emphasis or omission. Nonetheless, there are many people to thank for reading and for supplying figures. These include Bill Loomis, Rob Kay, Larry Blanton, Mark Grimson, Rex Chisholm, Margaret Clarke, Ted Steck, Ted Cox, Alan Kimmel, Steve Alexander, Richard Gomer, Adam Kuspa, Gad Shaulsky and Chris West. The initial chapters were written in the laboratory of Michel Véron and I thank him. Thomas Winckler hunted down the original reference of Brefeld and had photographs made. Carole Parent made invaluable changes in the chapter on aggregation. Herb Ennis, now working with us in what he thinks is semi-retirement, read and corrected the chapter on spore germination. John Bonner, who has much more experience at writing books than I do, read a number of chapters and gave great encouragement. He showed his enthusiasm for research and writing throughout.

The members of my laboratory, Jakob Franke, Herb Ennis, Stefan Pukatzki, Dee Dao, Grant Otto, Mary Wu, and Palma Volino, kept experiments going during this exercise and I thank them. They are probably not looking forward to having me back in the lab. My colleague Gregg Gundersen was a great help with the chapter on motility, as was John Condeelis of the Albert Einstein College of Medicine.

My hope is that this book will help many people use *Dictyostelium* as an experimental organism. I have included a Resources chapter to help with technical details, including media and recipes. The amoebae have a lot to offer and the tools that biologists love – genetics, genomics, biochemistry – are always improving. I have found, over the years, that the major benefit of the organism is the people who work on it. On the whole, we are a remarkably congenial crowd, and our annual meeting has something of the atmosphere of a camp meeting. We all look forward to it, wherever it is. At the last meeting I organized with Greg Podgorski, at Snowbird in Utah, the staff of the resort had recently hosted a conference of Prozac sales people. They thought we were much more fun, and no doubt they were right.

Richard Kessin
New York, August, 2000

1

A Brief Introduction to *Dictyostelium discoideum* and its Relatives

Dictyostelium discoideum is the most studied species of the social amoebae, which are also known as the cellular slime molds. All of these organisms live in the soil and feed on bacteria, living a solitary life until the bacteria are consumed. The onset of starvation forces a major revision in the life cycle, and the amoebae respond by collecting into aggregates which transform into an organism that undergoes cell differentiation and morphogenesis. The result is a fruiting body consisting of a ball of resistant spores suspended on a stalk. *D. discoideum* and similar species have evolved strategies to survive in the harsh environment of the soil. A close examination of these strategies raises questions at all levels of biology: How do the amoebae sense starvation and other stresses, and how do they respond? How do they communicate with each other and how do they move? What mechanisms of signal transduction do they use, and how do those resemble the mechanisms of more complex organisms? How did the extraordinary cooperativity of development evolve? Rather than forcing the reader who has no experience with these organisms into details immediately, this chapter will provide a short glossary of terms and an overview of development, first in *D. discoideum*, and then in a few related species.

The developmental cycle begins when the amoebae consume all of their prey. If they do nothing to protect themselves, they will die from starvation. In response to this pressure, three distinct responses can occur – the amoebae can form microcysts, macrocysts, or fruiting bodies. The last is by far the most studied because it exhibits the fundamentals of all developing organisms. The cells signal each other to insure their correct proportion and pattern and they regulate – creating two full organisms from the two halves of a severed one. Development in *D. discoideum*, and numerous organisms like it, is composed of two phases: an aggregative period during which cells assemble in response to a

chemotactic signal, and a complex fruiting body stage in which cells differentiate and rearrange themselves to form a mass of spores supported by a stalk.
The Dictyostelia have strictly separated growth and developmental stages. A
cell either consumes bacteria or other nutrients and divides mitotically, or it
develops in response to starvation. One of the useful consequences of this
separation is that genes that are induced during development are usually not
needed for mitotic growth and so they can be mutated without affecting the
viability of the growing organism. Starvation induces a variety of new genes
whose products are necessary for chemotaxis toward cAMP (or other chemoattractants). These will be presented in detail in future chapters, but for the
moment it is sufficient to realize that cAMP is the molecule that the amoebae
recognize during chemotaxis. Their ability to synthesize, release, detect, and
degrade cAMP is critical for aggregation and none of these capacities exists in
the growing amoebae – all are induced as the cells starve.

The amoebae are grown on lawns of bacteria or in a sterile liquid medium.
To begin development in the laboratory we remove the source of nutrients and
put the cells on a moist solid substratum. The substrates can be agar or filter
paper, as long as it is sufficiently moist. No nutrients are provided during
development – the amoebae aggregate and make their fruiting structures
entirely on metabolic reserves accumulated during the trophic phase. After
the washed amoebae are deposited on the substrate, there is a period of apparent inactivity as many of the genes required during growth are down-regulated,
or their proteins are degraded and new genes – whose products are essential for
aggregation – are induced. Gradually, after a period of hours, occasional
amoebae begin to release cAMP into the population and as the macromolecules that produce, detect, and modulate the cAMP signal are made in increasing amounts, the propagation of the signal becomes stronger and stronger.

The signals that *D. discoideum* amoebae use are relayed, which allows the
organism to collect cells from a wide area, so that an aggregate of 100,000 cells
can result. The relay system is shown diagrammatically in Fig. 1.1. A single cell
in a field of starving cells releases a pulse of cAMP, and other cells detect the
cAMP as it binds to their cell surface receptors. First the cells change their
shape and then respond in two ways – by releasing another pulse of cAMP, and
by moving up the gradient of cAMP released by the first cell. Thus there is an
outwardly propagated wave of cAMP and an inward movement of cells. The
cAMP propagation and the cell movement happen in steps – the central cells
release a pulse of cAMP about every 6 minutes and inwardly moving cells only
move as long as the slope of the gradient is positive, so that when the wave
decays, the amoebae stop. With dark-field illumination we can see the moving
cells because they appear lighter than non-moving cells that are awaiting
another wave of cAMP. Such results can be seen in Fig. 1.2. Amoebae can
move into aggregation centers from aggregation territories as much as 1 cm
across. In the soil, movement of cells into an aggregate would be from three
dimensions and over much tougher terrain. The circular patterns of aggregation often evolve into spirals and eventually, as aggregation progresses, the
concentric rings break down into long streams of cells, which still move in a

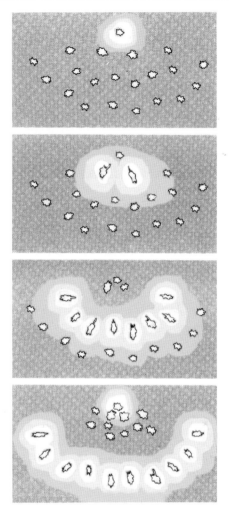

Figure 1.1 During starvation cells develop the capacity to synthesize, detect and destroy cAMP. When one cell releases a pulse of cAMP, neighboring cells detect it, move up its gradient toward higher concentrations and after a minute or so, release cAMP of their own. This attracts more outlying cells. The process is repeated about every 6 minutes under laboratory conditions. (Reprinted by permisson of *American Scientist*, magazine of Sigma Xi, The Scientific Research Society (Kessin and van Lookeren Campagne, 1992).)

periodic fashion into the center. As the cells move into the aggregate they become adhesive and capable of constructing a three-dimensional structure.

Once the cells have collected into a central point, a process that takes about 8 hours under most laboratory conditions, a stage is attained which is called the loose aggregate. This is shown in the scanning electron microscope figure produced by Grimson and Blanton (Fig. 1.3). The loose aggregate is relatively flat, with indistinct borders, but over the next few hours it changes to a

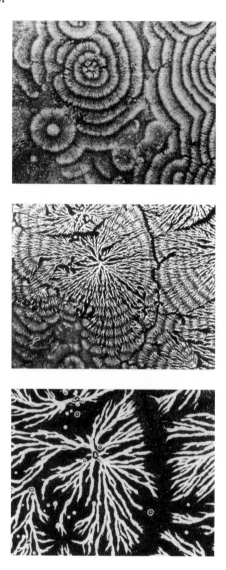

Figure 1.2 The patterns of aggregation can be seen by dark-field illumination because moving and stationary cells reflect light differently. In the region of the immobilized cells (dark bands) there is no longer a cAMP gradient, a situation that will change when the next wave of cAMP is propagated from the center. (Courtesy of Peter C. Newell, University of Oxford.)

hemispherical shape, shown as the second structure in Fig. 1.3, the tight aggregate, or mound. It is covered with a layer of mucopolysaccharide and cellulose that is called the sheath. The third structure in the figure, moving from lower right to left, has formed a critical new element, called the tip. The tip is the source of signals that organize the behavior of cells that are behind it. The tip is

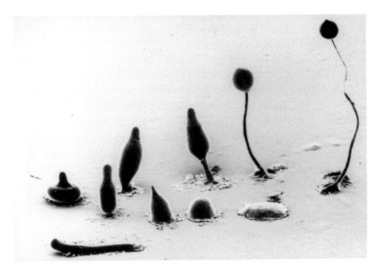

Figure 1.3 The post-aggregation stages of *Dictyostelium* development. Moving counter-clockwise, the stages encountered are the loose aggregate, the tight aggregate, the tipped aggregate, an elongated form called the finger, a slug, and then the stages of culmination leading from the Mexican hat stage to the fruiting body. (This scanning electron micrograph is by R. Lawrence Blanton and Mark Grimson, Texas Tech University.)

an important group of cells that controls development – cut it off and development stops until a new tip is formed. The tip functions like the Mangold/ Spemann organizer in vertebrates. Under the control of the tip, the aggregate elongates and makes a structure called a finger, or standing slug. At this point the organism may proceed directly to the production of fruiting bodies, or the finger may fall over and migrate in a structure called the slug or pseudoplasmodium. This is shown as the structure out of the progression in Fig. 1.3. The slug is wonderfully phototactic and is also capable of migrating up very shallow heat gradients. Once they have aggregated, *Dictyostelium* aggregates have many of the properties of an embryo – they have polarity, they have exquisite proportioning, they regulate, and they have an organizing center – the anterior tip.

The transition from migrating slug or standing slug to the fruiting body occurs by a process called culmination. The origins of cells that will make the prespore cells or the prestalk cells can be traced to the growing cells and depends to a certain extent, in a way we will describe later, on the position of a cell in the cell cycle when starvation is imposed. Prespore cells are those which have synthesized certain proteins in preparation for becoming spore cells, while prestalk cells are similarly identified by the expression of proteins that contribute to the stalk. As their names imply, each type is a progenitor of a terminally differentiated cell. Whether this separation into two precursor cell types occurs by a positional information mechanism or by a sorting mechanism is a question that we will defer. Neither of these cell types is fully differentiated or committed. Until very late in development, one type can convert to the other and given food, each type will revert to an amoeboid state.

By the time of aggregation, genes that are essential to the formation of the spore or the stalk have been expressed in the two precursor populations, and by the time of the tipped aggregate these two cells types have nearly sorted out, so that the cells that will make the stalk are on top and the cells that will form the spores are on the bottom. The topological problem faced by the developing cells is how to get the spores on top and the stalk on the bottom. The stalk is formed by a movement of cells from the tip through a collar at the apex of the aggregate. A special class of prestalk cells migrates through and down toward the substratum. During this period they become vacuolated as they make cellulose and die. Gradually the prespore cells are lifted up, as shown in the four last structures of Fig. 1.3. The left most structure in Fig. 1.3 is called a Mexican hat, for obvious reasons. It is the earliest of the culminating forms. The next stages (left to right) are called early to late culminants. As the spore mass is pulled up the stalk, the prespore cells undergo encapsulation, in which the contents of internal vesicles are released by exocytosis to form a layered cell wall of mucopolysaccharide and cellulose around each spore. Encapsulation spreads in a wave from the top of the developing spore mass to the bottom. How this is regulated so that it only happens at the very end of culmination will be described in a future chapter. The final structure, the fruiting body, contains a spore mass, supported by a slender stalk. The ratio of stalk to spore cells is about 1:4 and varies little between large fruiting bodies and small ones.

In *D. discoideum*, no mature stalk cells are present in the slug. Some species have a variant on this theme in which the stalk forms in the center of a migrating slug, which appears to have a rod down the middle. Such is the case with *D. mucoroides*, the first species isolated by Brefeld in 1869. As the slug migrates, it extends along the stalk which is left behind and can provide tensile strength, so that migration is in the air as well as on a surface. There are a variety of other species, some much more common than *D. discoideum*. These include *D. purpureum, D. giganteum, D. lacteum* and many others (Cavender, 1990; Hagiwara, 1989; Raper, 1984).

There are other variations – in the species *Polysphondylium violaceum* and *Polysphondylium pallidum*, culminating fruiting bodies develop delicate whorls, such that a series of secondary stalks and spore masses are produced along the main axis (Byrne and Cox, 1986; Harper, 1929). How this occurs so precisely, so that the secondary spore masses are evenly spaced, is a fundamental question of pattern formation. The various mature fruiting bodies are shown in Fig. 1.4. There is another difference among species, particularly the two species of *Polysphondylium*. Although these are all aggregative organisms, they do not all use cAMP as a chemotactic molecule. *P. violaceum* uses a dipeptide called glorin, which was identified in a tour de force of chemical analysis by Shimomura, Suthers, and Bonner (1982).

The significance of aggregation as a means of producing a mass of cells for differentiation is discussed in the next chapter, but the Dictyostelid species described above have more distant relatives (Cavender, 1990). These are

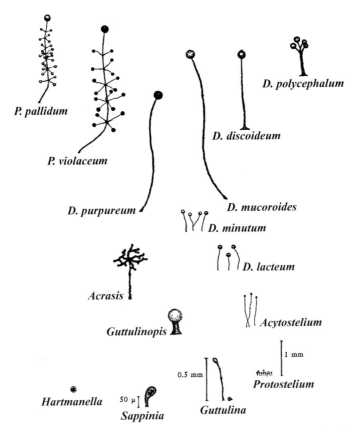

Figure 1.4 The various fruiting body structures of Dictyostelids. Note the distinction between *Dictyostelium* and *Polysphondylium* with its secondary stalks. Other much more distantly related organisms are also shown. These share an aggregative life style but may be of independent evolutionary origin. *Protostelium* forms a stalk and spore from a single cell (Bonner, 1967). (Reprinted by permission of Princeton University Press.)

grouped under the title of Acrasidae and Acytosteliaceae by Raper, whose 1984 book should be consulted for keys and other material for assigning unknown organisms to particular classes or species, as well as excellent descriptions (Raper, 1984). A series of aggregative organisms, all of which have an amoeboid trophic form and a fruiting structure that depends on aggregation, are known. These include species like *Acrasis rosea, Copromyxa protea,* and *Guttulina rosea* (Bonner, 1967; Raper, 1984). Most are little studied, especially from the cell biological or molecular point of view, but all share the capacity to assemble multiple cells to form a fruiting structure and all are inhabitants of the soil. Little is known about the details of their development or whether the chemotactic and developmental mechanisms they use resemble those of *D. discoideum*. There is no reason to believe that these organisms and *D. discoi-*

deum are monophyletic – the aggregative strategy for creating a larger organism could have evolved many times, and probably did (Blanton, 1990).

Since their discovery in 1869 by Brefeld, these organisms have been called slime molds. Phylogenetically they are not fungi, as we will learn in Chapter 3, nor are they slimy, as organisms go. For this reason, and because the title *slime mold* does not sound sufficiently elevated or convey an idea of the true evolutionary niche, many workers in the field have decided to use the designations *Dictyostelium* or social amoeba. Attempts to redirect a language are usually not successful, as the French Academy is no doubt aware, but in this book, we will use *Dictyostelium* or social amoeba.

2

A History of Research on *Dictyostelium discoideum*

The first description of *Dictyostelium*, by the mycologist Oskar Brefeld, is 130 years old (Brefeld, 1869). Many of the features of these organisms that modern workers assume to be obvious – the phagocytic nature of the amoebae, the separation of growth and developmental cycles, the absence of cell fusion in the aggregate – were not apparent to early workers. Culture systems had not been developed and the ease of manipulation that now makes these organisms so attractive would not be used until the 1930s by Kenneth Raper (Raper, 1937).

Brefeld (1869) first observed *Dictyostelium mucoroides* while examining the fungal flora in horse dung, and then grew purer cultures in rabbit dung. Even with this difficult culture method, Brefeld realized that the amoebae were the trophic (feeding) form, and that they aggregated to give rise to fructifications. He named the species *Dictyostelium* (Dicty means net-like and stelium means tower) because the aggregation territories he observed looked like nets (Fig. 2.1) and the fruiting bodies like towers (Fig. 2.2). He added the qualifier *mucoroides* because the new organism resembled the fungus *Mucor*. This was a misnomer because closer examination by Brefeld (1884) established that his new species did not have the same sporangial walls as the fungus, but instead the spores were suspended in a drop of liquid. The germination of the spores led not to hyphae and a mycelium but to distinctly amoeboid cells, which the microscopes of the time were quite capable of resolving. Brefeld correctly determined that the walls of stalk and spores contained cellulose.

It was left to Van Tieghem to show in 1880 that, far from fusing in the manner of *Physarum*, the cells of *D. mucoroides* and several other species retain their individual, if cooperative, character in the aggregate (Van Tieghem, 1880). Brefeld, realizing his earlier mistake, published a paper in 1884 which discerned the individual nature of the cells in the pseudoplasmodium (Brefeld,

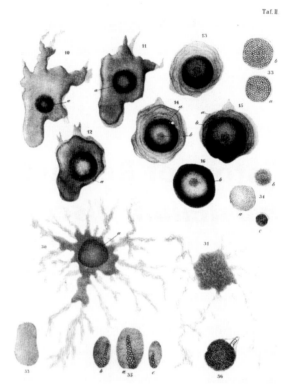

Taf. II

Figure 2.1 The original drawings of Brefeld show the aggregation fields of *D. mucoroides* and the aggregates. Stages that we now call tight aggregates are also shown. In 1869, Brefeld thought these were syncytial. On the upper right are drawings of what are probably macrocysts (Raper, 1984). (I am grateful to Thomas Winckler of the Johann Wolfgang Goethe University for finding the original paper and preparing these photographs.)

1884). Brefeld's 1884 paper was richly illustrated, which the Van Tieghem papers were not.

The nineteenth century investigators were constrained by difficult growth conditions based on concoctions of dung, sometimes stiffened with agar, sometimes not, but in all cases contaminated with bacteria upon which, we now know, the amoebae feed. Prior to the work of the great immunologist Elie Metchnikoff, phagocytosis was poorly recognized and there was no reason for early workers to believe that the amoebae obtained nourishment in a way different from that of the fungi – by extracellular digestion. That cells could surround and digest bacteria internally was a revolutionary idea. So it is not surprising that in 1899, when G. A. Nadson reported that *D. mucoroides* grew with a known species of bacteria – *Bacillus fluorescens liquifaciens* – he assumed that the two organisms were symbionts, rather than predator and prey (Nadson, 1899/1900). In 1902, Potts developed media more sophisticated than

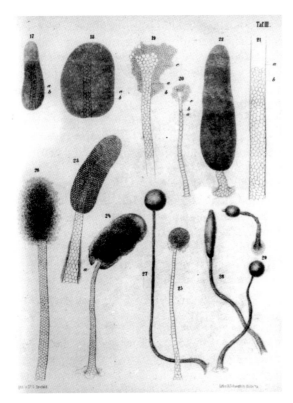

Figure 2.2 The fruiting stages of *D. mucoroides* show the structure of the stalk and the events of culmination. (These are the original drawings. I am grateful to Thomas Winckler of the Johann Wolfgang Goethe University for finding the original paper and preparing these photographs.)

simple dung and realized that the growth of the amoebae depended on the presence of bacteria (Potts, 1902). Potts was the first to maintain the amoebae in a persistent vegetative stage by providing them with fresh bacterial food sources, but he did not postulate that the bacteria were digested intracellularly. This was left to Vuillemin in 1903, who – possibly under the influence of Metchnikoff – realized that amoebae engulfed the bacteria, that digestion was intracellular, and that *D. mucoroides* and its bacterial food source were not symbionts (Vuillemin, 1903). Some of these early papers, including that of Vuillemin, lack figures or data and read almost as anecdotes. Vuillemin used the French verb *englober*, which means to engulf, so we give him the credit for discovering a fundamental property of these organisms.

Olive (1902), in his *Monograph on the Acrasieae*, made a number of shrewd observations, including the separation of growth and aggregative phases, and the fact that amoebae of different species sorted from each other. Ingestion of bacteria was observed, but Olive did not think this served the purposes of nutrition. His descriptions are precise and organize what was then known about the Acrasieae and their relationship to similar organisms such as the Mycetozoans, whose plasmodia do not resemble the pseudoplasmodia of the cellular slime molds. Olive suggested a relationship with the non-developing soil amoebae, an idea which will be explored in the next chapter.

What is missing from these earlier papers is any comparison of the elaborate development of the Acrasieae with an embryo or any remark on the extraordinary proportionality that the developing aggregates maintain, perhaps because the early workers were microbiologists and mycologists. There may be another historical reason – developmental biology before the turn of the twentieth century was an adjunct of evolutionary biology and had yet to take on the experimental nature that began with the insights of Wilhelm Roux and which came to fruition in the 1920s and 1930s with the work of Spemann, Waddington, and Needham, among many others. The social amoebae were not assumed to be of interest to experimental embryology until embryology moved out of the shadow of the evolutionary theories of Haeckel. The capacity of the developing cells to signal each other, to produce organizing centers, and to regulate would not be clarified for many years. One of the earliest workers to recognize the embryological nature of *Dictyostelium* development was Arndt (1937).

There was a period of inactivity that spanned the decades before and after World War I, when only a few studies were published. This drought was broken by three papers of Harper, that closely examined the morphogenesis of *Polysphondylium pallidum* and *D. mucoroides* (Harper, 1926, 1929, 1932). Harper's contribution was to provide accurate descriptions, to emphasize the proportionality problem that the development of these organisms presents, and to establish their light sensitivity. Harper also provides the direct link with modern studies because it was he who encouraged Raper as a young student to study the Acrasiales (Raper, 1984). In the late 1930s Arndt made films of the aggregating amoebae and realized their developmental interest (Arndt, 1937).

2.1 The classical experiments of Kenneth Raper

Raper started by isolating a previously unknown species, *Dictyostelium discoideum*, which comes from the hardwood forest of Little Butt's Gap, outside of Ashville, North Carolina (Raper, 1935). That strain is now called NC4 and its descendents (with identical RFLP patterns) can still be isolated from the site (Francis and Eisenberg, 1993). The species can be recovered worldwide, but is not the most common of the Dictyostelids. Raper recovered *D. discoideum* and other species from soil samples, whereas previously the organisms had been recovered from dung. *D. discoideum* possessed a number of singular characteristics. To use his words (Raper, 1984):

Foremost among these were a freely migrating pseudoplasmodial stage and the development under optimal conditions of sorocarps of essentially uniform construction and constant proportions irrespective of their size. The sorocarps consisted of beautifully

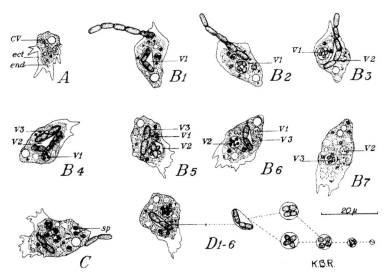

Figure 2.3 The amoebae are superb phagocytes. The phagocytic properties of the amoebae are shown in these camera lucida drawings of the ingestion of large chain-forming bacteria, in this case *Bacillus megatherium*. (The drawing is from Raper (1937), and is also presented in Raper's 1984 book. Reprinted with permission of Princeton University Press.)

tapered stalks (sorophores) which arose from flattened, cellular discs and bore at their apices unwalled, globose-to-slightly-citriform, white-to-yellowish spore masses (sori).

D. discoideum has become the type species. Its convenience of manipulation was a prerequisite to effective experimentation and Raper solved this problem first. Workers as late as Harper were still using dung, but Raper showed that nearly any species of bacteria, spread as a lawn, would support the luxuriant growth of *D. discoideum* or related species. He settled on a two-member system, employing *E. coli* or *Aerobacter* (now *Klebsiella*) *aerogenes*, established the media for growth and solved the problem of limited material. In the course of solving the growth problem Raper made elegant drawings of the amoebae ingesting a very large bacterium, *B. megatherium* (Fig. 2.3). He identified contractile and digestive vacuoles and settled the question of phagocytosis (Raper, 1937).

In the early 1940s, Raper and Thom published their studies of the developmental capacities of *Dictyostelium*. There were a number of developmentally interesting findings – the first concerned proportionality, which was maintained over a variety of aggregate and slug sizes (Raper, 1940b). No matter what the size of the aggregate or the slug that formed from it, the proportion of spores always was 80%, while the remaining 20% formed the stalk. Raper was the first to recognize that the proportionality problem is a critical issue in *Dictyostelium* development. But what were the precursors of the spore and stalk cells? Did particular regions of the slug give rise to spores or was there a mixed population? Today we have available all manner of marked strains

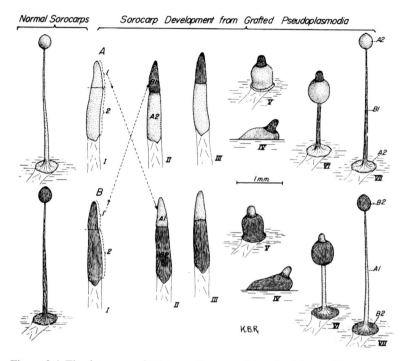

Normal Sorocarps   Sorocarp Development from Grafted Pseudoplasmodia

Figure 2.4 The fate map of *Dictyostelium* was determined by grafting experiments. Amoebae grown on *Serratia marcesans* are red and give rise to red fruiting structures. The red anterior of a slug grafted onto a colorless posterior gives rise to the red stalk. Red posterior segments give rise to the spore masses, as shown in this drawing (Raper, 1940a,b). The shading of the red segments has been increased to improve clarity. (Reprinted with permission of Princeton University Press.)

that express β-galactosidase or green fluorescent protein, but Raper had none of these and solved the problem by growing the amoebae on *Serratia marcesans*, which is red. The pigment of *Serratia* is not degraded by the amoebae and the resulting aggregates, slugs, and fruiting bodies are red. In an elegant series of transplantation experiments, Raper showed that the front of the slugs gave rise to the stalk and the rear to the spores. The original drawings that established the fate map are shown in Fig. 2.4.

The pseudoplasmodium was shown to be polarized and controlled by its tip. Further grafting experiments showed that fresh tips recruit cells from a slug and control their fate, much like an embryonic organizer. When tips are transplanted to a slug, a number of smaller slugs result, all of which migrate and then form normal fruiting bodies (Fig. 2.5). This result raised the possibility that a substance is secreted from the tip to control cells to the posterior. A variant of the question leads us to ask why each slug had only one tip, why not form tips everywhere? Not only is there a recruitment activity in the tip, but also an activity that suppresses secondary tip formation. Certain species form multiple tips – *P. pallidum* and *D. caveatum* are two examples. The capacity to

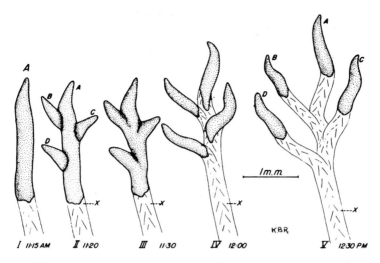

Figure 2.5 The tip controls events in the slug. These camera lucida drawings demonstrate the regulatory behavior of the tip. A normal pseudoplasmodium is shown on the left. The other structures represent apical fractions of foreign pseudoplasmodia grafted onto a normal slug. Each tip entrains a fraction of the cells, like an embryonic organizer (Raper, 1940a). (Reprinted by permission of Princeton University Press.)

control tip number is regulated and is crucial for the ultimate morphogenesis of the fruiting body. How this happens is still not known.

The nature of the commitment of the cells to development and the extraordinary regulative ability of the severed slug were first described by Raper and Thom (Raper, 1940a,b). When a slug is severed, such that its prespore and prestalk regions are separated, the fragments will still culminate but produce rather defective fruiting bodies. Front ends produce structures rich in stalk and poor in spores, violating the proportionality rule. Rear sections produce more normal fruiting bodies. We now know that there is a reserve of prestalk cells called anterior-like cells among the cells of the prespore population (Sternfeld and David, 1982). These migrate to the front of the severed slug and form the prestalk cells. The front fragment is the only one capable of motion and if it is allowed to migrate in the dark, it will eventually form a normal fruiting body as prestalk cells within this region transdifferentiate to make a new population of spore precursors.

Next, Raper and Thom (1941) determined that different species, when mixed, would segregate and form separate fruiting bodies. The different species ignored each other and apparently had non-compatible chemotactic and adhesion systems, a result that has been confirmed by Bonner and other workers (Bonner and Adams, 1958). An interesting exception is *D. caveatum*, which was not discovered until many years later. Instead of ignoring the other species, it preys on them, and we will consider this in the next chapter when we discuss evolution.

2.2 Chemotaxis and aggregation

How the cells of the aggregate assembled was a question that had been considered as early as 1902 by Olive and Potts, who raised the possibility of chemotaxis, but provided no evidence to support the hypothesis (Olive, 1902; Potts, 1902). The examination of chemotaxis – and eventually the determination of chemotactic molecules – was the work of John Tyler Bonner. Bonner had one precursor. In 1942, Runyon put amoebae on both sides of a semipermeable membrane and found that during aggregation, streams of amoebae moving toward aggregation centers were perfectly aligned (Runyon, 1942). The most likely explanation for this was that a small molecule penetrated the membrane, and controlled cell movement on both sides. While chemotaxis may have been the most likely explanation, there remained other possibilities, such as heat or electrical stimuli, which may now appear unlikely. Bonner (1947) solved these problems by the following experiment, which is shown diagrammatically in Fig. 2.6. He induced amoebae to begin aggregation under a layer of liquid and then, as the cells were streaming into the center of the territory, he forced a current of buffer across the aggregate. If the center is the source of a chemotactic molecule, cells on the upstream side of the aggregate will be deprived of the chemotactic agent by the current and should move randomly. Those cells downstream of the center, if it is the source of the agent, should be exposed and will move in a narrow stream. This was exactly what was observed. This substance was named acrasin, a generic name for any chemotactic molecule produced by a member of the order Acrasiales. In *D. discoideum* and a number of other species, the acrasin is cAMP (Konijn *et al.*, 1968), in *P. violaceum* the molecule is glorin (Shimomura *et al.*, 1982) and in other strains a variety of other molecules, including pterins, are the acrasins. The origin of the word acrasin comes from Acrasia, a witch in Canto I of Spenser's epic poem the Faerie Queene, who lured errant knights to: *The cursed land where many wend amis/And know it by the name, it hight the Bower of blis.*

Once the dependence of aggregation on acrasin was determined, the question of acrasin chemistry became paramount, and here the field was held up for nearly twenty years. What were the problems? The first was the instability of acrasin, as determined by Brian Shaffer (Shaffer, 1961, 1962). Shaffer developed chemotactic tests and showed that acrasin removed from colonies of aggregating cells was unstable and unable to attract amoebae after a few minutes at room temperature. He could attract cells if he constantly placed fresh droplets of liquid from aggregating colonies next to ones which had not aggregated, but had developed to the point where they were ready to do so. Shaffer concluded that an enzyme (now known to be a cyclic nucleotide phosphodiesterase) destroyed the acrasin. He could recover the substance in a bath underneath cells aggregating on a dialysis membrane. Separated from any macromolecules, acrasin was stable, confirming the prediction of enzymatic degradation.

The basis of chemotaxis is gradients, which can be analyzed even if the molecule in question is not known. The destruction of acrasin, Shaffer realized,

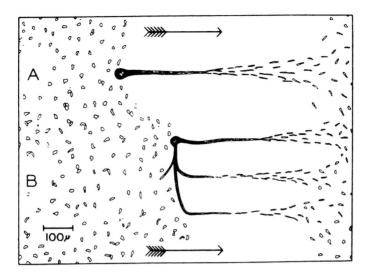

Figure 2.6 The discovery of chemotaxis: In the top part of the figure (A), water moved across a territory of amoebae that had starved long enough to be ready to aggregate. The direction of flow is shown by the arrow. Acrasin could only flow downstream. In the lower part of the figure (B), aggregation commenced in still water so that cells could stream into the center from every direction. A flow was then applied, wiping out any aggregation upstream of the source of acrasin (Bonner, 1947, 1967). (Courtesy J. T. Bonner. Reprinted by permission of Princeton University Press.)

was an important part of the process of chemotaxis. Without degradation, levels of the acrasin would rise, the gradient would become more shallow or be obliterated, and directed motion of the cells would stop. The aggregation territory was too big for a single gradient and so Shaffer proposed that the chemotactic agent was emitted in pulses, which diffused locally and stimulated nearby cells to release more acrasin. The hypothesis of pulsatile relayed signaling had the virtue that it explained the motion of cells in time-lapse films first made by Arndt in the 1930s and explained the rings of aggregating cells observed by Bonner, Raper, and others. Pulsatile signaling has an advantage in energy costs. If each signal must be greater than the last to maintain a gradient, the cells would consume huge amounts of ATP in the synthesis of cAMP. There were other requirements of cells that moved inward in a periodic way after relaying their signals – they had to be refractory to cells signaling behind them, and Shaffer realized this as well. His meticulous experimentation led to an elegant theory, which proved to be essentially correct.

The relay theory predicted a pace-maker consisting of a group of cells in the center of the aggregate. We still do not know the reason that signals occur within a certain period – initially about 6 minutes in *D. discoideum*. Ennis and Sussman (1958) believed that the initial signaling cell was genetically different and called it the I cell, suggesting that this cell was present in a population at a frequency of about 1 in 1,000. Shaffer disproved this by showing that populations of 100–200 cells aggregate. All that we know of the pace-maker cells is that they are the first among equals, generated by chance, and are not genetically distinct (Bonner, 1960). Other species may have distinct founder cells (Glazer and Newell, 1981; Newth et al., 1987; Shaffer, 1963).

Cyclic AMP was not known until 1956 when Sutherland and his colleagues, trying to explain the effect of epinephrine on glycogen degradation in liver cells, first realized that there was an intracellular messenger and determined its structure (Rall et al., 1957). Sutherland and Rall's paper was arguably the first contribution to the field of second messengers and signal transduction. Despite a major effort to isolate the chemotactic molecule, involving a number of scientific misadventures, the discovery of cAMP as the chemotactic agent rested on a guess. By 1967, cAMP was involved in any number of processes and its chemotactic effects were tried in the laboratory of John Bonner (Bonner et al., 1969; Konijn, 1968; Konijn et al., 1967, 1968). Its powerful effect on chemotaxis and sensitivity to Shaffer's acrasinase left no doubt that it was the chemotactic substance for D. discoideum.

2.3 Biochemistry and molecular biology

There were other major contributors to the establishment of Dictyostelium as an experimental organism. If the experiments of Raper, Bonner, and their students and colleagues were largely biological, Maurice Sussman realized that he could take advantage of the natural synchrony of development to study the biochemistry of development and to ask if enzymes in a eukaryotic organism were as inducible as those in bacteria – then the only examples known. Sussman began by determining the major components of the sorocarp and then asking what enzymes were necessary for their synthesis (Franke and Sussman, 1973; White and Sussman, 1961). The natural synchrony of development was enhanced by uniform deposition of starved cells on filters so that the appearance of enzyme activities could be analyzed. In this way the developmental regulation of enzymes that are important for polysaccharide synthesis was determined, and dependence on protein and mRNA synthesis was established. Although it may seem odd, or even quaint with the hindsight of more than 30 years, it was not obvious at the time or universally accepted that the increases in enzymes of polysaccharide biosynthesis were due to gene activation. Raquel and Maurice Sussman, steeped in the genetic approaches of Jacob, Lwoff, and Monod, were the protagonists of gene activation (Sussman, 1955; Sussman and Rayner, 1971). The other side, sustained by Barbara Wright, supported the idea that changes in activity were due to flux of substrates and stability of enzymes (Wright and Gustafson, 1972). The debate resulted in a victory for the gene inductionists.

A remarkable property of aggregated cells is that they become adhesive. The nature of this adhesion was pursued over a number of years by Günther Gerisch and his colleagues, who pioneered general methods for the study of cell adhesion that were later used in mammalian cells (Gerisch, 1968). The basis of the approach was immunological and was directed toward the isolation of antibodies that would block adhesion (Beug *et al.*, 1970). In this way, two elements of an adhesive system were detected. One of these was sensitive to calcium chelators, the other was insensitive. In the past decades, the details of these systems and others have been elaborated, but since they will have a chapter of their own, they will not be explained until then.

The period from the late 1960s to the mid-1980s saw an exponential increase in the number of research articles involving *D. discoideum*. Later chapters will describe these relatively recent results, but certain older papers – such as those discussed above – will find their way into the discussion. The establishment of the field depends on these elegant older experiments, which should not be forgotten as we exploit modern molecular approaches.

3

The Evolutionary Biology of *Dictyostelium*

During the construction of a fruiting body, the social amoebae show impressive cooperativity. One of the problems that we face is to explain how such cooperativity and the altruism of the stalk cells evolved. There are numerous questions to be asked: what advantage does multicellularity provide? What is unique about the evolution of an organism that increases its size by aggregation rather than feeding and growth? How can parasitism be avoided in the developmental cycle? Can we learn anything that can be exploited as we study the development of these or other aggregative organisms? What is the relationship of the Dictyostelids to other multicellular developing organisms? Before asking such questions, certain fundamentals are required, beginning with the fact that the social amoebae are capable of three developmental cycles – microcyst, macrocyst, and fruiting body formation. Each developmental option stems from a trophic, or feeding, amoeboid cell population. Formation of the microcyst, the macrocyst, and the fruiting body are each highly programmed events. Only the last one receives great attention, but knowing about the first two is essential, because the microcyst and macrocyst pathways may be evolutionary precursors of the fruiting body.

3.1   A digression into ecology

The various species of social amoebae were first found in dung and later were recognized to be ubiquitous in forest soils. Soil as an environment has been thoroughly studied, and one of the results of this body of work shows that bacteria, the prey of *Dictyostelium*, are dispersed in patches in the soil. Amoebae are also dispersed and may be dormant as microcysts, macrocysts, or spores, awaiting the appearance of bacteria. There is direct evidence for the patchy distribution of *Dictyostelium* amoebae in the soil thanks to sampling methods by Francis and Eisenberg (1993) and also by Buss and Ellison (Buss, 1982; Ellison and Buss, 1983). The variety of soil bacteria is large, and amoebae ingest a variety of bacterial species (Depraitere and Darmon, 1978; Raper and Smith, 1939). Amoebae of *D. discoideum* are remarkably omnivorous and feed on a variety of bacteria and even yeast. Whether *Dictyostelium* can feed on important colonial bacteria in the soil such as the Streptomycetes, is not known. The mass of an amoeba is about 1,000 times that of a bacterium and therefore ingestion of this many bacteria must occur before a cell can divide. Amoebae contain, in their chemotactic repertoire, an ability to detect folate gradients and it is likely that this is a food detection mechanism (Blusch and Nellen, 1994; Pan *et al.*, 1972). Unfortunately, it is difficult to prove that what they can do in the laboratory, they do in nature, and it is possible that they locate prey by additional chemotactic mechanisms.

A critical element for survival in the soil is dispersal and according to John Bonner (1982), this is the adaptive value of the stalk. The loose ball of spores extends into crevices in the soil where they are picked up by the carapaces of a variety of micro-arthropods or by annelids (Huss, 1989). The stalk, according to this theory, is an instrument of dispersal and is therefore critical for survival in the patchy environment of the soil. Many organisms, including Gram-positive bacteria like *Bacillus subtilis* or *Bacillus anthracis*, or eukaryotes such as yeast, form spores in isolation. The social amoebae also do so as microcysts, but in addition, they have invented an elaborate multicellular life cycle that projects the spores onto a long stalk of dead cells. A curious feature of the maturing fruiting body, noted by Bonner and Dodd (1962), may also have adaptive significance related to dispersal. Mature fruiting bodies project into open spaces away from surfaces, thus putting themselves in the way of passing denizens of the soil.

There are other necessities for a life in the soil. The most severe environmental challenge, other than starvation, is osmotic. Although the floor of hardwood forests is likely to be buffered from extremes of dampness or aridity, there will still be a large range of osmotic conditions. The amoebae have evolved ways to cope with changes in osmolarity by efficiently expelling water through a dedicated vacuolar system. Contractile vacuoles were observed by Raper in 1937 and were shown in the previous chapter. These organelles are widespread among soil amoebae and other protists, and we will return to them in Chapter 5.

3.2 Soil amoebae have predators

No environment is free of competitors and predators, and to understand how *D. discoideum* and other cellular slime molds evolved we must ask what these might be. The primary grazers of soil bacteria, other than a large variety of amoebae, are nematodes. It is virtually impossible to isolate amoebae from the soil without at the same time isolating nematodes – often more than one species. Nematodes are competitors for scarce food, but they also feed on amoebae. The amoebae are not defenseless. When the choice is between feeding on the soil bacterium *Klebsiella aerogenes* or *D. purpureum* amoebae, a small nematode, isolated from the same soil sample as the amoebae, prefers the bacteria (Kessin *et al.*, 1996). We infer from the chemotaxis assays shown in Fig. 3.1, that these amoebae secrete a substance that repels the nematodes. When spread on agar plates, nematodes wander. Given a choice between *K. aerogenes* and *D. purpureum*, they chose the bacterium. The lower panel shows that even in the absence of the bacterium, *D. purpureum* repels the nematodes. Only when there is no choice and they are spread evenly on a lawn of the amoebae, will the nematode eat them. A population of *Caenorhabditis elegans* can maintain itself on a lawn of amoebae growing on an otherwise sterile medium of proteose peptone and yeast extract.

The relationship of nematodes and *D. discoideum* is more complex than ingestion or repulsion. Once amoebae aggregate, they form a sheath of mucopolysaccharide and cellulose around the collective of cells. When this happens, nematodes no longer penetrate the aggregate, which soon transforms into a migrating slug. The slug convoys cells toward light and up heat gradients, presumably to the surface of the soil, all the while protecting them from nematodes. When the slug transforms into a fruiting body, it rises above nematodes of all larval stages, but one form of nematode – the dauerlarva – climbs up the stalk and into the spore mass where it is trapped by surface tension (Fig. 3.2). Dauerlarvae are resting forms of the nematode that do not feed, but have a tendency to climb, presumably seeking dispersal. The spore mass is held aloft only by surface tension, and the writhing of the dauerlarvae gradually causes the spore mass to slide down the stalk to the substratum. There the adult nematodes (in this case, *C. elegans*) ingest the spores. Although the mature nematodes destroy the amoebae by ripping out patches of membrane, the spores pass harmlessly through the gut and are dispersed as much as 5 or 6 cm by the highly motile worms, which simultaneously distribute bacteria.

3.3 The amoebae can respond to starvation in three ways

Amoebae have three choices open to them when confronted with starvation. First, at least some species can form microcysts, in which each cell elaborates a two-layered cellulose coat and undergoes other changes to prepare it for dormancy. Second, amoebae can form macrocysts, which requires the presence of two mating types, and is a sexual cycle (Raper, 1984). These structures have a three-layered cellulose coat at maturity, and their formation is favored by wet conditions. The third option is the most complex – aggregation and fruiting body formation with the production of spores, which also have a three-layered cellulose perimeter.

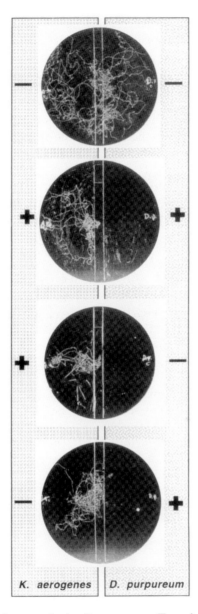

Figure 3.1 Repulsion of nematodes by *D. purpureum*. Ten adult hermaphrodites were deposited on the center strip, allowed to migrate for 1 hour, and then killed with chloroform vapor. Their tracks were photographed. See Kessin *et al.* (1996) for details. (Reprinted with permission of the National Academy of Sciences.)

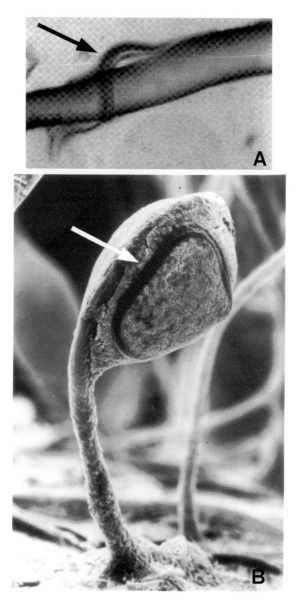

Figure 3.2 Nematodes do not penetrate slugs. Panel A: Migrating slugs were mixed with nematodes – the worms crawl under the slugs but do not penetrate the sheath. Panel B: Nematode dauerlarvae (a resting form), climb the fruiting bodies and disrupt the spore masses (Kessin *et al.*, 1996). Arrows indicate the nematode (*C. elegans*). (Reprinted with permission of the National Academy of Sciences.)

3.3.1 The microcyst

Encystment is a process common to amoebae, such as *Hartmanella* and *Entamoeba*. Other soil amoebae have the capacity to encyst and some, the testate amoebae, carry a pre-made coating from which part of the amoeba projects (Ekelund and Ronn, 1994). Although not known in *D. discoideum*, encystment has been studied in the related species *Polysphondylium pallidum*. The laboratory conditions for the formation of microcysts have not been thoroughly established, but there is a requirement for starvation, and the percentage of microcysts in a culture appears to respond to osmotic pressure and perhaps to ammonia (Lonski, 1976). Several of the sets of enzymes that are induced during microcyst production are also used during the more elaborate developmental cycles. In addition to cellulose synthesis, cysteine proteinases and glycosidases appear to be developmentally regulated during all of the developmental cycles (North and Cotter, 1991; O'Day, 1973). Encystment is a solitary form of development and involves no chemotaxis, no multicellularity, and no cell-type proportioning. At least superficially, the encystment process resembles that of many other free-living soil amoebae (Raper, 1984). The microcyst form of development is little studied, yet it is part of the developmental repertoire of these simple organisms and the evolutionary innovations of the microcyst, such as cellulose synthesis, are maintained in the other cycles.

3.3.2 The macrocyst

During the macrocyst cycle, cells of two mating types fuse under certain well-defined environmental conditions. There is a requirement for submersion and starvation, as well as low ionic strength. The fused cells form a giant cell, which can have many nuclei or just two. This structure attracts other amoebae by chemotaxis to cAMP. These cells are engulfed to become endocytes, and eventually the giant cells produce meiotic offspring (Fig. 3.3). The sexual cycle is complicated by the fact that selfing occurs. If a few cells of the NC4 (*matA*) mating type are cultured with a much greater number of the V12 mating type (*mata*), macrocysts form (Fig. 3.4). One cell type may not have to fuse with another but rather provides a signal for the V12 cells to form macrocysts (Bozzone and Bonner, 1982; Lewis and O'Day, 1977). Several authors postulate that the extensive cellulose walls of the macrocyst provide a protective function (Bozzone and Bonner, 1982). What has not been developed from macrocyst studies is a working system of sexual genetics, although recently some progress has been made using strains that are from a different source (Francis, 1998).

As in the case of the microcyst, several biosynthetic pathways, including cellulose biosynthesis, occur in the giant cell after the fusion events of macrocyst formation. As we will see in a later chapter, cellulose biosynthesis and extrusion is a complicated process and it is not likely that it evolved twice. The macrocyst capability of greatest evolutionary and developmental interest is the

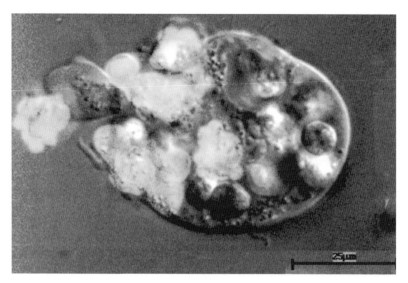

Figure 3.3 Giant cell of NC4 and V12. A giant cell, formed from the fusion of starving amoebae of two mating types, attracts and then engulfs peripheral cells. One mating type has been marked with green fluorescent protein (GFP). Some cells have been ingested and others are attached to phagocytic cups in the process of ingestion. Ingested cells are called endocytes. This is a fluorescence image superimposed on a Nomarski image. The lack of coincidence of the two images is due to the fact that the amoebae move between the two photographs. (This image was taken in the author's laboratory by Irina Paley and Theresa Swayne.)

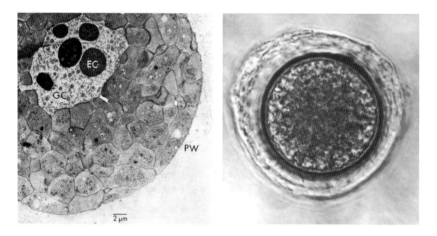

Figure 3.4 The structure of forming and mature macrocysts. An electron micrograph is shown on the left; a phase image of a mature macrocyst is shown on the right. Each endocyte is about 6–8 μm in diameter. GC = giant cell; EC = endocyte; PW = primary cell wall. The macrocyst is about 50 μm in diameter. Note the boundary of the giant cell on the left. The endocytes are pycnotic. The mechanism by which ingested cells are killed is not known. According to Raper, the walls of the macrocyst are produced by the giant cell after it has ingested all of the peripheral cells (Raper, 1984). (The left panel is reprinted from Erdos *et al.* (1972), with permission of *The European Journal of Cell Biology*.)

collection of cells by chemotaxis. Chemotaxis is toward cAMP and involves some of the same enzymes, as in fruiting body formation. Bozzone has shown that there is functional overlap between the stages of aggregation to form a fruiting body and the process of forming a macrocyst (Bozzone, 1983). Cells in the aggregation phase of fruiting body construction form macrocysts more rapidly than vegetative cells, if the second mating type is provided. Thus, chemotaxis may have evolved as an adaptation of a sexual stage, which also serves a dormancy function, much like the asci of fungi. The zygote of the macrocyst feeds on the endocytes so that maturation of the macrocyst after fusion may not depend on starvation the way fruiting body development does. A curious dependence on ethylene has been noted in macrocyst formation in *D. mucoroides* and *D. discoideum* (Amagai, 1992; Amagai and Maeda, 1992). Ethylene is a potent plant hormone, and this finding may reflect a common ancestor of plants and *Dictyostelium*. The macrocyst achieves aggregation, but does not have elaborate mechanisms of slug formation and movement, or of cell-type proportioning.

3.3.3 *Fruiting bodies*

The fruiting body is formed through complex and polarized cell movements that are more elaborate than those of the macrocyst. A form of cell–cell recognition has evolved so that aggregated cells are not engulfed to form endocytes. Yet, cellulose synthesis and chemotaxis occur. Chemotaxis in the formation of the fruiting body of *D. discoideum* is more elaborate and involves a relay mechanism, which is either suppressed or does not exist during the aggregation to form a macrocyst. Macrocysts are formed from hundreds of cells, including endocytes, while fruiting bodies can spring from aggregates of 100,000 cells. What is different in fruiting body construction is that cells are adhesive, move among each other, and sort to sources of cAMP or other molecules, both during and after aggregation. The slug that forms under the control of the tip is thermotactic and phototactic. Within the slug, the proportions of the two cell types are stabilized, implying a signaling system between them. In contrast to the macrocyst, where most of the members of the population die as fodder for the giant cells, only 20% of the cells are lost in the formation of the stalk of the fruiting body. This leaves 80% of the cells as spores, which may increase the chances that a resistant cell will find a favorable environment and creates a selective pressure for the formation of a more efficient dormancy alternative.

3.4 The forms of development evolved in a sequence

Evolutionary biology makes the argument that not all forms can evolve at once, and therefore there has to have been a sequence. Since complex biochemical mechanisms such as cellulose synthesis and deposition appear in all three forms, a conservative (cladistic) argument is that all forms had a common origin. If we postulate that the order of evolution is amoeba → microcyst → macrocyst → fruiting body, we are suggesting that an ancestor of existing species made only microcysts and macrocysts. This ancestor then adapted chemotaxis to formation of an aggregate and a primordial fruiting body. The diversification of the fruiting bodies of the various species could have taken place subsequently. Many species are known to form macrocysts, which as sexual structures would also serve to increase the genetic combinations that are essential for selection of the elaborate properties of the fruiting body (Raper, 1984).

I suggest that relatively simple developmental programs, such as those that lead to microcysts and macrocysts, evolved before the fruiting body, but after the elaboration of a successful trophic feeding phase. This argument is based on several ideas. The first idea is that complex innovations arise from simple ones, which are retained. Major events in the evolution of *Dictyostelium* were the ability to protect themselves as single cells (microcysts), the ability to survive as aggregates in the form of a macrocyst, and to evolve the motile slug and the individual spore, both of which provide advantages of protection from predators and of dispersal (Bonner, 1982; Kessin *et al.*, 1996). The second idea that suggests the evolutionary sequence microcyst, macrocyst and then fruiting body, is that the fruiting body is much less costly in terms of lost cells than the macrocyst. This is part of its selective advantage. Each stage incorporates critical innovations that evolved in the less complex stage without (at least in the case of *P. pallidum*) destroying the program for microcyst or macrocyst development. A third argument for the suggested order of evolution is that in order to produce anything as complicated as the development of fruiting bodies, natural selection needs diversity upon which to work. The existence of a prior sexual stage would provide new gene combinations and greater diversity.

The suggested ontogeny of the developmental forms is shown in Fig. 3.5. A corollary of this is that at the time the ancestor of the Dictyostelids diverged from other eukaryotic cells, it may have had none of the developmental features that we find so fascinating. The time of this divergence is discussed below – but suffice it to say that all methods put the origin of these species several hundred million years prior to the establishment of terrestrial life in the early Silurian period, some 430 million years ago (Raff, 1996).

3.5   Genetic heterogeneity in wild populations

Any discussion of evolution should ask what genetic heterogeneity exists in wild populations of *Dictyostelium*. Is there a pool of variants to be selected? Francis and Eisenberg have shown that there is genetic variability within a limited area at the original site of isolation (Francis and Eisenberg, 1993). *Dictyostelium* and other social amoebae are widely dispersed by birds (Suthers, 1985), nematodes (Kessin *et al.*, 1996), and small arthropods (Huss, 1989). The vast numbers of isolated populations have accumulated a huge pool

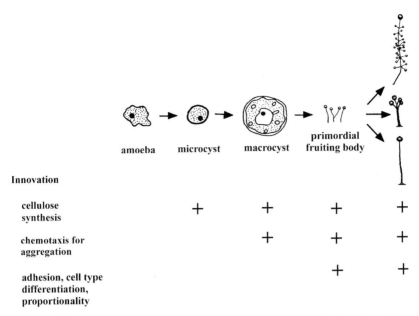

amoeba   microcyst   macrocyst   primordial
fruiting body

Innovation

cellulose
synthesis   +   +   +   +

chemotaxis for   +   +   +
aggregation

adhesion, cell type   +   +
differentiation,
proportionality

Figure 3.5 One view of the evolution of the life cycles. The complexity of the life cycles increases from left to right, but in this view, innovations of the simplest life style are retained in the most complex one. Microcysts have not been observed in *D. discoideum*, but because of their existence in *P. pallidum* they are assumed to be an ancestral form that has been lost in certain species. Diversification of fruiting body types might occur from a primordial stalk and spore producing aggregate. The examples shown here are *P. violaceum*, *D. polycephalum*, and *D. discoideum*.

of genetic diversity, which can be acted upon if these separated cells are ever brought together. There may be considerable spread of variant strains and competition among the variants to form spores rather than stalk.

3.6 The evolution of cooperativity

How does a cooperative assembly of cells evolve? During fruiting body formation, 20% of the cells are lost to form stalk and in the case of the macrocyst, an even greater percentage is lost as endocytes. One idea that explains such altruism is kin selection, which derives from work on social insects. This theory states that an organism (in this case a cell) could be lost as long as the genes that are preserved as a result of its death are closely related to those of the dead cell. It was elaborated by Hamilton (1964) and discussed for the particular case of *Dictyostelium* by Armstrong (1984). Thus in populations in which all cells or animals have a high coefficient of relatedness, whether in social insects or in social amoebae, one expects the evolution of cooperativity. As Armstrong has pointed out, if the survival of differentiating cells is greater than that of non-differentiating cells during the same time period, division of labor will be selected for – if the differentiation is independent of genotype. If we imagine a situation in which a gene evolves that promotes stalk formation, but at the expense of the cell that contains it, the gene would spread in the population if identical genes were contained in the spores.

The concept of kin selection has been discussed by Gadagkar and Bonner (1994) for the specific cases of the social amoebae and social wasps, and by many other authors with respect to other social organisms. A problem arises in the generation of cooperativity, even by kin selection, because parasites that do not contribute to the stalk cell population are likely to arise. One case of this is discussed below. One could make the argument that *Dictyostelium* and other amoebae are rapidly dispersed and that most populations are essentially clonal and thus prime candidates for kin selection mechanisms and free of parasitic mutants. If this is so, then we are at a loss to explain the source of the variation that allowed the evolution of the fruiting body from an earlier form. The interesting suggestion has been made that there are active mechanisms to thwart parasitism (Armstrong, 1984; Frank, 1995).

3.7   The cells in a *Dictyostelium* aggregate compete to form spores

Most organisms studied by developmental biologists – sea urchins, nematodes, *Drosophila*, or *Xenopus* – result from a zygote and a blastocyst. All cells in the body are genetically identical, barring somatic mutation, the immune system, and post-meiotic cells. Because all reproduction passes through a germ line, a cell achieves no advantage by dividing faster or developing better. After meiosis occurs, genomes are no longer identical and there is a potential selective advantage. This is presumably the source of the phenomenon of meiotic drive, which promotes the selective non-Mendelian inheritance of certain chromosomes (Ardlie, 1998; Lyttle, 1993).

Despite the complexity of its development, *Dictyostelium* does not develop from a zygote or a blastocyst – during development most Dictyostelids assemble cells from a relatively large area, and the resulting aggregate may or may not be genetically homogeneous. As Francis and Eisenberg established, there is a pool of genetic diversity. This natural chimerism has consequences when viewed through the lens of evolutionary theory. Although natural chimerism does not occur frequently among the animals, it does occur in other developing organisms such as the myxobacteria (Kaiser, 1986, 1993; Dao *et al.*, 2000). Natural chimerism influences the evolution of a number of phyla as described by Buss (1982), who pointed out the evolutionary consequences of chimerism and the lack of a germ line.

The consequences of chimeric development stem from the fact that in the macrocyst cycle, and in the construction of the fruiting body, many of the cells die, either as endocytes during macrocyst formation or as stalk cells during fruiting body construction. If a single cell in a population carries a mutation that allows it to avoid the endocyte or stalk cell fate – to form the giant cell of the macrocyst or to form the spores of the fruiting body – the frequency of this cell's progeny will increase in a genetically heterogeneous population. The mutant forms are parasites and eventually, they will subvert normal development. These organisms are usually haploid, and therefore mutant phenotypes do not have to be dominant to have an immediate effect. Such variant cells, when grown and developed as clonal populations, may be impaired in their ability to make a fruiting body, and in the wild there would be selection against them because they cannot form a protective cell and because mechanisms of dispersal depend on a normal fruiting body.

There have been two reports indicating that in freshly isolated uncloned populations of cells, variants preferentially make spores in chimeras with normal fruiting cells. The first report was by Filosa (1962), working with *D. mucoroides*. Starting with an uncloned population, Filosa found individual colonies with variant morphologies. One, which formed slugs but no fruiting bodies, preferentially made spores when mixed with wild-type cells. The second observation of parasitism was by Buss (1982), also with *D. mucoroides*. Buss found a single soil sample which, when plated clonally, gave rise to normal colonies and colonies that only formed spores. The latter, in chimeras with the fruiting strain, could suppress the wild-type if their initial concentration was high enough. If their initial proportion in the chimeric population was less than 5%, these cells maintained a stable proportion with the wild-type. The mechanisms that these cells, which we call *cheaters*, use to avoid the stalk cell fate in a chimera are probably central to development itself. One might imagine that they have mutated to secrete greater amounts of a factor that causes other cells to differentiate into stalk. One such compound, called differentiation inducing factor (DIF), has been described and we will come back to it in due course. *Cheaters* are, by definition, parasites, and if such strains can be found again or produced in the laboratory, the molecular basis of the parasitism can be studied. In insect societies, parasitism is based on stealing or distorting signals, and it is possible that the *cheater* mutants of *Dictyostelium* do the same. Avoiding parasitism is a classical problem at all levels of evolution. One way that organisms might do this is to make the chemotactic system of aggregation more specific. *Dictyostelium*, like other societies, must have evolved mechanisms to police itself (Armstrong, 1984; Frank, 1995).

The Buss and Filosa strains have been lost, but in our own laboratory we have created a *cheater* mutant of *D. discoideum* (Ennis *et al.*, 2000; Dao *et al.*, 2000). This was done by insertional mutagenesis methods that will be described in the next chapter. A population of 20,000 cells, each with a different mutation, was allowed to form spores and then the spores were harvested and treated with detergent to kill any cell that was not a spore. The spores were then plated, allowed to grow, aggregate, and form spores again. The process was repeated 20 times and at the end, one mutant made up 3% of the population. At the beginning of the experiment it was so rare that it went undetected. Mutation of the *chtA* gene creates a mutant with a very long slug that is defective in culmination. When developed alone it is severely disadvantaged. However, in a chimera with wild-type cells it forms spores preferentially. This

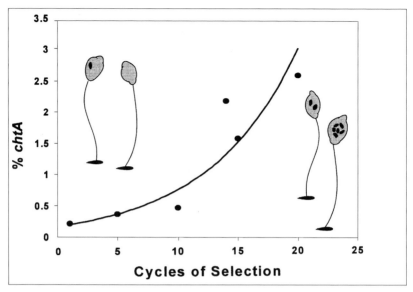

Figure 3.6 In the competition to form spores rather than stalk, some mutants can cheat. The increase in frequency of the *chtA* mutation is shown. After six to eight cell divisions, amoebae aggregated and formed fruiting bodies. Spores were recovered and the process was repeated. Ultimately a morphologically defective cheater mutant was found to increase in the population (Ennis *et al.*, 2000; Dao *et al.*, 2000). Wild-type populations also compete such that one genetically distinct population may take advantage of another (Joan Strassman, personal communication).

can be seen if the mutant cells are marked with a gene that expresses β-galactosidase. In slugs formed from chimeras, the mutant cells always populate the zones that are fated to give rise to spores (Fig. 3.6). The affected gene has been recovered and codes for a protein that acts in proteolysis by binding target proteins to the proteolytic machinery. This causes ubiquitination of the proteins and leads to their degradation.

Competition among amoebae during growth and development need not lead to a cell that preferentially forms a spore because it ignores signals to be stalk. If in the aggregate those with more developed chemotactic mechanisms or better motility reach a zone from which spores are more likely to arise, this strain will have a selective advantage in propagation. Thus knockout mutants, for example of certain small myosins (Novak *et al.*, 1995) may have minimal observable phenotypes when cultured alone, but may have major disadvantages in a chimera with an isogenic parent. The haploid nature of the organism and the rather severe difference in the prospects of a spore and a stalk cell will lead to the rapid spread of mutations that improve fitness. In an experimental context, any mutant that lacks an obvious visible phenotype should be tested for its behavior in chimeras, where it may be at a selective disadvantage in the race to be a spore rather than a stalk. Ponte, Bozzaro and their colleagues have recently shown that many mutant strains that have no visible phenotype are at a selective disadvantage in chimeras or when forced to aggregate on soil, an environment that is more arduous than smooth agar (Ponte *et al.*, 1998). The case of *chtA* and the detection of subtle phenotypes by Ponte *et al.* suggest that in *Dictyostelium*, evolutionary theory can lead to experimental results.

3.8 Size limitations in an aggregative organism

One of the major innovations of evolution has been to combine feeding and differentiation. Animals grow and differentiate at the same time, allowing them to attain large sizes and to fill an enormous number of ecological niches. Feeding and developing at the same time requires a mechanism to maintain the differentiated stage across the cell division cycle. *Dictyostelium* does not have this capacity – it either divides or it differentiates and all developmental progress is wiped out when we let the cells divide by feeding them. This process has been called erasure by Soll (Finney *et al.*, 1979). It takes about an hour after nutrients are supplied, and is dependent on protein synthesis. The inability to feed and develop at the same time leads to size limitations that depend in part on the density of food bacteria in the soil and the distance that cells can migrate. The separation of growth and development is an enormous advantage to the experimenter, who can easily maintain the many mutants that are affected in development.

3.9 The extraordinary parasitism of *D. caveatum*

The urge to find new species of *Dictyostelium* has led people into strange places – none more odd than the bat guano-filled caves of Blanchard Springs, Arkansas. The temperature and humidity in these caves is constant, and a sample yielded a number of species of social amoebae, among which was one which had an extraordinary ability to parasitize other species (Waddell, 1982). *D. caveatum* parasitizes the development of *D. discoideum*, *D. purpureum*, *D. polycephalum*, *D. rosarium*, *D. minutum*, *P. pallidum*, *P. violaceum*, *Acytostelium leptosum*, and others. It is not choosy. Alone, the amoebae of *D. caveatum* aggregate without streams and undergo a normal morphogenesis leading to a fruiting body with multiple stalks spring from a single base. The reason this species is extraordinary is that when mixed with any of the species listed above – even in ratios of 1:1,000 or 1:10,000 – the only species to emerge as spores is *D. caveatum*. *D. caveatum* amoebae co-aggregate with their peers and at first everything is normal – the host species progresses into development. Then, at around the tight aggregate stage, there is a massive conversion to *D. caveatum* because these cells, while developing with the others, eat them by phagocytosis of little pieces at a time (Fig. 3.7) (Waddell and Vogel, 1985). This process is called nibbling, and is distinguished from another case where there is a loss of self recognition – the phagocytosis carried out by giant cells during macrocyst formation. In this case, the victims are eaten whole. For every *D.*

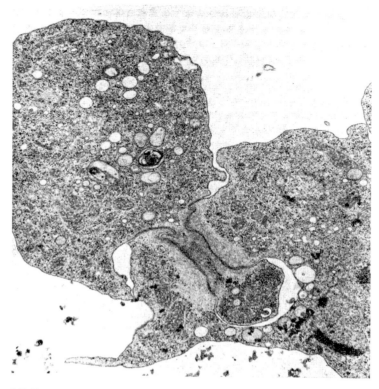

Figure 3.7 *D. caveatum* eats its neighbors in small pieces. The darker pinched-off cell is the prey (Waddell and Vogel, 1985). (Reprinted with permission from *Experimental Cell Research*.)

caveatum cell that is gained, two or three of the host cells are lost. Eventually, when none of the host cells is left, *D. caveatum* makes its own fruiting body. If few cells are added, say one *D. caveatum* to 10,000 *D. discoideum*, the host cells get farther into development, to the slug stage, but then the slug stops migrating and the tip structure is lost as the *D. caveatum* cells cannibalize the slug. The rare *D. caveatum* cell acts to arrest the host cells' development, giving *D. caveatum* more time to feed. The means that *D. caveatum* uses to create a stage-specific block are of interest to normal development. One possibility to explain developmental inhibition stems from a discovery by Mizutani and Yanigasawa (1990), who showed that certain killer strains of *P. pallidum* secrete an inhibitor of growth that affects other strains and species. Perhaps *D. caveatum* is using such an interesting compound.

There are several intriguing aspects to *D. caveatum*. First, it does more than eat the host cells – it inhibits their development, even when very few cells are present. Second, how is it that in the original sample of bat guano, *D. caveatum* was rare? It should have spread and eaten everything else unless the prey species have defense mechanisms. How do *D. caveatum* amoebae get into the aggregates of the prey? Are they universally chemotactic, capable of aggregation toward cAMP, glorin and other chemotactically active molecules, or are they tracking a common molecule? Third, the self–non-self recognition mechanism has broken down. Normally, the various species of the social amoebae do not ingest one another. *D. caveatum* has enormous phagocytic capacity, and this is apparently directed at molecules on the surface of other species. *D. caveatum* does not consume itself, except in a cannibalistic mutant isolated by Waddell and Duffy in which, when bacteria are consumed, the amoebae start devouring each other and do not form fruiting bodies, possibly because they cannot starve and trigger the developmental pathways (Waddell, 1986). In theory, new mutagenesis methods, which introduce insertions and have become available since this work was done in the early 1980s, could be employed to isolate receptors for phagocytosis (see Chapter 5). If *D. caveatum* recognizes a particular molecule on the surface of the prey cell, a library of restriction enzyme-mediated integration (REMI) mutants should contain some cells which are mutated in this receptor and therefore escape the voracious *D. caveatum* amoebae.

3.10 Molecular phylogeny

The phylogenetic position of *Dictyostelium* has been a source of confusion. The classical work on the taxonomy of the cellular slime molds is that of Raper (1984), who describes them as a class of primitive fungi. Briefly, Raper put them in the kingdom Mycetae (fungi), in the division Myxomycota and the class Acrasiomycetes (cellular slime molds). Other organisms that are related at the class level are the Protostelids and the true (or plasmodial) slime molds, such as *Physarum polycephalum*. The Acrasiomycetes were divided into the subclass Acrasidae, of which there are various genera (for example, *Acrasis rosea*) and the subclass Dictyostelidae. The Dictyostelids contain the most studied species, including *D. mucoroides*, *P. pallidum*, and *D. discoideum*.

Since the advent of molecular phylogeny, there is no longer any reason to postulate that the *Dictyostelium* and its immediate relatives have a close relation to the fungi. The tendency to put organisms into kingdoms has given way to other concepts, which stem from our ability to analyze sequences and to ask, by various algorithms, which organisms diverged earlier. In this view, life is partitioned among the bacteria, the archaea, and the eukaryota. At the tip of the eukaryotes, the animals, plants, fungi and several other groups radiate. These organisms have been called the Crown Group. A long series of simple eukaryotes, the diplomonads, the microsporidia, and the trichomonads, which are amitochondrial, are descendants of the earliest divergence of eukaryotes. After them, the ciliates and the flagellates diverge. Some of these organisms are free living – many are noticed because they are parasites. The ensemble is usually called the Protista, a term which dates from Haeckel in the nineteenth century, who was the first to propose that single-celled eukaryotes gave rise to multicellular organisms. The Protista are best viewed as having long, independent lines of descent with an ancient divergence. One question for those of us who work with *Dictyostelium* is: does this organism represent one of the most complex of the pre-Crown Group organisms, as the ribosomal sequence data indicate? Or, do the social amoebae form part of the plant/animal/fungal group, perhaps even along the animal lineage with a common origin of multi-cellularity, as Loomis and Smith (1995) have suggested?

In peering so far back, it is not surprising that the view is a little hazy. We are not blessed with a fossil record – a fact that is not likely to change – and therefore we will have to rely on sequence comparisons of ribosomal RNAs or conserved proteins. Another tactic we can use is to ask whether major innovations, such as the G_1 portion of the cell cycle, specific signal transduction pathways, common sequences in mitochondrial genomes, or membrane sterol biosynthesis are consistently present in one group of organisms, but not another. For example, if *Dictyostelium* lacked particular innovations present in plants, animals and fungi, we would be more justified in placing their divergence before the Crown Group radiation and saying that plants, animals, and fungi shared a more recent common ancestor than *Dictyostelium*.

All phylogenetic inference makes certain assumptions, and each method has problems that are peculiar to it. The basic assumption is that nucleotide and amino acid sequences diverge with time, and that by the application of a number of increasingly powerful and mathematically different methods, relationships among lineages can be determined. By grouping closely related organisms we can produce a tree that gives us an idea of their order of divergence. We assume that there is no lateral gene transfer – that when we say a species diverged, we really mean that it stopped exchanging genetic material with other species. We know that this is not always true. A particular example concerns the cyclic nucleotide phosphodiesterase that *Dictyostelium* uses to control cAMP levels during aggregation in the formation of macrocysts and fruiting bodies. It has no known homologues in plants or animals, but shares homology with enzymes from *Saccharomyces cerevisiae* and *Schizo-saccharomyces pombe*. This is not a surprise since both groups of organisms, according to Fig. 3.8, shared an ancestor. What is surprising is that the *Dictyostelium* cyclic nucleotide phosphodiesterase shares distant homology with a secreted phosphodiesterase of *Vibrio fisherii*, a bacterium (Callahan *et al.*, 1995). This homology can be best explained by lateral gene transfer.

A second difficulty, especially of protein sequence analysis, is that of paralogues – genes that share common ancestry due to duplication. When drawing phylogenetic inferences, it is important not to compare sequences that are the result of past duplications and have evolved since under different selective pressures. If an organism like *Naegleria* has multiple α-tubulins, it is difficult to choose one to compare with the single *Dictyostelium* α-tubulin. This problem has been overcome by using genes that code for elongation factors which fall into two clearly defined groups, have no paralogues, and for which sequence information exists from many organisms (Baldauf and Doolittle, 1997). The other strategy is to compare large numbers of protein sequences, being careful not to compare proteins coded by paralogous genes (Kuma *et al.*, 1995).

Much phylogenetic inference has been carried out with ribosomal RNA sequences. Data from more than 1,000 species are currently available. Ribosomal RNA sequences within one organism are identical and do not suffer from problems of divergence of paralogous genes, yet phylogenetic inference may be complicated by wide divergences of nucleotide frequency. How do we compare the GC-rich genome of *Giardia lamblia* with the AT-rich genome of *Dictyostelium*? Phylogenetic comparisons analyze sequences within known structural motifs where differences in base composition vary from the larger molecule and base sequence divergence is reduced. Certain regions of ribosomal RNA are highly conserved and vary slowly so that they are used to measure ancient events. Others vary more widely (McCarroll *et al.*, 1983). Variations in base frequency may be legitimate indications of evolutionary divergence. Using carefully aligned sub-sequences that form known ribosomal structures also minimizes the base divergence problem. Given the problems both with ribosomal methods and with methods based on protein sequence, the most conservative approach is to assume that a particular phylogeny is robust if it can be derived from ribosomal and protein sequence data and by more than one method of comparison. The ribosomal analysis gives an early divergence of *Dictyostelium* before the evolution of the plants, animals, and fungi. The results of most protein analyses give a divergence time that is still early, but more or less at the time of the crown group organisms. Figure 3.8 shows a recent positioning of the Mycetozoa based on a comparison of elongation factor 1α (EF-1α) sequences (Baldauf and Doolittle, 1997). Note that by this analysis, the Dictyostelids, *Physarum*, and the Protostelids (which make a fruiting structure with a spore and stalk out of one cell) are all thought to share a common origin, as Raper originally postulated.

Evidence from sterol biosynthesis pathways dissociates *Dictyostelium* from the fungi and puts it as an ancient derivative of photosynthetic algae (Nes *et al.*, 1990). Other evidence suggesting a protistan or an algal origin is the near identity of the *Dictyostelium* and the *Acanthamoeba* mitochondrial genomes (Angata *et al.*, 1995). These authors have now sequenced the entire mitochondrial genome of *Dictyostelium*, and their own phylogenetic inference suggests a closer relationship to green plants than to animals and fungi. At least one gene that is normally present in the nucleus of higher eukaryotes, is mitochondrial in *Dictyostelium* (Cole *et al.*, 1995).

There is also evidence for a later divergence closer to the animals. Loomis and Smith (1995) found a number of open reading frames coding for known proteins of *Dictyostelium* that are more closely related to animal proteins than to the corresponding yeast proteins. This would mean that yeast diverged earlier than *Dictyostelium*, or that the rate of substitution is greater in fungi than in other lineages. These studies are weakened by the lack of comparison with the genes of other protists. They are important in that even if the rapid evolution of yeast is responsible for the apparent close relation of *Dictyostelium* sequences to corresponding animal sequences, the fact remains that the

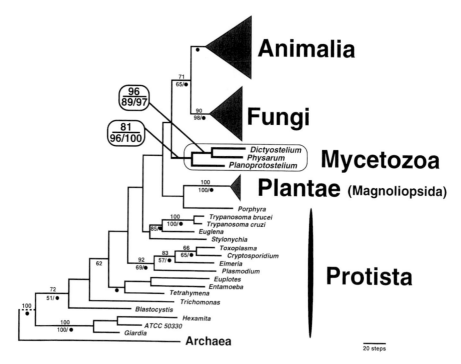

Figure 3.8 EF-1α phylogeny showing the relative time of divergence of the Mycetozoans, including *Dictyostelium, Physarum,* and a protostelid (see Fig. 1.4). Figures above and below the nodes indicate bootstrap values, which are an indication of reliability. Several methods of analysis were used to produce this tree. Phylogenies derived from the sequences of other proteins, notably α-tubulin, β-tubulin, and actin support this tree (Baldauf and Doolittle, 1997). (Courtesy of Sandra Baldauf, York University. Reprinted with permission of *American Scientist.*)

Dictyostelium proteins are more similar to mammalian proteins than are those of yeast. *Dictyostelium* genes and their products may give a better approximation to the structure and regulation of mammalian proteins than do the sequences derived from budding yeast.

A number of findings do not fit the idea of independent origin of the multicellularity of *Dictyostelium.* Certain associated functions such as the use of glycogen synthase kinase (GSK3) in cell fate determinations are conserved in *Dictyostelium* and in higher organisms (Harwood *et al.,* 1995). This result may indicate that the development of *Dictyostelium* is not independent of development in higher organisms. Rather, the cell fate-determining mechanism that employs GSK3 may have evolved in an ancient ancestor of eukaryotic cells. Junctional complexes discovered in culminating structures by Blanton and Grimson are consistent with a common origin of developmental mechanisms in animals and *Dictyostelium* (see Chapter 11). It will be interesting to determine whether *Dictyostelium* contains signal transduction pathways that exist in animal cells, but not in fungi or plants. *Dictyostelium* has STAT proteins, which budding yeast lack (Kawata *et al.*, 1997). STAT stands for Signal Transducers and Activators of Transcription. These factors are associated with several mammalian growth factors and cytokine-induced signal transduction pathways. The structure of the pyrimidine biosynthetic genes is consistent with a phylogenetic divergence after the precursor to the animals diverged (Boy-Marcotte *et al.*, 1984). The organization of the pyrimidine biosynthetic genes resembles the mammalian organization, rather than that of the fungi.

To conclude – *Dictyostelium* and its relatives diverged long ago. Precisely when this occurred is difficult to know with certainty, and depends to a great extent on whether one prefers comparisons of protein or ribosomal sequences. The interesting evolutionary questions raised by these organisms – how did cooperativity evolve; how was parasitism avoided; what was the order of evolution of the various developmental options – are independent of a complete understanding of the time of divergence. An evolutionary analysis helps us to design new selection procedures and to design tests for the fitness of mutants that previously were not known to have a phenotype.

4

The Genome and Genetics

4.1 The genome is relatively small and will be sequenced soon

The chromosomes of *Dictyostelium* are at the limit of light microscopic detection, and although seven are generally observed, the physical and genetic maps of the genome reveal only six chromosomes. Originally, the genome size was determined to be about 50 megabases (Mb) (Firtel and Bonner, 1972; Sussman and Rayner, 1971), about 1% of the size of the human genome, and about 11–12 times that of *E. coli*. More recent estimates yield a genome size of 34 Mb (Cox *et al.*, 1990; Kuspa and Loomis, 1996).

One of the curiosities of the genome is its extreme AT-richness, especially in intergenic regions and introns. The stop codon is nearly always TAA. Coding sequences are skewed to an AT bias – when the third position of a codon can be an A or a T, it usually is. Overall, the base composition is 77% AT, while in coding sequences A and T constitute about 65% (Firtel and Bonner, 1972; Kimmel and Firtel, 1982; Sussman and Rayner, 1971). Other organisms such as *Plasmodia falciparum* are also exceptionally AT-rich, though what led to this AT-richness is unknown.

There is another strange element in the genome – coding sequences tend to have long repeats of the triplet AAC (Kimmel and Firtel, 1985; Shaw *et al.*, 1989). The AAC repeats are present in all three reading frames, leading to long stretches of asparagine, glutamine, or threonine in the deduced proteins (Shaw *et al.*, 1989). Their role, if any, in the function of the proteins is unknown, nor is it known how they arose. Surveys of the databases to locate homologies are not successful unless the homopolymer stretches are removed. They can also create problems if hybridization probes are not well designed.

Introns are present in most genes but are generally short – about 100 base pairs. The splice junctions known from higher organisms are conserved, indicating that the splicing mechanism evolved in a common ancestor of amoebae and animals (Wu and Franke, 1990). The genome is relatively compact – a number of genes have been found that flank rather short promoter regions. Calculations, assuming about 3,000 base pairs for a gene with introns and regulatory sequences, indicate that slightly more than 10,000 genes are coded by the organism. With this repertoire it runs its metabolism (*Dictyostelium* makes everything it needs except a few vitamins, eight amino acids, and a carbon source; Franke and Kessin, 1977) and two developmental programs – for the creation of macrocysts, and fruiting bodies. The organism presumably has many genes that help it cope with adverse environmental conditions and with the large number of different species of bacteria that it is capable of consuming.

Early experiments employed hybridization kinetics to learn what percentage of the genome was expressed, and when (Blumberg and Lodish, 1981; Firtel and Jacobson, 1977; Jacquet *et al.*, 1981). More recently, cDNAs have been made from transcripts and these have been used for large-scale sequencing projects. Two such efforts have been made by T. Morio *et al.* (personal communication) by creating libraries from the slug stage of development. These cDNAs can be searched at http://www.csm.biol.tsukuba.ac.jp/Series.html or through the *Dictyostelium* home page. Recently, the cDNA sequencing project has expanded to vegetative cDNAs. A high percentage of newly sequenced cDNAs have no homologues in the databases. As we will see below, *Dictyostelium*'s genetic capacity offers a route to determining the function of such genes.

A genome sequence project is underway and the full genome sequence should be known within the next few years. The currently available sequences are already a great assistance in sequencing projects. The material for starting such a process is in place because of an excellent physical map that has been correlated with the genetic map. The physical map of the *Dictyostelium* genome has been pioneered by Kuspa, Loomis, and colleagues (Kuspa and Loomis, 1996; Kuspa *et al.*, 1992; Loomis and Kuspa, 1997). The map is critical to the localization of new genes and to genomic sequencing. The relatively small genome has been represented with about 5-fold redundancy in 1,000 yeast artificial chromosomes (YACs), each containing about 200 kb of sequence. These large fragments have been separated by pulse-field gel electrophoresis and displayed on membranes. Each new sequence should hybridize to approximately five YACs. Closely linked sequences will hybridize to an overlapping set of YACs, and since not all of these YACs have the same ends – being staggered in relationship to one another – they can be aligned and then extended into even larger units called *contigs*. The resolution of the map has been increased by introducing a series of insertions with restriction enzyme-mediated integration (see below). A known sequence with a rare restriction site was introduced into 147 different strains so that restriction fragment length polymorphisms could be generated (Kuspa and Loomis, 1994a). These were used to confirm the map created from overlapping YAC sequences. Nearly all of the genome is currently mapped in a series of ordered YACs and contigs. Sequenced genes

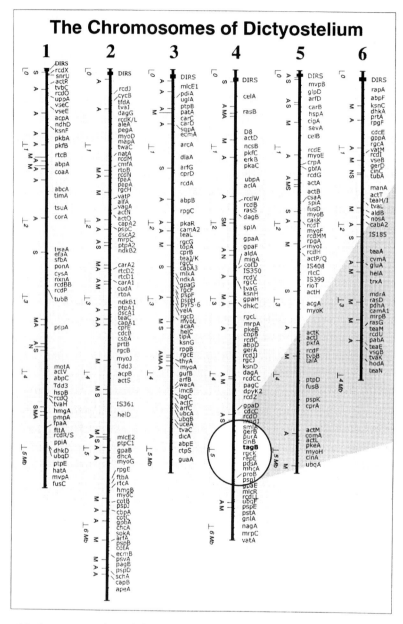

Figure 4.1 A representation of the six chromosomes. The area around the genes coding for *tagB*, *C*, and *D* is shown. These genes are involved in early events in stalk formation and are described in Chapter 11, Section 11. See Kuspa and Loomis (1996) and Loomis *et al.* (1995) for a list of gene names, or consult the database described in Chapter 13. (This figure was prepared by Richard Sucgang of the Baylor College of Medicine.)

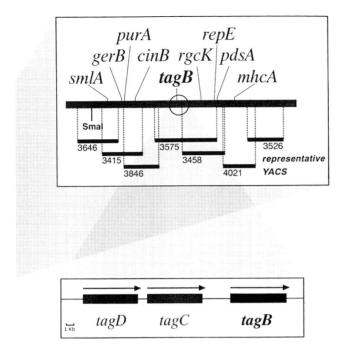

Figure 4.1 (*continued*)

are located, on average, every 100 kilobase pairs along the genomic DNA, although this will rapidly change.

The correlation of the genetic and the physical maps has been accomplished by probing the chromosomes of *Dictyostelium* with sequences that are located on known linkage groups (Kuspa and Loomis, 1994a). A sequence introduced into the genome can be mapped to a linkage group by standard methods of parasexual analysis, and then used to probe YACs or whole chromosomes separated by pulse-field gel electrophoresis. In this way chromosomes have been identified with linkage groups (Loomis *et al.*, 1995; Welker *et al.*, 1986). (See Chapter 13 for a variety of linked Websites. Workers can immediately see a listing of YACs and the map position of an increasing number of genes.) The format used is shown in Fig. 4.1. One of the values of the system has been to act as a central information source so that when laboratories recover new genes they can immediately ask whether the gene has been studied before and by whom, so that redundant effort is minimized.

4.2 The ribosomal genes are coded in an extra-chromosomal palindrome

There are other known landmarks of the nuclear genome. The ribosomal rRNA genes are coded in a large extra-chromosomal palindromic sequence.

The dimeric sequence is linear, and occurs in 100 copies per genome. The ribosomal genes were first noticed as DNA fragments with a buoyant density that was greater than the rest of the nuclear DNA (Cockburn *et al.*, 1976; Firtel and Bonner, 1972; Firtel and Jacobson, 1977). The ribosomal palindrome has been partially sequenced (Ozaki *et al.*, 1984; A. Kuspa, personal communication). The 88-kb palindrome contains no open reading frames, but does contain uninterrupted sequences that code for 17S, 26S, and 5S ribosomal RNAs. As in other organisms, the 17S and 26S sequences are first transcribed as a large precursor, which is then processed (Cockburn *et al.*, 1978). The structural RNAs occupy a small part of the palindrome. No role for the remaining sequence is known. Because of the high copy number of the ribosomal palindrome, restriction enzyme digestion of nuclear DNA produces a number of visible bands on ethidium bromide-stained gels. The nucleolus, the site of rRNA synthesis, appears attached to the nuclear membrane, so that when viewed in the microscope, the nuclei sometimes have the appearance of a three-cornered hat. Despite the fact that most growth-specific genes are inactivated during development, the ribosomal genes are transcriptionally active (Kessin, 1973).

Dictyostelium has many tRNA genes, and some of these have been characterized by Dingermann and his colleagues (Dingermann *et al.*, 1986, 1987; Marschalek *et al.*, 1989). One of the most useful results of this work is to create nonsense suppressors (Dingermann *et al.*, 1990). Suppression can be used to control the expression of a gene product or for other purposes (Dingermann *et al.*, 1992).

4.3 *Dictyostelium* species contain several families of replicating plasmids

In addition to the ribosomal palindrome, *Dictyostelium* species have a variety of extra-chromosomal plasmids, some of which have been exploited as transformation vectors (Firtel *et al.*, 1985; Hughes *et al.*, 1988; Leiting *et al.*, 1990; Slade *et al.*, 1990). About 20% of wild isolates harbor nuclear plasmids, which can be divided into four families, based on sequence and structural similarities. Plasmids in the Ddp1 and the Ddp2 families are the best studied (Gonzales *et al.*, 1999). Plasmids Dpp1 and Dpp3 from *D. purpureum* define the other two families, based on sequence and structural homology (Kiyosawa *et al.*, 1993). Families are defined by sequence homology and do not connote incompatibility groups.

The Ddp1 plasmids contain a number of genes expressed during growth and development, in addition to an origin of replication. All of the plasmid-carried genes expressed during growth are essential for long-term maintenance, while deletion of the Ddp1 genes expressed during development had no detectable effect on long-term maintenance (Hughes *et al.*, 1994; Kiyosawa *et al.*, 1995). The origin of replication of Ddp1 has been localized to a 543-bp region (Kiyosawa *et al.*, 1995). The Ddp1 elements necessary for extra-chromosomal replication have been utilized to drive expression of heterologous genes. Most recently, a tetracycline-regulatable promoter has been introduced into a version of Ddp1, providing a useful inducible system for *Dictyostelium* (Blaauw *et al.*, 2000).

The Ddp2 family of plasmids contains a conserved family of *rep* genes that code for a protein that is required for plasmid replication (Slade *et al.*, 1990). The Rep protein binds DNA (Shammar and Welker, 1999). The *rep* gene and a characterized inverted repeat are necessary for plasmid maintenance (Shammat *et al.*, 1998; Rieben *et al.*, 1998). This gene should not be confused with the DNA repair genes, also called *rep*, which are discussed in Chapter 4, section 6.

There are sequence relationships among these families, but there is no relationship to other nuclear plasmids, such as the budding yeast 2 μm circle. Different plasmids can coexist in the same nucleus, indicating that there are different replication systems preventing incompatibility (Kiyosawa *et al.*, 1994). The *Dictyostelium* plasmids show a clear species specificity (Hughes *et al.*, 1988). The plasmids share the AT-richness of their hosts and are packaged in nucleosomes. Since these plasmids can constitute several percent of the nuclear DNA, they could present a burden on the cells. The complete plasmids are stable over time in the absence of any selection (Hughes and Welker, 1989).

In addition to the Ddp1 and Ddp2 families and the two families known from *D. purpureum*, a number of plasmids have been isolated from other species. The isolates come from *D. gigantium* and *D. firmibasis* (Gonzales *et al.*, 1999) and *D. mucoroides* (Kiyosawa *et al.*, 1994). The *D. mucoroides* plasmids fall into the Ddp1 family, and the *D. giganteum* and *D. firmibasis* plasmids are related to Ddp2 (Gonzales *et al.*, 1999).

4.4   The genome is littered with transposable elements

All *Dictyostelium* species contain several types of transposable elements which are relics of events in the earlier history of the species (Geier *et al.*, 1996; Leng *et al.*, 1998). Mobile genetic elements can be divided into retrotransposons with identical long terminal repeats (LTR) or into non-LTR elements, which are often defined as "longer-interspersed nuclear-elements" (LINEs). LINEs contain non-identical terminal structures and characteristic oligo-dT sequences at their 3′ ends. *D. discoideum* contains both retrotransposons with long terminal repeats, and LINEs. These elements occupy several percent of the genome. This is not uncommon – the human genome is equally littered with molecular parasites. In human and many other genomes, genomic DNA is often modified by methylation, which is thought to suppress transposition of various molecular parasites (Yoder *et al.*, 1997). No such modifications have been found in *Dictyostelium* genomic DNA (Smith and Ratner, 1991).

Table 4.1. *The mobile elements of the* Dictyostelium *genome*

Element	Copy number	Length (kb)	Fraction of genome (%)	Reference
Tdd2	20–30	4.7	0.2–0.3	Poole and Firtel (1984)
Tdd3	20–30	4.7	0.2–0.3	Poole and Firtel (1984)
DIRS-1	40	4.7	0.4	Cappello *et al.* (1985a);
(Tdd1)				Cappello *et al.* (1985b)
DREa	80–100	5.7	1.1	Marschalek *et al.* (1992b)
DREb	80–100	2.4	0.5	Marschalek *et al.* (1992a)
Skipper	15–20	7	0.3	Leng *et al.* (1998)

Note: The table was compiled by Leng, Steck and colleagues (Leng *et al.*, 1998). In addition to those noted here, there are several hundred DIRS-1- and DRE-related sequences in the genome.

4.4.1   Tdd-2 and Tdd-3

The Tdd-2 element was found in a region of genetic instability that flanks a discoidin gene. The Tdd-2 element is inserted in different sites in the several laboratory strains of *D. discoideum*, which indicates that there has been movement since the strains began to be used. The Tdd-2 elements have frequently been targets of a secondary integration event by Tdd-3, which is better characterized than Tdd-2. Insertion generates a 9- or 10-bp repeat, as do other transposable elements, due to a staggered cut in the DNA into which the new element inserts.

Tdd-3 is a 5.2-kb element that encodes two overlapping reading frames flanked by non-redundant regions that are not coding sequences. The deduced amino acid sequence of Tdd-3 is homologous to the reverse transcriptase of polyA transposons, and may also code for an endonuclease. The element is transcribed and polyadenylated during growth, and seems to have preferred insertion sites that are about 100 bp downstream of some tRNA genes. It is curious that neither element is found in *D. mucoroides* or *D. purpureum*. Either the invasion by these elements occurred after the speciation of *D. discoideum* or the sequences could have evolved rapidly in *D. mucoroides* and *D. purpureum* so that they are no longer recognizable by hybridization.

4.4.2   DIRS-1

DIRS-1, also called Tdd-1, is a 4.7-kb repetitive element that is present in about 40 copies in the genome (Rosen *et al.*, 1983; Zuker *et al.*, 1984). The physical mapping of the genome reveals multiple DIRS sequences near the telomere of each chromosome. One speculation is that they serve as centromeres, but this has not been confirmed. DIRS-1 contains inverted terminal repeats of 330 bp. Terminal repeats can also be located throughout the genome, independently of the complete 4.7-kb element. The sites of incorporation vary in the different laboratory stocks of *D. discoideum*, although no analysis has been done among the various Dictyostelid species. DIRS-1 is transcribed into what appear to be three open reading frames, and probing Northern blots with the internal sequence detects heterodisperse, polyA-containing RNAs that are found on polysomes. One of these has homology to reverse transcriptase. The promoters, which reside in the long terminal repeat-like elements, are heat shock and stress responsive and contain the consensus heat-shock element of budding yeast heat-shock promoters (Zuker *et al.*, 1984). Development also induces these promoters.

4.4.3   DRE

The *Dictyostelium* repetitive element, DRE, is a LINE-like element that integrates about 50 bases upstream of tRNA genes. There are about 800 tRNA genes in the genome and a large number of integrated DRE elements, making this the most abundant inserted element in the genome. About half of the DRE elements are truncated at their 5′ end (Marschalek *et al.*, 1992a,b). The DRE element contains two open reading frames that code for reverse transcriptase and integrase. Non-identical terminal elements contain transcription start sites. Apparently only the C module (Fig. 4.2) is necessary for correct transposition to the neighborhood of tRNA genes. An activity that binds to the transcriptional start site in the C module has been purified, but models to explain site-specific integration are not well established. In addition to the DRE elements, somewhat different elements, Tdd-2 and Tdd-3, also bind near tRNA genes (Marschalek *et al.*, 1990). None, so far, has been found to affect a coding sequence. The genome contains so many copies of the DRE elements that Southern blots cannot be used to detect single fragments, making it difficult to determine whether they transpose frequently.

4.4.4   Skipper

The first retrotransposon with long terminal repeats has recently been described, and is of the same class exemplified by Ty3 in budding yeast and *gypsy* in *Drosophila* (Leng *et al.*, 1998). Skipper is transcribed at high levels and contains three open reading frames that code for putative gal, pol, and a potential protease, in an arrangement that is reminiscent of certain mammalian retroviruses. It does not integrate near tRNA genes, but rather in novel segments of the genome and in several cases, next to the DIR-1 elements described above.

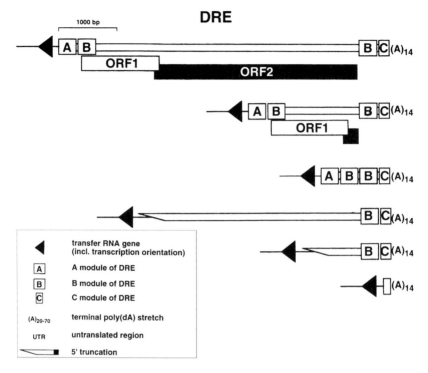

Figure 4.2 The structure of DRE and some of its deleted relatives. Many copies litter the genome. The arrow at the left represents a tRNA gene. A, B, and C are characteristic modules of DRE elements. All of the elements have polyA stretches at their 3′ ends. A variety of truncations are known among the many copies of the DRE element in the genome. These lack open reading frames (ORF), or modules, but remain associated with a tRNA gene. (This drawing was provided by Thomas Winckler of the Johann Wolfgang Goethe-University.)

4.5 The mitochondrial genome

The mitochondrial genome of *D. discoideum* is 55.5 kb, which is larger than animal mitochondrial genomes. The physical and genetic organization has been described (Fukuhara, 1982). Amoebae carry a large number of mitochondria such that as much as 35% of the total DNA is mitochondrial. This means that each cell carries about 200 mitochondrial genomes. The sequence of the mitochondrial genome has been completed. The genome contains 55,564 bp and can be found in Genbank with the following accession number: AB000109 (Y. Tanaka, personal communication). In addition to ribosomal genes, there are seven tRNAs and a number of open reading frames (Angata *et al.*, 1995; Cole and Williams, 1994). The AT-richness of the nuclear genome is shared by the mitochondrial genomes. The A + T content is 70.8% overall, with coding sequences being 73.6% and the ribosomal sequences being 71.5%. All open reading frames use the universal code.

The *Dictyostelium* mitochondrial genome codes for at least one electron transport enzyme that is nuclear in plants and animals (Angata *et al.*, 1995; Cole *et al.*, 1995). This result suggests that *Dictyostelium* diverged at an early date, before the immediate ancestor that gave rise to plant and animal mitochondria. The occurrence of this gene can be considered a conserved primitive feature that is not retained in plant or animal sequences, thus making the *Dictyostelium* a model for studying the endosymbiotic origins of mitochondria. *Acanthamoeba* shares many features of the *Dictyostelium* mitochondrial sequence (Komori *et al.*, 1997a,b; Ogawa *et al.*, 1997).

In *P. pallidum*, mitochondrial DNA polymorphisms have been detected and the behavior of the mitochondrial genome during sexual and asexual genetics has been studied. This species has a workable meiotic system based on macrocyst formation and germination. When two strains with restriction-fragment-length polymorphisms are mated, only one mitochondrial genome survives, indicating that one genome is selected and one excluded (Mirfakhrai *et al.*, 1990). No mechanism has been suggested. During asexual development, most DNA synthesis is shut down as the cells starve – yet there is clear synthesis of mitochondrial DNA in prespore cells. Thus initiation of mitochondrial DNA synthesis is developmentally regulated and cell-type specific (Shaulsky and Loomis, 1995).

4.6 Maintaining the genome – the DNA repair mechanisms of *Dictyostelium*

Amoebae are extremely resistant to DNA-damaging reagents, including ionizing radiation, ultraviolet irradiation, and alkylating agents (Deering, 1968; Deering *et al.*, 1970; Guialis and Deering, 1976). *Dictyostelium* is something of a champion when it comes to resistance to irradiation with ^{60}Co, requiring more than 300 krad to achieve 90% killing (Deering, 1988a). This is far in excess of what a mammalian cell could withstand. The basis for this resistance is exceptionally efficient DNA repair mechanisms. Deering and his colleagues have shown that a variety of mutants can be obtained which reduce the resistance dramatically. These fall into a number of complementation groups, as shown by parasexual analysis (Welker and Deering, 1978). Some of the enzymatic activities involved in repair have been characterized, but these mutants predate insertional mutagenesis techniques and therefore the genes involved have not generally been cloned (Deering, 1988b; Freeland *et al.*, 1996; Guyer *et al.*, 1985, 1986). One question that arises is why *Dictyostelium* amoebae should have been selected to be so proficient at repair. Deering makes the interesting suggestion that it is not ionizing or ultraviolet radiation that selected repair proficiency, but rather co-habiting with soil organisms that secrete a variety of DNA-modifying antibiotics, such as bleomycin (Deering *et al.*, 1996).

Xeroderma pigmentosum is a disease of humans that results in increased radiation sensitivity and a variety of developmental disorders. The homologous genes have been recovered from *Dictyostelium* by Stephen Alexander and his colleagues (Alexander *et al.*, 1996; Lee *et al.*, 1997b, 1998). These authors showed that three *rep* genes, *repB*, *repD*, and *repE*, are developmentally regulated and induced by ultraviolet light. These are not the same genes as described in Section 4.3 above. Deletion of *repB* causes an increased sensitivity to ultraviolet light, but not to the extent of some of the Deering mutants. Despite its developmental regulation, the loss of RepB causes no obvious developmental defect (Lee *et al.*, 1998). The *rep* genes may participate in developmental events, in addition to being deployed in the repair of the genome, as is apparently the case in xeroderma pigmentosum. Recently, *Dictyostelium* mutants that are resistant to the chemotherapeutic agent cisplatin have been isolated and are being used to study the basis of clinical resistance to this anti-metabolite (Li *et al.*, in press). *Dictyostelium* is a potentially valuable tool in the study of the disease, and of DNA repair in general.

4.7 Molecular genetics

The range of available genetic techniques is shown in Fig. 4.3. To study any biological process, it is essential to be able to introduce DNA into cells and to observe its expression. DNA can be introduced into *Dictyostelium* amoebae either as a calcium phosphate precipitate or by electroporation (Firtel *et al.*, 1985; Nellen *et al.*, 1984). The first transformations employed vectors that contained a gene providing resistance to the drug G418, which kills amoebae. This resistance gene was regulated by a promoter from an endogenous actin gene. The frequency of transformation was low, and required integration of the plasmid into the genome (Hirth *et al.*, 1982). Plasmids that confer resistance to G418 are usually introduced in multiple copies. Other selections depend on complementation of auxotrophic mutations that create requirements for uracil or thymidine. More recently, resistance to blasticidin has been used, and this has the advantage that apparently only a single copy of the gene that confers blasticidin resistance is required (Adachi *et al.*, 1994). Other vectors employ bleomycin and hygromycin resistance. For further references, see Kuspa and Loomis (1994b).

There is a choice of vectors, both integrating and extra-chromosomal. The latter tend to be more efficient because transformation does not require an integration step. These vectors are derived from the endogenous plasmids discussed above. The integrating vectors have been used for gene replacement and gene knockout, both in a linear or in a circular context. Marker genes, such as β-galactosidase, β-glucuronidase, or green fluorescent protein, function well in *Dictyostelium* when driven by a variety of promoters (Dingermann *et al.*, 1989). Initially there was worry that the skewed base composition and the constellation of tRNAs from *Dictyostelium* would not efficiently translate exogenous genes, though this seems not to be a major problem.

4.8 Mutagenesis

Although it has been possible to isolate mutants of *Dictyostelium* since it was first done by Sussman in 1953,[1] it was impossible until recently to recover the affected genes (Sussman and Sussman, 1953). Even after transformation became possible, the mutations could not be complemented by transformation with libraries of genomic DNA, and therefore the affected genes could not be recovered in a manner that was available in bacteria or later became a powerful tool in the study of yeast. This is still the case, and the reasons for our inability to do so (at least consistently) rest on several points. The frequency of transformation is the product of the efficiency of introduction of DNA and the probability of replication of the DNA. For integrating vectors, the latter is a rate-limiting step. The transformation frequency of extra-chromosomal vectors is higher because they do not require a low efficiency integration event into a chromosome in order to be propagated. Yet efficient complementation of chemically induced mutants is still being developed. In order to transform with a cDNA library and expect complementation of a gene whose mRNA represents 0.01% of the mRNA population, one would need $10,000 \times 6$ sequences. This assumes that the cDNAs are relatively full length and not directionally inserted, and that only one plasmid enters per cell during transformation. Thus many tens of thousands of transformants would be required to have a chance of complementing a mutation in a gene with a low abundance mRNA. Recovery of this many independent transformants is not yet routine in *Dictyostelium*. Construction of unidirectional full-length stage-specific cDNA libraries in extra-chromosomal vectors, and better methods of DNA introduction, may solve this problem soon. The alternative strategy – to use genomic DNA in large fragments, thus requiring fewer transformants – has not been successful (for reasons that are not clear), but which are often assumed to be due to the instability of the AT-rich *Dictyostelium* sequences during amplification steps in *E. coli*.

---

[1] The genetic nomenclature is as follows. Genes are designated by three lower case letters in italics, followed by an upper case suffix for different loci whose mutation gives similar phenotypes – e.g. *ecmA*. These are also in italics. Gene products are designated with the same symbol, but the first letter is in upper case and there are no italics – EcmA, for example. In this text, the phrase "*ecmA* gene" is redundant because *ecmA* already indicates the gene. When *ecmA* is a mutant allele, this will be stated. Chromosomes are designated by Arabic numerals, e.g., 1–6. Plasmids native to *Dictyostelium* species are designated by a prefix indicating the genus and species followed by a p for plasmid and a plasmid specific number. For example: Ddp1 or Dpp1. The former would be a plasmid recovered from *D. discoideum*, the latter from *D. purpureum*. Strains are labeled with two or three letter upper-case designations, followed by a serial number: for example: HM44, FR17, HTY217. See the following Web site for further information: http://dicty.cmb.nwu.edu/dicty/GeneNames.html.

Genetic approaches available in *Dictyostelium* amoebae

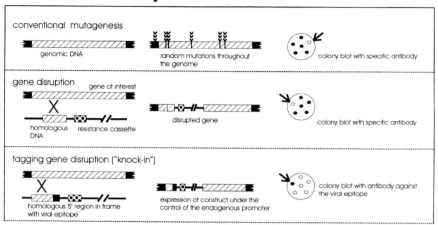

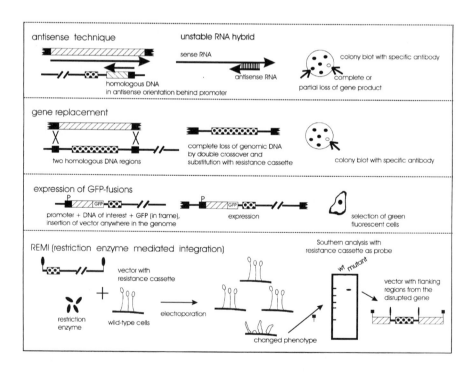

4.9 Restriction enzyme-mediated integration

Until complementation with prepared libraries becomes a reality, there is an alternative for gene recovery. This method, called restriction enzyme-mediated integration (REMI), has been introduced from yeast and is a form of insertional mutagenesis that permits the cloning of affected genes (Kuspa and Loomis, 1992). REMI employs a small shuttle plasmid with selectable markers for growth in *E. coli* and in *Dictyostelium*. The plasmid is cut with a single enzyme and then introduced into growth phase amoebae by electroporation. The difference between REMI and simple transformation is that at the time of electroporation, a restriction enzyme is also included. For example, if the plasmid is linearized with *Bam*H1 to generate single-stranded ends of GATC, *Dpn*II (which also generates GATC ends) is included at the moment that the linear plasmid is electroporated into the cell. *Dpn*II is used because it recognizes more sites in the genome than *Bam*H1. Apparently, the *Dpn*II enters the cell and recognizes appropriate sequences in the genomic DNA. Occasionally, the compatible ends of the plasmid are introduced at the restriction sites and ligated into place such that an insertion is created. Other combinations of restriction enzyme can also be used. The frequency of integrating transformation is elevated by the presence of the restriction enzyme. Transformants are then selected by growth in axenic medium in the presence of selective agents such as the drug blasticidin. One might expect that randomly cutting the nuclear DNA with a restriction enzyme might be mutagenic in its own right, but no elevated level of morphological mutants or lethality has been detected when the enzyme is electroporated into cells without plasmid DNA. There is no evidence that the genomic DNA actually suffers a double strand cut. *Dictyostelium* has exceptionally powerful DNA repair mechanisms (see Section 4.6) which may prevent damage from restriction enzymes (Deering, 1998b).

After resistant cells have been recovered, it is a simple matter to plate the cells on bacterial lawns where each cell forms a colony. Among these are

---

Figure 4.3 Molecular genetic manipulations in *D. discoideum*. Amoebae can be subjected to a variety of genetic treatments. Conventional mutagenesis is shown in the top panel, and was used before other techniques displaced it. In the second panel, targeted gene disruption is shown. The circular vector aligns with the endogenous gene and inserts by homologous recombination. A variant is the knock-in, shown in the third panel. In this variant, an epitope coding sequence is inserted in genomic DNA. Panel 4 shows that antisense constructs can also be used. These are sometimes effective when a product is coded by a multigene family and a single knockout is inadequate or if the knockout is lethal. In addition, standard gene replacements and GFP constructs can be introduced into the genome. The REMI procedure that is described in the text is shown in the last panel. (This summary of genetic techniques was prepared by Ludwig Eichinger, Soo Sim Lee, and Michael Schleicher. Reproduced from *Microscopy Research and Technique*. Published by permission of Wiley-Liss, Inc., a subsidiary of John Wiley & Sons, Inc.)

mutants that are affected in one aspect or another of morphogenesis because critical genes carry the plasmid as an insertion. Recovery of the sequences that flank the insertion site is accomplished by cutting genomic DNA with a restriction enzyme that does not cut the plasmid. Instead, flanking sequences are cut. This restriction digest is then ligated and used to transform *E. coli*. Only the fragment containing plasmid and flanking sequences will be recovered in transformed bacteria selected for resistance to ampicillin. The usual experience is that this works efficiently, but there are cases when the flanking sequences are rearranged or deleted.

Once the flanking fragments have been recovered, the entire plasmid can be used to create new mutants by homologous recombination in the wild-type strain. If the gene in question codes for a critical protein whose elimination is responsible for the phenotype, the mutant phenotype should be observed at high frequency after homologous recombination disrupts the gene in the parental strain. Recapitulation of the phenotype is a critical step. Spontaneous mutants may occur in the population of amoebae, and these can confound the analysis. It is critical that freshly cloned populations of amoebae be used for REMI mutagenesis to avoid adventitious mutations.

An increasing number of REMI mutants of great interest has been recovered, and every month produces new mutants that affect fundamental processes. Loomis and his colleagues have systematically recovered genes from REMI mutants and these have led to a number of important genes which are being used by the community. While these will be discussed in their proper developmental context, it is curious that often several laboratories have isolated the same mutant, which raises the question of randomness of REMI insertions (Dynes *et al.*, 1994). The only compelling evidence on the subject comes from Kuspa and Loomis (1994a), who used REMI to isolate insertions at 147 random sites in the genome. At least to a first approximation, there do not seem to be limitations on the sites of REMI insertions. To saturate the genome with mutations, it is essential that REMI experiments yield a large number of mutants. Since its introduction, the efficiency of REMI has increased so that one can now isolate many thousands of independent mutants. This increase in efficiency is useful in that being able to recover as many as 10^5 REMI mutants permits us to create suppressors, which is a powerful genetic method to examine pathways of regulation. This has been used to good effect in examining the role of the Tag genes, whose absence blocks development in the tight aggregate stage (Shaulsky *et al.*, 1996), and in the examination of proteolytic mechanisms that are employed during development (Pukatzki *et al.*, 1998). Other examples are in the suppression of the effects of the mutant phosphatase coded by the Spalten gene (L. Aubry, personal communication) and the suppression of a phenotype caused by mutation of *carC* (Bear *et al.*, 1998).

As a technique, REMI has certain limitations: it creates null mutants because in general it disrupts a gene. Thus many kinds of suppressors cannot be found. One would employ it preferentially to remove inhibitory genes that control a pathway and thus create second site suppression. It may be possible to make the technique more sophisticated and to bias insertions into promoter regions or introns by using enzymes that cut AT-rich sequences. The separation of growth and developmental cycles is what permits the effective use of the technique for the recovery of genes that are important for development. The other problem is that REMI mutants affected during the growth cycle are not recovered by current methods. Thus defects in genes that are critical to pinocytosis, phagocytosis, or the cell cycle are not available. This limitation would be overcome if stable diploids could be mutated. If an insertion can be found in a diploid, but not in the haploid products of the diploid, novel genes that are essential for growth could be recovered (R. Sucgang, personal communication).

4.10   Parasexual and sexual genetic manipulations

The haploid nature of *Dictyostelium* permits isolation of a rich variety of mutants. These can generally be divided into those that affect growth, such as those that confer drug resistance, temperature sensitivity, or auxotrophy, and those which affect various aspects of development. The study of the latter is facilitated by the fact that development is gratuitous – defects in it do not affect growth. One of the gaps that exist in genetic manipulation of *Dictyostelium* is the absence of a workable system of sexual genetic analysis. It is likely that the macrocysts that were described in Section 3.3.2 are a sexual stage, but there have been problems of germination, or in the cases where germination has occurred, no recombination took place so that only the genotype of one of the parents emerged. The macrocysts have already been encountered as an intermediate in the evolution of the fruiting body and for the problems it presents in kin selection. Further details of their formation and structure, as well as some speculation as to why the macrocyst has failed so far to provide a system of genetic analysis, will be described below.

4.10.1   Parasexual genetics

There is a system of parasexual analysis that fulfills some of the same research functions. *Dictyostelium* amoebae fuse to form diploids, and the diploids go through the same development that the normally haploid amoebae do. Although the fusion process is infrequent, occurring at a frequency of 1 in 10^5, it is a simple matter to select the diploids by using haploid partners with different thermosensitive mutations. If two such partners are incubated together they form aggregates, and in these aggregates cells occasionally fuse (Loomis and Ashworth, 1968). The rare diploids can be selected by incubating them on lawns of bacteria at the restrictive temperature, which is $27°C$ (Katz and Sussman, 1972; Loomis, 1969). By using two strains to make mutants that affect a fundamental process such as aggregation, culmination, or phagocytosis, one can do complementation tests and determine whether two mutations affect the same gene. Thus, if two aggregationless mutants affect different genes and both are recessive, the diploid containing both of them should be normal in its development. Spores formed by diploids are conveniently banana-shaped, whereas the haploid spores are smaller and elliptical.

The parasexual system has been exploited to examine the complexity of a developmental process. Williams and colleagues (Williams and Newell, 1976) and Coukell (1975), using *Dictyostelium*, and Warren *et al.* (1975), using *Polysphondylium pallidum*, created a series of aggregateless mutants in two different thermosensitive strains. They then made all pairwise crosses to form diploids. If the number of genes required for aggregation is very large, it would be unlikely that in crossing two groups of 20 mutants one would find a pair that did not complement. If the number of genes that are required for aggregation to take place is less, then the chances of independent isolation and identification by failure to complement becomes higher. The distribution of non-complementing classes allows an estimate of the number of genes. While these estimates are necessarily inexact, all of the estimates are in the range of 50 to 100 genes required for aggregation. Since the *Dictyostelium* genome is capable of coding for many times this number of genes, the process is not as complex as it might be, and there is a hope of understanding it in detail. In the years since these experiments were performed, a number of the genes involved in aggregation have been cloned and studied, and although the number does not yet approach 100, the original estimates of complexity seem to have been reasonably accurate. The parasexual system permits us to ask whether older chemically induced mutations are in the same complementation group as newer REMI mutants. A recent outstanding example of this type of analysis is by Chang *et al.* (1998), who showed that one of the classic rapid development mutants, *rdeA*, is in the same complementation group as a mutant made by REMI. *RdeA* codes for a phosphorelay protein, and will be discussed below.

The parasexual system permits linkage assignment because the diploids are not completely stable and segregate haploids, which inherit chromosomes randomly from either parent. The frequency of haploid segregation can be increased by the use of anti-microtubule agents that disrupt the mitotic spindle. Thus if a chromosome that carries a mutation in a gene coding for cycloheximide resistance always segregates with a mutation that results in an inability to aggregate, then one can be sure that the two mutants are linked. If the cycloheximide resistance and the aggregation gene lie on homologues provided by the two partners, they are said to be in trans, so that when cycloheximide-resistant haploids are selected, none of them will be aggregateless. If the two mutations are unlinked, then cycloheximide-resistant strains will segregate aggregateless or wild-type markers in equal numbers (Kessin *et al.*, 1974; Williams *et al.*, 1974b).

Using a variety of markers, which include drug resistance, color, spore shape, enzymatic activities, auxotrophs, and morphological defects, a genetic map has been constructed. More than 100 genes have been mapped. By including restriction-fragment-length polymorphisms, this genetic map has been linked to the physical map of the 34-Mb genome (Loomis *et al.*, 1995). The *Dictyostelium* karyotype has been seen to include seven physical chromosomes (Bonner and Frascella, 1952; Robson and Williams, 1977; Sussman, 1961; Zada-Hames, 1977), but only six of these have been assigned genetic markers. No markers have been assigned to linkage group V, with the possible exception of a phototaxis mutation (Darcy *et al.*, 1993).

The system of parasexual analysis has one further capability; it can be used to order markers on chromosomes, although the existence of physical mapping methods has rendered this unnecessary. Mitotic recombination occurs in *Dictyostelium*, and has been used to infer the order of markers in a centromere to telomere direction (Katz and Kao, 1974; Wallace and Newell, 1982; Welker and Williams, 1982). Centromere proximal markers undergo less recombination than distant ones. The efficiency of this system is limited.

4.10.2 *Sexual recombination*

The sexual cycle has been mentioned in the context of evolution. Cells of two mating types fuse to form the giant cells described in Section 3.3.2, and then there is a massive ingestion of solitary amoebae to form endocytes. A cellulose structure is constructed, and eventually these germinate to release recombinants. Though cumbersome, this system has been used with several species including *D. mucoroides* (MacInnes and Francis, 1974) and *P. pallidum* (Francis, 1975). The type species, *D. discoideum*, has not been successfully used for meiotic analysis except in non-standard strains (Wallace and Raper, 1979; Francis, 1998). The traditional strains seem to have an incompatibility – even by parasexual analysis it is impossible to form diploids of the standard NC4 and V12 strains (Robson and Williams, 1979).

The strains that appear to mate are independent isolates from Wisconsin soils, recovered many years ago by Raper (Wallace and Raper, 1979). These strains were used by Francis to show that high-frequency recombination takes place. Francis used polymorphisms of various linked genes that had been located on the physical map. The system has been shown to work, but takes six weeks in its current conception and does not yet have selectable markers incorporated into it (Francis, 1998). There has been little effort to mark the sexually competent strains with dominant selectable or visual markers.

The formation of the macrocyst was described in Chapter 3, Section 2. The macrocyst, independently of its genetic potential, displays a number of interesting features, including cell fusion, loss of self recognition, and a voracious phagocytosis until one cell expands 100-fold by engulfing the surrounding cells. A giant cell that is 100 times bigger than a normal amoeba can function with a DNA content that is only 2-fold greater. These phenomena have been studied and are reviewed by Urushihara and her colleagues (Aiba *et al.*, 1997; Higuchi *et al.*, 1995; Urushihara, 1997). These authors showed that a particular gene is essential for cell fusion, but that it is present in multiple copies and therefore deleting one does not block cell fusion. Other aspects of the biology of the macrocyst remain to be studied.

The full sequencing of the genome, which should take place over the next few years, will be an enormous asset to research in the field. The genomics projects being contemplated include more than the sequence and extend to more elaborate databases, cDNA libraries of full-length sequences that are arrayed for display technologies, and a variety of services to make research on the organism easier, faster and more productive.

5

Membranes and Organelles of *Dictyostelium*

Embryonic cells are spared some of the tasks that confront soil amoebae. Multicellular organisms are protected from direct contact with the environment because they are part of a large mass with specialized external epithelia. The plasma membrane of soil amoebae offers none of the protection of a multicellular organism. Every time it rains, the osmotic shock to the cells will be severe, because rainwater is essentially distilled water and like other single-celled organisms, *D. discoideum* and its relatives must be equipped to handle sudden changes in osmolarity. Other conditions, for example hyperosmotic mud, lead to hypertonic shock, for which the cells also have adaptive mechanisms. The internal membranes and organelles resemble those of higher organisms. Many of the membrane compartments of the amoebae can be marked by fusing the green fluorescent protein (GFP) gene to genes that code for appropriately targeted proteins. This useful property is illustrated in Plate 1.

5.1   The plasma membrane

The plasma membrane must be capable of movement on a variety of surfaces that constitute the matrix of the soil, whether the cellulose of decayed leaves, minerals, decaying vegetable matter, or films of bacterial growth. These cells must have substrate adhesion systems that are more versatile than those of embryonic cells, which move on a defined extracellular matrix. The amoebae eat all manner of bacteria and yeasts, and even undertake the occasional act of cannibalism, so several mechanisms of cell recognition must be deployed on the membranes. During growth and development, the plasma membrane must accommodate the molecules that mediate chemotaxis and other forms of signaling and cell-to-cell adhesion, as well as the fusion and phagocytosis that occur during the sexual cycle. The soil is a noxious environment, and the cells must cope with various anti-metabolites and poisons.

The plasma membrane can be isolated by several methods which have been described in detail by a number of workers and result in at least a 15-fold purification of membrane proteins. These results suggest that the plasma membrane bears less that 7% of the cell's protein (Goodloe-Holland and Luna, 1987; Murray, 1982; Steck and Lavasa, 1994). The lipid composition is not remarkably different from that of mammalian cells with phosphatidylserine, choline, and ethanolamine being major constituents in approximately equal amounts. Several novel unsaturated fatty acid species have been detected (Murray, 1982; Weeks and Herring, 1980). The phospholipid content does not change dramatically during development, nor are there observed changes in membrane fluidity. Plasma membranes are sometimes isolated bound to elements of the actin cytoskeleton and contain a number of integral membrane proteins that are involved in maintaining osmotic balance, in motility, in phagocytosis and pinocytosis, in adhesion to substrates or other cells, and in cell-to-cell signaling.

The sterol compositions of amoeboid membranes are different from those of higher organisms. The major sterol is δ-22-stigmasten-3β-ol (Murray, 1982). Sterols are the targets of polyene antibiotics, and *Dictyostelium* mutants that are resistant to nystatin and other polyene antibiotics have been recovered (Kasbekar *et al.*, 1983, 1988; Nes *et al.*, 1990). Initially, these mutants were created by chemical or spontaneous mutagenesis, but recently, with the advent of REMI, a mutant gene that leads to steroid defects and polyene resistance has been recovered (E. R. Katz, personal communication). This gene codes for an oxysterol-binding protein, the absence of which leads to resistance to the polyene antibiotic pimaricin.

Several proteins involved in cell-to-cell signaling – the cAMP receptor cAR1, the membrane-bound phosphodiesterase, and adenylyl cyclase – have been shown to localize to micro-domains which constitute about 5% of isolated plasma membranes (Xiao and Devreotes, 1997). About 20 other proteins are associated with this micro-domain, which includes one sterol, probably δ-22-stigmasten-3β-ol, the principal steroid in the bulk membrane. Adenylyl cyclase and cAR1 are integral membrane proteins, but several other integral membrane proteins are absent in the micro-domains, as are the three major phospholipids. One peripheral protein, the membrane-bound phosphodiesterase, is found in this fraction, located in close proximity to the cAMP receptor. It is salt-extractable, and may be bound to a separate integral membrane protein that serves as a transmembrane anchor. Thus, *Dictyostelium* amoebae may have a micro-domain that traps proteins that are important to signaling within a small patch of sterol. This introduces the idea that the plasma membranes of these amoebae deviate from a fluid mosaic model of protein diffusion. The micro-domains must be evenly and frequently dispersed over the membrane because immunocytochemistry with antibodies to the cAMP receptor cAR1 reveal an even distribution over the surface of fixed cells (Xiao and Devreotes, 1997).

5.2 Channels and pumps of the plasma membrane

Biochemical, electrophysiological, and molecular studies have revealed that the plasma membrane has several known channels, including a proton exporter (Gross *et al.*, 1988; Pogge-von Strandmann *et al.*, 1984; van Duijn and Vogelzang, 1989), a K^+ channel (Müller and Hartung, 1990; Müller *et al.*, 1986), and at least one important calcium channel. The calcium channel is regulated by multiple receptors (Milne and Coukell, 1991). Upon stimulation with cAMP, cells rapidly secrete protons and K^+ (Aeckerle and Malchow, 1989; Aeckerle *et al.*, 1985; Malchow *et al.*, 1978a,b). The plasma membrane contains a P-type H^+-ATPase, which serves to pump protons out of the cell. This pump is essential for survival in acidic conditions, which are commonly faced by free-living soil amoebae (Coukell *et al.*, 1997). It must be one of several such pumps, because its deletion by homologous recombination does not eliminate the hydrogen ion secretion from the cells. Calcium entry is stimulated by chemoattractants, including cAMP (Nebl and Fisher, 1997; Verkerke-van Wijk and Schaap, 1997). Curiously, this occurs without the involvement of GTP- binding proteins, despite the fact that the cAR receptors are G protein-coupled receptors (Milne and Coukell, 1991). Calcium is also released from intracellular stores and evokes a number of responses, which will be described in their developmental context. Calcium is gradually removed from the cell by the action of a Ca^{++}-ATPase to restore resting levels (Böhme *et al.*, 1987).

Another membrane pump of interest to tumor cell biology, which has been detected in *D. discoideum* is TagB, a homologue of the multidrug resistance transporter gene (MDR) (Shaulsky *et al.*, 1995). TagB, which plays a critical role during development (see below), was discovered by a REMI screening procedure (Shaulsky *et al.*, 1995). Its structure identified it as potential molecular transporter, although to date its transport ability has not been directly demonstrated. We will describe TagB further when we discuss one of the last events in development, culmination, where it plays a critical role.

5.3 Membrane systems that transiently connect to the plasma membrane

Water regulation is critical to survival of soil amoebae, and *D. discoideum* has a rich system of contractile vacuoles which it uses to control its osmotic status (Heuser *et al.*, 1993; Nolta and Steck, 1994; Steck *et al.*, 1997). There are subsurface cisternae with radiating tubules, as shown in Fig. 5.1 by Heuser and colleagues (Clarke and Heuser, 1997; Heuser *et al.*, 1993). The tubules are postulated to fill with water and swell (see Fig. 5.1B). The appearance of the tubules depends on the osmotic conditions – in a hyperosmotic environment, the cisternae are flattened and have many narrow tubular extensions. Under hypo-osmotic conditions, the cisternae become swollen and are incorporated into a growing vacuole, which then fuses with the plasma membrane to discharge excess fluid. The system of vacuoles and tubules is continuous and is not – as has previously been postulated – the result of vesicle fusion (Nolta and Steck, 1994). As the micrographs in Fig. 5.1 show, the central vacuole and the radiating tubules are studded with many 15-nm pegs, which are proton pumps – the vacuolar H^+-ATPase (Bracco et al., 1997; Clarke and Heuser, 1997; Heuser et al., 1993; Liu and Clarke, 1996). The ATPase is homologous to those in higher organisms, and reacts with sera prepared against the mammalian form of the enzyme.

Heuser et al. reported the presence of carbonic anhydrase in cytoplasmic extracts, and they suggested a mechanism in which HCO_3^- is the counter-ion during the acidification of the vacuoles, generating osmotically active carbonic acid and its dissociation products, which could draw water into the vacuoles (Clarke and Heuser, 1997; Heuser et al., 1993). This hypothesis is supported by the discovery of a bicarbonate transporter (Giglione and Gross, 1995). Steck et al. (1997) demonstrated that certain amino acids are excreted by the cells in response to hypotonicity. The vacuole system also stains with antibodies to calmodulin (Zhu and Clarke, 1992) and has a P-type-Ca^{++}-ATPase in contractile vacuoles, indicating that it may function in calcium homeostasis as well as water homeostasis (Coukell et al., 1997; Moniakis et al., 1995). An excellent view of amoebae filling their vesicles with water and expelling it can be found on the homepage of the Heuser laboratory (www.heuserlab.wustl.edu). This process is shown in real time with labeled cells.

The amoebae also respond to osmotic stress by mobilizing their cytoskeletons (Insall, 1996; Kuwayama et al., 1996; Kuwayama and van Haastert, 1998; Rivero et al., 1996b). Hyperosmotic stress causes D. discoideum cells to reduce their volume and to round up, as if in a defensive posture, this occurring within a few minutes. Actin and myosin are redistributed in this process to line the cell cortex, but there is no dramatic synthesis of new proteins observed on two-dimensional gels (S. C. Schuster, personal communication). Hisactophilin, a pH-sensitive actin-binding protein, changes its distribution as the cells rearrange their cytoskeletons (Stoeckelhuber et al., 1996; S. C. Schuster, personal communication). Guanylyl cyclase is stimulated by hyperosmolar conditions, and the resulting cGMP production is known to mobilize the cytoskeleton, particularly by phosphorylation of myosin II (Kuwayama et al., 1996; Newell, 1995b). Thus increased rigidity of the cells may also have evolved as a protective measure. Several proteins have been described which play a critical role, in addition to guanylyl cyclase. One of these is DokA, which is a homologue of a bacterial histidine kinase family. These signal transducing elements are common in D. discoideum, bacteria, fungi, and plants, but have not been described in animals. The hybrid kinases contain a sensor domain, a histidine kinase catalytic domain, and a response-regulator domain. In these proteins there is an intramolecular autophosphorylation on histidine and transfer to the aspar-

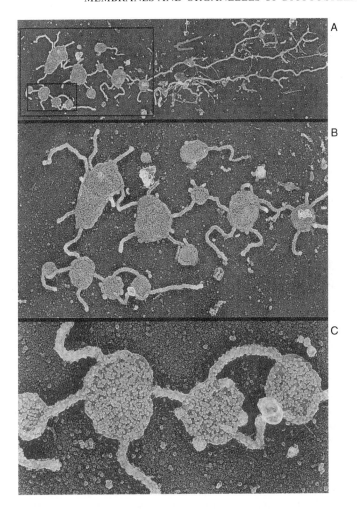

Figure 5.1 A deep etch view of the contractile vacuole system. In this figure, the entire contractile system has been squeezed out of an axenic amoeba. Panel A shows that there are flattened cisternae connected by narrow tubules (see also Plate 1). The entire panel stretches over 15 μm and serves to illustrate that the system is an extensive network, not merely one or two vesicles. Panel B shows that larger cisterns are connected to tubules by continuous membranes. These tubules blow up into full-fledged vacuoles under hypotonic conditions. The connected vesicles in panel B span 7 μm. Panel C shows that vesicles and tubules are studded with pegs that represent the proton pumps (Clarke and Heuser, 1997; Heuser *et al.*, 1993). (Illustrations courtesy of John Heuser.)

tyl residue of the response regulator part of the molecule. This then activates a variety of effectors (Loomis *et al.*, 1998). Absence or overproduction of DokA causes osmosensitivity (Schuster *et al.*, 1996).

Cells are also protected from sudden osmotic shock when they are encapsulated as spores. As we will see in a later chapter, amoebae emerge from their spore coats with a certain amount of caution. Among the elements of the environment that they sense are the osmotic conditions, with high osmolarity functioning as a signal to maintain dormancy. In the spore mass itself the conditions are extremely hyperosmolar. The maintenance of dormancy is controlled in part by a unique adenylyl cyclase. Mutants of the ACG (adenylyl cyclase of germination) gene germinate in hyperosmolar conditions, when the wild-type does not (van Es *et al.*, 1996).

5.4 Axenic cells feed by macropinocytosis

The amoebae are extraordinarily efficient phagocytes, with most strains capable of consuming a variety of bacteria (Depraitère and Darmon, 1978; Raper, 1937; Raper and Smith, 1939). For reasons of convenience, starting about 30 years ago, strains were created which were capable of growing on an axenic medium free of bacteria or particles (Sussman and Sussman, 1967; Watts and Ashworth, 1970; Franke and Kessin, 1977). While all strains can grow by phagocytosis of bacteria, only a few mutants can grow in the axenic or defined media, and they do so by fluid-phase uptake (Hacker *et al.*, 1997). This ability is conferred by several mutations in the original non-axenic strain (Williams *et al.*, 1974a). Addition of labeled fluid-phase markers, such as fluorescent dextrans, can be followed through the various compartments of the cells in pulse–chase experiments (see Plate 2). The compartments can be isolated by magnetic retention after feeding colloidal iron to the cells (Adessi *et al.*, 1995; Nolta *et al.*, 1994).

The compartments isolated after very short labeling periods present a slightly confusing picture, largely because there may be more than one means of initial uptake. An established uptake pathway involves clathrin-coated pits, which are 0.1 μm in diameter. Magnetic isolation studies by Nolta *et al.* (1994) found that vesicles loaded for 3 minutes were 0.1–0.2 μm in diameter. These turned over rapidly – the life of a primary pinosome was calculated to be less than 1 second – in contrast to animal cells where the coated vesicles are thought to persist for minutes, although there may be coated vesicles with different lifetimes in some mammalian cells. This presents a problem because the number of pinosomes observed by electron microscopy is not more than a few hundred. Given known rates of fluid uptake, one would have to predict many thousands of coated vesicles per cell. The idea that clathrin-coated pits mediate most of fluid-phase uptake depends on a singular role for clathrin in the formation of coated pits. In clathrin heavy chain-defective mutants of *D. discoideum*, the rate of fluid-phase uptake is reduced by about 80% (O'Halloran and Anderson, 1992). The large reduction may be due to defects in other clathrin-mediated events, which impede the vesicular economy of the cell. Thus, some of the 80% reduction could be indirect (Hacker *et al.*, 1997).

These authors proposed a solution to this paradox by postulating that most of fluid-phase uptake is mediated by larger structures, termed macropinosomes, which are 0.6 μm in diameter (de Hostos *et al.*, 1991; Hacker *et al.*,

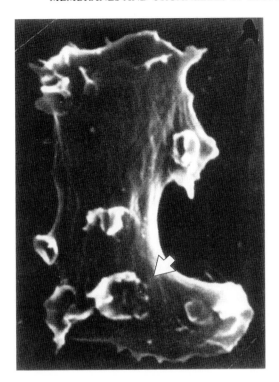

Figure 5.2 Ruffles transformed into macropinosomes. These scanning electron micrographs show crowns on the surface of a cell. See also Plate 2. Fluid is taken up in the closing crown structure. The actin-binding protein coronin is located at the tips of the crowns, and null mutants lacking coronin are defective in fluid uptake by macropinocytosis. Other cytoskeletal proteins also accumulate in the crowns (Hacker *et al.*, 1997). (Reprinted with permission of the *Journal of Cell Science* and the Company of Biologists, Ltd.)

1997; Thilo and Vogel, 1980). The genesis of this vacuole is clathrin-independent, and depends on an actin ruffle which protrudes from the cell and fuses at the tips to form a vesicle. These structures, called crowns, are shown in Fig. 5.2 and are observed in the axenic cells. They depend on actin, myosin I, and the cytoskeletal protein coronin (see Chapter 6). Because they are so much larger than pinosomes, the formation of macropinosomes from only a few crowns per minute would be enough to account for all of fluid-phase uptake. Since the macropinosomes are few in number, they could be missed in magnetic fractionation studies, thus accounting for the discrepancies. Endocytosis in *D. discoideum* has recently been reviewed (Aubry *et al.*, 1997). The events that generate crown structures constitute a pathway that is known to share components with phagocytosis, such as a requirement for the participation of coronin in the generation of the crown.

A number of mutants that affect endocytosis or phagocytosis have been recovered by elegant selection procedures (Bacon *et al.*, 1994; Cohen *et al.*,

1994; Labrousse and Satre, 1997). One procedure employed a photoactivatable dye, which kills amoebae that take it up, leaving mutant amoebae to survive. These endocytosis mutants fall into a limited number of complementation groups, but since they were prepared by chemical mutagenesis, their genes have not been recovered (Labrousse and Satre, 1997). The other method employed fluorescent bacteria which were fed to the amoebae. Those amoebae that did not take up the bacteria were not labeled and were recovered after cell sorting. These mutant strains could still grow by macropinocytosis, which makes the mutations conditional and therefore they could be recovered after REMI mutagenesis – which would allow access to the affected genes.

There is a curious effect of bisphosphonate drugs – the pharmacological agents used to treat osteoporosis – on the growth of *D. discoideum* amoebae, potentially affecting macropinocytosis. The cells that degrade bone are called osteoclasts and, except for their extraordinary phagocytic activity and motility, they might seem to have little in common with *D. discoideum* amoebae. Yet the series of bisphosphonate analogues each has an identical potency in *D. discoideum* and in osteoclasts. The modifications of these pyrophosphate analogues that inhibit *D. discoideum* growth at low concentrations are the most effective at inhibition of osteoclasts (Brown *et al.*, 1998; Rogers *et al.*, 1995). The site of action of these drugs is not known, but the pharmacological results suggest that it is the same in the amoebae as in the osteoclasts. The use of REMI methods could yield this target.

5.5 Phagocytosis

Amoebae ingest a variety of prey by coordinating surface recognition, signal transduction, and mobilization of the cytoskeleton. A comparison with pinocytosis is shown in Plate 2. A number of mutants that are affected in phagocytosis have been isolated, and these have been used to discern some of the properties (but not yet any details) of the mechanism (Bacon *et al.*, 1994; Cohen *et al.*, 1994; Vogel *et al.*, 1980). From results with mutants defective in phagocytosis, Vogel and colleagues have postulated that there are at least two types of receptor that initiate phagocytosis. One is non-specific, and the other is specific for glucose residues on any extracellular particle, including bacteria. The specific receptor has been called a lectin-type receptor because of its ability to recognize terminal sugars. The non-specific receptor also recognizes bacteria, but can be mutated. The mutants rely on the lectin-type receptor, and no phagocytosis occurs in the presence of glucose.

No genes that code for phagocytosis receptors have been cloned, despite the mutant collections used above. It should be possible to use REMI mutagenesis to recover such genes, if a way can be found to allow conditional growth of the amoebae. The fact that there are apparently two receptors makes this possible. A problem in such work is that a multiplicity of genes is involved in phagocytosis. In addition to receptors, phagocytosis also depends on the cytoskeleton. A variety of cytoskeletal mutants affect phagocytosis without necessarily blocking it (Chadwick *et al.*, 1984; Chia, 1996; Cox *et al.*, 1996; Fukui and Imamoto, 1983; Furukawa *et al.*, 1992; Maniak *et al.*, 1995; Simmonds *et al.*, 1987). The cytoskeleton is discussed in Chapter 6.

During phagocytosis, the amount of membrane and the total volume of ingested liquid must be returned to the cell surface. As Steck and his colleagues have pointed out, in animal cells such restoration occurs at a pre-lysosomal phase of the endocytic pathway (Padh *et al.*, 1993). Post-lysosomal excretion in mammalian cells (other than macrophages) may not occur at a high rate, with indigestible material remaining in the cell. *D. discoideum* is faced with the problem of undegradable cell wall material from a variety of species of bacteria. These husks are returned to the surface through a distinctive class of large and internally neutral post-lysosomal vacuoles. Similar mechanisms operate in the protists *Acanthamoeba* and *Entamoeba* and the ciliates *Tetrahymena* and *Paramecium* (Padh *et al.*, 1993).

5.6 Lysosomes

Whatever their origin or size, the early vesicles – labeled for as long as 3 minutes – had not yet been converted to lysosomes because these bodies contained no acid phosphatases, no intrinsic acidity, and no vacuolar H^+-ATPase. The maturation process continues with the binding of adaptins and the conversion of pre-lysosomal vesicles to lysosomes between 3 and 15 minutes after labeling (Nolta *et al.*, 1994). After 15 minutes, the labeled compartments are acidic, with a high activity of vacuolar H^+-ATPase. The lysosomal compartment has a unique lipid called BMP (bis(monoacylglycerol)phosphate), which also marks mammalian lysosomes (Rodriguez-Paris *et al.*, 1993). Using magnetic techniques, a number of lysosomal proteins have been isolated and sequenced (Adessi *et al.*, 1995; Rodriguez-Paris *et al.*, 1993; Temesvari *et al.*, 1994). In addition, the membrane components of lysosomes have been heavily enriched and shown to contain several Rab-like small GTPases (Temesvari *et al.*, 1994). Other lysosomal constituents, including various proteinases, a mannosidase and a glucuronidase, have previously been isolated and their genes have been cloned (Bush *et al.*, 1994). When colloidal iron techniques are used with short-term loading, lysosomes were found to be physically homogeneous, between 0.4 and 1.1 μm in diameter, and frequently bound to one another. With long-term loading, it has been determined that the cell commits about 3.3% of cell protein and 11% of lipid to the endocytic pathway (Rodriguez-Paris *et al.*, 1993).

The targeting of enzymes to the lysosome follows a standard route, but with certain variations. In higher organisms, lysosomal enzymes are targeted by mannose-6-phosphate (Freeze and Wolgast, 1986). *D. discoideum* lysosomal enzymes are bound by the mammalian mannose-6-phosphate receptor *in vitro*, but the cells apparently do not have such a receptor itself. The lysosomal enzymes of these organisms have other modifications, and recent work has shown that in addition to the asparagine-linked oligosaccharides, there is a class of lysosomal enzyme which contains N-acetylglucosamine-1-phosphate linked to serine. Remarkably, confocal microscopy has revealed that the two sets of enzymes lie in distinct structures (Freeze, 1997; Souza *et al.*, 1997). Vesicles with N-acetylglucosamine-1-phosphate fuse with bacteria-loaded phagosomes less than 3 minutes after ingestion, whereas the fusion of mannose-6-phosphate-containing vesicles does not occur until 15 minutes later. The fusion occurs with separate phagosomes, so that there appear to be at least two types of acidic digestive compartments (Souza *et al.*, 1997).

A number of mutants of the glycosylation pathways exist, and these result in N-linked oligosaccharides that are under-sulfated or under-phosphorylated. Freeze, Cardelli and their colleagues have used these mutants to ask whether α-mannosidase or β-glucosidase are properly targeted to lysosomes (Freeze *et al.*, 1989). Targeting is not affected, but is less efficient with the altered N-linked oligosaccharides. A review by Freeze (1997) gives an excellent summary of the status of glycosylation and lysosomal protein targeting. The type of sugars added to N-linked oligosaccharides and their further modification by sulfation and phosphorylation changes during development (Ivatt *et al.*, 1984).

5.7 Endoplasmic reticulum, Golgi, and nuclei

D. discoideum has an endoplasmic reticulum (ER) that is morphologically similar to that of other organisms (Monnat *et al.*, 1997). It contains several of the expected enzymes, including protein disulfide isomerase (Monnat *et al.*, 1997; Sarkar and Steck, 1997) and enzymes involved in glycosylation (Freeze, 1997). Curiously, the protein disulfide isomerase does not contain an ER signal KDEL, raising questions about how the enzyme is retained. Mutations in the genes that code for lysosomal enzymes or the gene products involved in glycosylation cause a retention in the ER (Woychik and Dimond, 1987). The organism also contains an active Golgi apparatus. Perhaps this is best seen in another cellular slime mold *Fonticula alba*, where it participates in the massive synthesis and regulated exocytosis that occurs during the formation of the spore coat (Deasey and Olive, 1981). In *D. discoideum* the Golgi apparatus is best visualized by staining for comitin, a protein which resides on the outside of the Golgi vesicles and also binds F-actin, perhaps providing an exo-skeleton for the organization of Golgi vesicles. See Plate 1. Immunofluorescence staining reveals a distinct peri-nuclear Golgi apparatus (Weiner *et al.*, 1993).

The nuclei are small, with pronounced nucleoli. A nucleosomal structure is maintained in chromatin, which has basic histones that are related (but not identical) to those of higher eukaryotes (Arakane and Maeda, 1997; Bukenberger *et al.*, 1997; Coukell and Walker, 1973; Garside and MacLean, 1987; Hauser *et al.*, 1995; Heads *et al.*, 1992). No structural studies have been done of nuclear pores or of the nuclear membrane, although pores are observed (R. Chisholm, personal communication).

5.8   Mitochondria and peroxisomes

D. discoideum is an aerobic organism and contains a large number of mitochondria. The mitochondrial genome has been completely sequenced, and the
details have been discussed in Section 4.4. The regulation of the nuclear-
encoded cytochrome C oxidase has been studied during growth and development (Sandona *et al.*, 1995). It is structurally the most simple of the eukaryotic
cytochrome C oxidases. One of the subunits – VII – varies with oxygen tension.
Subunit VIIe is replaced by a larger subunit under hypoxic conditions, where
development is blocked (Sandona *et al.*, 1995). The gene coding for a topoisomerase localized to mitochondria has been cloned (Komori *et al.*, 1997a). It
contains an N-terminal mitochondrial targeting signal. Deletion of the signal
prevents the enzyme from reaching the lumen of the mitochondria, but this
does not appear to have an effect on viability or development (Komori *et al.*,
1997b). The *cluA* gene encodes a 150-kDa protein that is required for the
dispersion of mitochondria (Zhu *et al.*, 1997). It is a functional homologue
of the yeast *CLU1* gene (Fields *et al.*, 1998). An actin-related protein has been
identified that binds to mitochondria, but its function is unknown (Murgia *et
al.*, 1995). Disruption of the large ribosomal rRNA gene in a subpopulation of
the mitochondria impairs signal transduction in slugs (Wilczynska *et al.*, 1997).

   D. discoideum has been reported to contain peroxisomes which contain
catalase and urate oxidase, but these have been little studied (Hayashi and
Suga, 1978; Parish, 1975).

5.9   The autophagic vacuole

During the development of spore and stalk cells, each acquires a characteristic
membrane organelle. In the prespore cells, prespore vesicles form and their
contents form the spore wall. We will discuss these when we come to culmination and spores. In the prestalk cells, there is an autophagic vacuole that digests
cellular contents and is specific to prestalk cells. This acidic organelle accumulates the often used dye neutral red, which was the original marker for prestalk
cells (Bonner, 1952b; Yamamoto and Takeuchi, 1983). In a study that
employed light and electron microscopy, Yamamoto and colleagues found
that during the loose aggregate stage of development, food vacuoles are lost
and autophagic vacuoles increase in number, initially in all cells, but after tip
formation, only in the prestalk cells. The autophagic vacuoles of prestalk cells
then become larger (De Chastellier and Ryter, 1977, 1980; Yamamoto *et al.*,
1981).

   The cell and structural biology of *D. discoideum* is a nearly unlimited area of
research which should be enriched by the forthcoming genome sequence, much
of which will reveal genes dedicated to the survival of vegetative cells. The
development of complementation techniques will be a powerful spur to the
study of vegetative functions.

6

Cell Motility and the Cytoskeleton

If *Dictyostelium* amoebae are not the Ferraris of moving cells, they are at least a respectable entrant in the cell motility Grand Prix. Amoebae can move as fast as 10–15 µm/min (Varnum and Soll, 1984). They do not do this in a random walk, but move directionally, up gradients of cAMP or folate. *D. discoideum* responds to gradients that vary by as little as 2% from the front to the back of a cell, as we will describe in Chapter 8. This astonishing chemical sensitivity raises a number of questions. How are signal transduction pathways connected to the cytoskeleton such that stimulation of chemotactic receptors leads to movement up a chemical gradient? How are the components of the cytoskeleton organized to promote sudden movement of cells? How does the mobilization of the cytoskeleton differ in the various protrusions, such as pseudopodia and filopodia? How are responses terminated? In this chapter we will review how the actin-based cytoskeleton is mobilized during cell movement, while in Chapter 8 we will consider how it reorganizes to drive the cells up a gradient of cAMP, toward centers of aggregation.

Dictyostelium is one of the few organisms with impressive motility and tractable genetics. The cells move toward folate during growth and toward cAMP during development. The amoebae are useful for optical observation, so that with a few tricks, the movements of macromolecules within the cells can be observed by a variety of microscopic methods. Above all, the cells can be manipulated genetically and the mutational analysis of components of the actin and myosin cytoskeleton has produced several surprises. The first surprise was that there is only one myosin II gene, and the cells can do without it. Amoebae without myosin II live as long as they are bound to a substratum where motility can lead to cytofission, which overcomes their defect in cytokinesis. They become multinuclear and die in suspension cultures. Thus the mutant is conditional – a circumstance that has been beautifully exploited in a dissection of the properties of the myosin molecule.

The second surprise was that the elements of the cytoskeleton, whether actin-binding proteins or unconventional myosins, have overlapping functions. The cells survive without individual proteins. Their fitness may be reduced and, with careful imaging analysis, differences between mutant cells and their isogenic parents may be apparent, but in many cases, mutant cells grow and develop under laboratory conditions. The cytoskeleton forms from many elements and these are dynamic, constantly being redeployed in the formation of macropinosomes, phagosomes, filopodia, pseudopodia, ruffles, substrate contacts, the mitotic spindle, or the contractile ring of cytokinesis. Many proteins are reconfigured to do different tasks, and this presents a problem for genetic analysis because the effects of a mutation may be pleiotropic – affecting several processes. We have seen this problem in Chapter 4 in the discussion of clathrin mutants. Despite partial redundancy and pleiotropy, the *Dictyostelium* cytoskeleton presents opportunities that have been exploited by many laboratories, so that the organism has become one of the most used in studies of motility. All of the cytoskeletal elements of *Dictyostelium* are also found in mammalian cells, and it is often illuminating to study the same protein in *Dictyostelium* and in mammalian cells (Zigmond *et al.*, 1997).

The study of motility and the complexity of the components that drive cell movement have been revolutionized by the advent of green fluorescent protein (GFP)-labeling methods, which permit an assessment of subcellular localization in living cells in real time. This technology has been widely introduced in *Dictyostelium* and in mammalian cells, and in many cases is being used to repeat experiments that were previously done by immunocytochemistry on fixed cells. The result is a greater appreciation of the dynamism of the cytoskeleton and the transient localization of some proteins.

6.1 Actin and its binding proteins

An organized approach to the actin cytoskeleton must start with the two main players – actin and the myosins – because most other proteins play supporting roles, as handlers and manipulators of actin or myosin. Actin represents about 8% of cell protein and its concentration is about 0.1 mM (Podolski and Steck, 1990) – the result of the concerted action of 17–20 actin genes (McKeown *et al.*, 1978; Romans and Firtel, 1985). These genes may be differentially transcribed, but the actin molecules that result are all nearly identical in sequence. Posttranslational modifications include tyrosine phosphorylation, which has been correlated with cell shape change, and acylation, the significance of which is not known.

Actin-binding proteins bind monomers, attach the actin filaments to each other, bind to their ends, sever them, or tie them to the membrane (Schafer and Cooper, 1995). Actin is in a dynamic equilibrium between the monomeric G-actin and the filamentous F-actin. The filaments of actin have polarity and the addition of monomeric G-actin to a filament depends on two events – generation of free barbed ends and increasing the free monomer pool. The barbed ends grow rapidly and are the site of actin assembly *in vivo*. The concentration of monomeric G-actin is also controlled by a series of sequestering proteins. If it were not for the restraining action of these proteins, almost all actin would exist as filaments. The G- and F-actin are found in approximately equal amounts in resting cells, but in chemotactically stimulated cells, the amount of F-actin increases by 50–60% within a few seconds (McRobbie and Newell, 1985a) (see Chapter 8). It is the generation of barbed ends that is thought to be the most critical event.

6.1.1 G-actin-binding proteins

There are a number of regulators of actin assembly and disassembly. We can sort actin-binding proteins on the basis of whether they interact with G-actin or with F-actin, as the summary shown in Fig. 6.1 indicates. For reviews, see Fukui (1993), Furukawa and Fechheimer (1997), Karakesisoglou *et al.* (1999), Noegel *et al.* (1995), and Noegel and Luna (1995). Molecules that sequester G-actin prevent the growth of actin filaments and in *Dictyostelium*, the two best known are profilin I and profilin II, which are conserved in other organisms (Haugwitz *et al.*, 1991). In *Dictyostelium*, deletion of both profilin genes leads to defects in growth, motility, and cytokinesis that are likely to be the result of altered G/F-actin ratios (Haugwitz *et al.*, 1994). Among the other G-actin-sequestering proteins is CAP (cyclase associated protein), which is localized in the leading fronts of cells (Gottwald *et al.*, 1996) and may be involved with the mobilization of microfilaments near the plasma membrane. The profilins and CAP bind to phosphatidylinositol 4,5-bisphosphate (PIP_2), but the role of PIP_2 as a modulator of the actin cytoskeleton is not well established in *Dictyostelium* (Gottwald *et al.*, 1996; Haugwitz *et al.*, 1994). We do not know whether CAP has other functions. Figure 6.1 shows a third G-actin-binding protein, cofilin, but this protein may bind both forms of actin (Aizawa *et al.*, 1997). It participates in pseudopod formation following cAMP stimulation, as shown by GFP–cofilin fusions and since no knockout mutant has been produced, it may be required for survival (Aizawa *et al.*, 1995). Cofilin is part of a large family of conserved proteins.

6.1.2 F-actin-binding proteins

A large array of proteins acts on the filaments of F-actin, either to cross-link them or to cap and sever them. Following Fig. 6.1, we note that several proteins are dedicated to the capping and severing of actin filaments. Cap32/34 is a ubiquitous actin-binding protein that caps the barbed ends of actin filaments (Eddy *et al.*, 1996; Haus *et al.*, 1993). Capping, as in the case of cofilin or

Actin-binding proteins in *Dictyostelium*

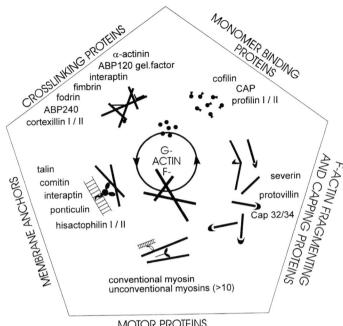

Figure 6.1 The actin-binding proteins. The molar ratios of F- and G-actins are approximately 1:1 and are kept under regulation by monomer binding proteins as well as by proteins that can cap or sever the actin filaments, exposing barbed ends. Capping proteins sequester the barbed ends and prevent expansion. Single F-actin filaments are weak, and a variety of cross-linking proteins bind the filaments, with an increase in viscosity and elasticity. These binding proteins have complex domain structures that are critical for actin binding, subcellular localization, and rapid regulation. The micro-filament networks produced may vary. Membrane anchoring proteins tie actin to the plasma membrane or the membranes of vesicles. Myosin II and perhaps the myosin Is act as motors. (Courtesy of Michael Schleicher, Ludwig-Maximilians-Universität, Munich. Reproduced from *Microscopy Research and Technique*. Published by permission of Wiley-Liss, Inc., a subsidiary of John Wiley & Sons, Inc.)

profilin, is sensitive to PIP_2 levels (Haus *et al.*, 1991). The apparent function of Cap32/34 is to keep filaments short and develop a meshwork of actin fibers.

Binding of agonists to the cell surface results in a very rapid rearrangement of the cytoskeleton, as demonstrated in Chapter 8. One potential participant in these reformulations of the cell's architecture is severin, an F-actin-severing protein and a homologue of human gelsolin (Eichinger *et al.*, 1998; Eichinger and Schleicher, 1992). These calcium-sensitive proteins sever actin filaments at low calcium concentrations, not by covalent bond disruption, but rather by severing the non-covalent bonds between monomers. After the severing, the proteins bind the barbed end of the actin filament and block elongation and recruitment from the G-actin pool. The resulting increase in short filaments leads to a dramatic decrease in viscosity. Phosphorylation of severin may participate in the rapid rearrangement of the cytoskeleton, converting localized gelled actin into a soluble state. Disruption of the severin gene does not produce cells with a dramatic phenotype (André *et al.*, 1989). Perhaps other severing proteins, such as cofilin (Aizawa *et al.*, 1997) compensate for the loss of severin (Hofmann *et al.*, 1993). Alternatively, the phenotype of a severin mutant may be subtle and may have escaped notice. By careful microscopic analysis (Soll and Wessels, 1998), or by assays in which the development of the mutant is made more challenging, severin mutants may reveal a phenotype (Ponte *et al.*, 1998). For a full treatment of the various actin-binding proteins, the reader should consult the reviews quoted above, but one particularly interesting protein, previously thought to act only in translation, is EF-1α, which serves to localize polysomes to the cytoskeleton (Liu *et al.*, 1996).

6.1.3 *Cross-linking proteins*

Because of the low tensile strength of an individual actin filament, many of the effects of actin are mediated by the creation of a three-dimensional cortical network of many filaments. The proteins that mediate this architecture are cross-linkers and bundling proteins. These proteins can bind actin filaments into orthogonal or into parallel arrays. By definition, a cross-linking protein must have two actin-binding sites, and these can be located in tandem on one polypeptide or created by dimerization in parallel or anti-parallel fashion (Eichinger *et al.*, 1999). The actin cross-linking proteins have a modular structure with a domain of about 250 amino acids implicated in actin binding and other domains that are required for multimerization, membrane binding, or regulation by sudden changes in calcium concentration. A variety of actin cross-linking proteins are listed in Fig. 6.1. These are described more fully in the following reviews, which contain the appropriate references (Furukawa and Fechheimer, 1997; Karakesisoglou *et al.*, 1999; Noegel *et al.*, 1995; Noegel and Luna, 1995).

One of the best characterized of the actin cross-linking proteins is ABP-120, a homodimer, with each subunit arranged in an anti-parallel orientation. ABP-120 is a member of the filamin family, and is in part responsible for maintaining orthogonal arrays of actin filaments. ABP-120 has been eliminated by homologous recombination, with a resulting non-lethal defect in motility (Cox *et al.*, 1992). In addition, there are cytoskeletal and phagocytosis defects – all of which are corrected when the gene encoding ABP-120 is restored to the cells (Cox *et al.*, 1996).

Taking advantage of the gene knockout technology, a number of mutants have been created to eliminate various other actin-binding proteins, including severin and α-actinin (Noegel and Witke, 1988). The effects of single mutants have been subtle, at least with the techniques used to characterize phenotypes (Noegel *et al.*, 1997; Rivero *et al.*, 1996a). The ABP-120 work stands out because of the elaborate biochemical analysis of actin networks, the phenotypic analysis, and the fact that all defects were restored by complementation with the wild-type gene. The advanced microscopy and image analysis system developed by Soll and colleagues has been invaluable in detecting differences in speed and pseudopod extension rates, among numerous other parameters, in ABP-120 and other mutants, including those affecting the small myosins (Soll and Wessels, 1998; Wessels *et al.*, 1998). Geneticists are trained to want mutants that have dramatic phenotypes. Evolution is more patient, and works with an increase in reproductive fitness of 0.001/generation or much less. A slight increase in fitness of a cell due to expansion of a multigene family is sufficient to explain the presence of the gene and its product. Molecules like the various actin-binding proteins or the small myosins probably produce an incremental increase in fitness. Our efforts at mutational deconstruction may return the cells to a non-lethal *status quo ante*.

6.1.4 *Attaching to the membrane*

In addition to the molecules that control the length and cross-linking of the actin filaments, a number of proteins secure the filaments of the cortical actin network to the cell membrane. These include ponticulin, talin, comitin, interaptin, and hisactophilin, all of which are important regulators of cortical–membrane interactions. Hisactophilin, which has two isoforms, binds actin filaments to the plasma membrane in a pH-dependent manner (Scheel *et al.*, 1989). A current hypothesis is that hisactophilin serves as a pH sensor by reversibly connecting the cortical actin network to the plasma membrane in response to changes in the proton concentration in very localized areas of the cell (Hanakam *et al.*, 1996).

Ponticulin is an integral membrane protein that serves to nucleate actin assembly and may be a major link between the actin cytoskeleton and the plasma membrane (Hitt *et al.*, 1994). Ponticulin mutants are affected in actin binding to the membrane, as observed by microscopy and membrane purification, but the loss of the ponticulin gene is not lethal. Rather, the role of ponticulin may be the stabilization of pseudopods (Shutt *et al.*, 1995). The *Dictyostelium* homologue of talin binds the membrane and appears to be enriched in filopodia, as is F-actin (Kreitmeier *et al.*, 1995). This enrichment occurs at the migrating fronts of cells within 30 seconds of chemotactic stimulation. The signal transduction pathways leading from the cell surface cAMP receptor to the relocalization of talin and other cytoskeletal molecules are now subjects of intense investigation. Talin's multiple roles are illustrated by a mutant that is defective in phagocytosis and in cell–substrate adhesion (Niewohner *et al.*, 1997). Actin-binding proteins connect membranes other than the plasma membrane to the cytoskeleton. Comitin binds the actin cytoskeleton to intracellular membranes (Weiner *et al.*, 1993), while interaptin binds the cytoskeleton to the endoplasmic reticulum and the cytoskeleton (Rivero *et al.*, 1998).

6.2 Myosin motors – conventional myosin

Dictyostelium has one conventional myosin (myosin II) gene and as many as ten unconventional myosins, all of which convert chemical into mechanical energy. Deleting the conventional myosin leads to a genetically satisfying cellular calamity, but pin-pointing the roles of the unconventional myosins has proved to be more difficult.

Myosin II is one of the more thoroughly studied molecules in biology (Fig. 6.2). It is a hexamer consisting of paired myosin heavy chains, two regulatory light chains, and two essential light chains. The light chains bind to the head of myosin and play structural and signal reception roles. The head of the molecule is responsible for the movement stroke that turns the chemical energy of ATP into the mechanical energy that leads to the movement of the myosin head along an actin filament (Ruppel and Spudich, 1996a). The three-dimensional structure of the *Dictyostelium* myosin head has been solved, leading to a precise model of the ATP and actin-binding sites, and the location of the light chains. The structure of *Dictyostelium* myosin is very similar to that of mammalian myosin. The parts of the molecule that act as a lever, pulling it along the actin filaments, have been located (Bauer *et al.*, 1997; Rayment *et al.*, 1996; Ruppel and Spudich, 1996b; Spudich *et al.*, 1995; Uyeda *et al.*, 1996).

The tail of the myosin II molecule is a coiled-coil dimer, and is responsible for its assembly into arrays. While these are not as extensive in amoebae as they are in skeletal or smooth muscle, the tails still perform this function (Moores and Spudich, 1998). Its assembly is dependent on phosphorylation by the action of a specific myosin heavy chain kinase (Egelhoff *et al.*, 1993). The phosphorylation of the heavy chain tail is necessary to maintain the polarity of the cells, and also suppresses turning during chemotaxis, so that the cells migrate directly up gradients (Stites *et al.*, 1998). When cells are stimulated with cAMP, myosin concentrates in the cell cortex around the entire periphery of the cell and causes the cell to change shape, rounding up in what is called the cringe response (Futrelle *et al.*, 1982), which soon ends as the cells begin to extend pseudopods. Despite its role in motility, myosin II is not localized in extending pseudopods.

The myosin II gene was the first *Dictyostelium* gene to be inactivated by homologous recombination and astonishingly, although there is only one conventional myosin gene in *Dictyostelium*, inactivation did not kill the cells (De Lozanne and Spudich, 1987; Knecht and Loomis, 1987; Manstein *et al.*, 1989). Deletion of the gene leads to defects in cytokinesis such that multinucleate cells appear. This defect causes a failure of growth when cells are grown in the normal way, shaking in an axenic medium. What prevents this mutation from being lethal is the fact that when the cells grow on a plastic substratum, they acquire enough traction to pull apart (Spudich, 1989).

The myosin II mutants have been useful for many studies, but the initial revelation was that conventional myosin is not necessary for the formation of

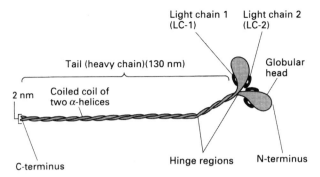

Figure 6.2 The structure of myosin II. The globular head domain acts as a lever arm. It binds regulatory and essential light chains, as well as actin. The rod-like tail is the site of interaction with other myosin II molecules, the assembly of which is regulated by phosphorylation (Warrick and Spudich, 1987). (Reprinted with permission of Cold Spring Harbor Laboratory Press.)

pseudopods. The multinucleate cells extend pseudopods and aggregate. They are not robust in this activity and do not move as well as normal cells, but there does not seem to be an absolute requirement for conventional myosin in the complex process of pseudopod formation or retraction. The critical role is in cytokinesis.

The multinucleate cells of the myosin II mutant can be rescued with exogenous plasmids coding for the myosin molecule – such transformants grow in shaking culture and carry out normal cytokinesis (Egelhoff *et al.*, 1990). Having established that complementation of the myosin II mutant occurs, it became possible to do detailed structure–function studies and to determine which residues are necessary for the lever arm action of myosin II. The effects of amino acid substitutions in the molecule can be dissected with a knowledge of the three-dimensional structure (Spudich *et al.*, 1995). The organism lends itself to structure–function studies of the domains of critical proteins, as we will see in Chapter 8, when we consider the structure of the cAMP receptor, cAR1.

The biochemical facility offered by *Dictyostelium* allows isolation of mutant myosins and their examination as motors in elegant *in vitro* tests. Thus, various mutations have been created that are cold-sensitive or define sites of importance within the head or tail of the myosin II molecule (Patterson and Spudich, 1995; Ruppel and Spudich, 1996b). The lever arm of the myosin head has been shortened, and the length of the movement step that is produced along actin filaments has been tested by a combination of biochemistry and computer-based image analysis. The speed of translocation depends on the length of the lever arm (Ruppel and Spudich, 1995; Spudich *et al.*, 1995). The elegant use of the genetics, biochemistry, and optical techniques has produced a detailed picture of a critical movement event – the progress of myosin along an actin filament.

6.3 Myosin motors – the unconventional myosins

Dictyostelium contains a number of unconventional myosins that lack the coiled-coil domain of conventional myosin and therefore do not associate in higher-order filaments. These are well-known molecules, first isolated from *Acanthamoeba*, and some at least are thought to act as motors. The globular domains of myosin I molecules have a high degree of homology with the motor domain of myosin II, but divergence occurs in the carboxy-terminal regions, which have a variety of recognizable motifs. There are at least 10 known myosin I molecules in *D. discoideum*. These have been grouped into large and small variants, and some of their genes have been knocked out by homologous recombination. As with actin-binding protein mutants, initial thoughts that the mutations of the various myosin I genes would be simple to interpret, were not correct. Their analysis has required complex image analysis (Soll and Wessels, 1998). For a review of the unconventional myosins from *Dictyostelium*, see Uyeda and Titus (1997), while for an assessment of these molecules in higher systems, see Mermall *et al.* (1998).

Image analysis has shown that movement velocities are reduced by about 50% in *myoA* and *myoB* mutants, but there is no other dramatic phenotype. Creation of double mutants, for example of *myoA/B*, *myoB/C* or *myoB/D*, does not make motility worse than the reduction observed in the single mutants (Jung *et al.*, 1996; Novak *et al.*, 1995; Soll and Wessels, 1998; Uyeda and Titus, 1997). Using image analysis, Wessels *et al.* found that *myoB*-null cells extended lateral pseudopods more frequently than wild-type cells, and this reduced chemotactic efficiency (Wessels *et al.*, 1991). There are effects on fluid-phase pinocytosis that appear to be additive. These subtle phenotypes make the mutants and the roles of the type I myosins difficult to study. For example, the kinds of structure–function studies that illuminated the subdomains of the myosin II molecule are not possible with the small myosins.

Overexpression of MyoB causes more profound defects in motility and manipulation of the actin cytoskeleton (Novak and Titus, 1997). These studies are difficult to interpret because the excess MyoB may simply tie up actin. A more convincing, if subtle, role of the myosin Is comes from cell localization studies. The localization of myosin I to the leading edge of chemotactic cells, filopodia, macropinocytic crowns, and the cortex suggests that the force of myosin I-like molecules can be exerted at any point in the cell. The molecules are important, even if no single myosin I is crucial for survival under laboratory conditions (Jung *et al.*, 1993; Morita *et al.*, 1996; Novak *et al.*, 1995).

A new structure containing a small myosin has recently been detected in migrating *Dictyostelium* amoebae. These are called "knobby feet" or eupodia, and are small projections under the leading edge of cells, which bind to a substrate. The movement of these feet can be observed by differential interference video microscopy. They contain MyoB distributed in a ring around each foot, but no myosin II. These structures are not adhesion plaques because they contain no stress fibers (Fukui and Inoue, 1997). These fine details can be detected by special techniques to flatten *Dictyostelium* cells and to make them more optically favorable (Fukui *et al.*, 1987).

6.4   Building and retracting the pseudopod

The study of pseudopod extension and retraction has a long history, which has been summarized by Condeelis (1992). The earliest idea was that the rear of the cell would contract due to the actions of myosin II and this would force the extension of a pseudopod at the opposite end of the cell. The presence of myosin II at the tail of the polarized cell is consistent with this idea (Rubino *et al.*, 1984; Yumura *et al.*, 1984). However, the tail contraction model suffered a fatal test when myosin II was eliminated by homologous recombination, and pseudopods were still extended (De Lozanne and Spudich, 1987; Knecht and Loomis, 1987).

Two additional models for producing protrusive force are depicted in Fig. 6.3. The first of these is the frontal contraction model, which supposes that the force of pseudopod extension is provided by an interaction with actin and unconventional myosins. At least one of the myosin I isoforms is present in newly created pseudopods (Titus *et al.*, 1993). Deletion of *myoB* does not block the formation of pseudopods. Instead, the pseudopods form more frequently than in the parental cells, suggesting a role for MyoB in restraining pseudopod formation (Wessels *et al.*, 1996). Since there remain a number of other myosin I isoforms, one cannot say that these do not participate. The other alternative is that the force of pseudopod extension may be driven by actin polymerization and cross-linking, without the critical participation of myosin I. Such a mechanism, called the cortical expansion model, is shown in Fig. 6.4 and has received support from experiments in a number of systems, including *Dictyostelium* (Condeelis, 1992, 1998; Soll and Wessels, 1998).

The increase in actin filament length occurs within a second after a cAMP stimulation, and therefore any model to explain local pseudopod extension must accommodate rapid signaling from receptors on the cell surface. Under *in vivo* conditions, actin filaments require barbed ends to grow, and it is the barbed ends that are the primary effectors of the actin cytoskeleton. Growth from the barbed ends of actin filaments can be stimulated in four ways – by freeing G-actin; by removing capping proteins from F-actin; by severing existing capped filaments; or by *de novo* nucleation (Condeelis, 1992, 1998). Most capping proteins, including the profilins described above, bind to PIP_2, and the generation of this phospholipid is a way to remove the capping proteins. The biochemistry of phosphoinositide synthesis is well worked out in *Dictyostelium* (van Haastert and van Dijken, 1997). However, the role of PIP_2 in sequestering capping proteins in *Dictyostelium* is not yet clearly demonstrated *in vivo*. Deletion of a phospholipase C gene has no major effects on cell movement (Drayer *et al.*, 1994).

Uncapping of F-actin does not seem to be a major source of new filament formation (Eddy *et al.*, 1997). Rather, capping is used to terminate a response

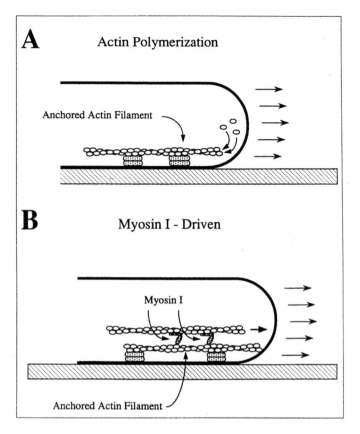

Figure 6.3 Two mechanisms of expanding a pseudopod. In panel A, actin monomers are added to anchored actin filaments. These would be cross-linked to increase strength, and osmotic swelling would also occur. Membrane lipids would have to diffuse into the expanding structure to accommodate the shape change. In panel B, the expansive force involves a motor in the form of one or more myosin I molecules. (Reprinted from Egelhoff and Spudich (1991), with permission from Elsevier Science.)

in the seconds after a chemotactic signal, as described in Chapter 8. If the uncapping of short filaments were the major mechanism for exposing barbed ends, one would expect that there would be a reduction in capping proteins in cytoskeletons prepared after chemotactic stimulation. Uncapping of F-actin does not appear to be extensive enough to account for the explosive growth of actin filaments; for a review, see Condeelis (1998).

A third way for the actin cytoskeleton to produce barbed ends is by severing. Potential severing proteins are listed in Fig. 6.1. The activity of these molecules is regulated by pH and by phosphorylation, providing them with rapid communication with events at the plasma membrane. Severing appears to be the main mechanism for creation of barbed F-actin filaments (Condeelis, 1998).

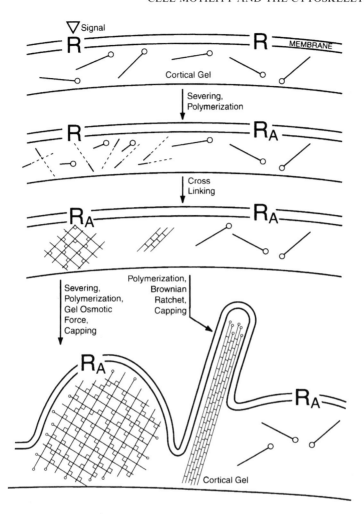

Figure 6.4 The cortical expansion model. This model proposes that extracellular signals lead to localized exposure of actin filament barbed ends, which then expand by incorporation of G-actin and are cross-linked by combinations of the actin-binding proteins listed in Fig. 6.1. R indicates a receptor and R_A indicates an adapted receptor, as described in Chapter 8. Open circles indicate capping proteins. Lines indicate actin filaments and dotted lines indicate expanding actin filaments. Depending on the type of cross-linking and local increases in osmotic pressure, pseudopods, filopods, or other protrusive structures can be formed. The ratchet mechanism refers to the likelihood that the membrane vibrates by Brownian motion, letting actin monomers be interposed. (Courtesy of John Condeelis. Reproduced from *Motion Analysis of Living Cells.* Reprinted with permission of Wiley-Liss, Inc., a division of John Wiley & Sons, Inc.)

The fourth mechanism, *de novo* nucleation, begins with a nucleating complex containing ponticulin (Schwartz and Luna, 1988; Shariff and Luna, 1990). Ponticulin appears to serve a function in stabilizing pseudopods, and does not appear to perform a nucleating activity in their initial formation (Shutt *et al.*, 1995). Various models combine the uncapping, severing and *de novo* nucleation mechanisms. The fact remains that the rapid local polymerization of actin is essential for generation of protrusive force.

6.5 Strengthening the filaments of actin

The next step in the generation of a protrusive structure is cross-linking. Cross-linking is necessary because single actin filaments are not sufficiently rigid to push against the plasma membrane, while cross-linked filaments are stiff and can generate a force. The actin meshwork in pseudopods and lamellapods of motile cells has an orthogonal geometry with the filaments linked at right-angles. As we have seen, one of the major proteins involved in this cross-linking is ABP-120, but there are a number of others, as indicated in Fig. 6.1.

The events shown in Fig. 6.4 occur over the course of a few seconds. First, the amoebae stimulate the formation of barbed ends through signal transduction pathways that remain poorly known (see Section 6.2). Second, by the mechanisms just described, actin polymerizes. Third, a meshwork is formed through the actions of cross-linking proteins like ABP-120 or the other proteins listed in Fig. 6.1. The meshwork may contain actin bound orthogonally or bundled in parallel fibers, and this determines the protrusive structures that are made.

The creation of actin filaments and cross-linking do not suffice to explain protrusion. Force must be exerted against the plasma membrane. The expansion of the pseudopod could employ myosin I molecules, which have been localized there. So far, it has not been possible to show that any of the myosin I molecules is essential for generation of a pseudopod. Alternatively, the energy associated with actin polymerization could be adequate to provide a protrusive force – as in pollen tube formation, *Listeria* propulsion, or acrosome filament extension. One mechanism by which this could happen has been called a ratchet, in which the plasma membrane vibrates by Brownian motion and then actin monomers are interposed between the growing filament and the membrane. Protrusion rates of 0.5 μm/minute can be accommodated with this model (Peskin *et al.*, 1993).

6.6 Moving the cell

Extension of a pseudopod is only the first step in cell movement. The cell must adhere at the tip of the protrusion and detach from the uropod in the rear. The centroid of the cell must be displaced. The general process is shown in Fig. 6.5, as drawn for mammalian cells by John Condeelis. Pseudopods of *Dictyostelium*

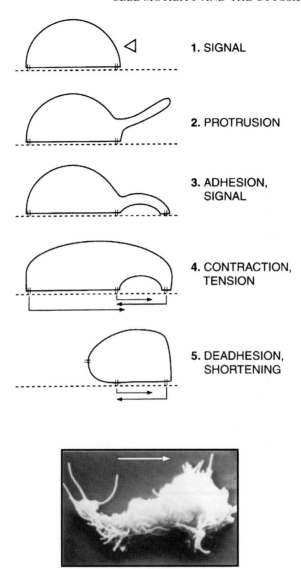

Figure 6.5 The essential steps in cell motility. Upper panels: Pseudopods protrude in response to local signals and then make contact with the substratum. Tension in the front of the cell promotes contraction, and then the cell must release the substratum in the rear and shorten for translocation to occur (Condeelis, 1998). (Courtesy of John Condeelis. Reproduced from *Motion Analysis of Living Cells*. Reprinted with permission of Wiley-Liss, Inc., a division of John Wiley and Sons, Inc.) The lower panel shows a scanning electron micrograph of an amoeba. The arrow indicates the direction of movement before fixation. Note the uropod extending to the left, and the many fine filopods. (Reprinted from Egelhoff and Spudich (1991), with permission from Elsevier Science.)

may project above the substratum or on it. Those which bind the substrate promote turning, those that do not are generally retracted (Luna *et al.*, 1998; Shutt *et al.*, 1995). Adhesion to the substratum is a subject in which studies of *Dictyostelium* lag behind those of mammalian cells. We do not yet know of any integrin-like molecules, nor of stress fibers extending from focal adhesion plaques. The newly described eupodia may fulfill some of these functions. Alternatively, the genome project may reveal new genes involved in the attachment to the substrate. It is clear that the variety of substrates to which the amoebae bind is large. The steps shown in Fig. 6.5 include the timely detachment of the rear of the cell and the contraction toward the front. In contrast to pseudopod extension, retraction of the uropod may be mediated by myosin II.

6.7   Signaling to the cytoskeleton

During chemotactic aggregation the cells perceive increasing and decreasing gradients of cAMP and respond by rapid mobilization of their cytoskeletons (Schleicher and Noegel, 1992). Because signaling can occur in suspension it is possible to synchronize the response and to examine the signaling pathways that lead to mobilization. The pathways that lead to actin polymerization and (equally important) to its depolymerization will be discussed in Chapter 8. Briefly, cGMP is produced on chemotactic stimulation. This happens over the course of a few seconds and is essential for causing the incorporation of myosin II into the cortex of the cytoskeleton. The signaling pathway that causes myosin II to incorporate into the actin cytoskeleton acts through the phosphorylation of myosin light chain kinase (Silveira *et al.*, 1998). The cGMP does not do this directly – a kinase is required. This may be a cGMP-dependent kinase, but this enzyme has not been well characterized from *Dictyostelium*.

The proteins that generate barbed ends of actin filaments are responsive to pH, calcium and other messengers that change on chemotactic stimulation. Severin, as already described, is phosphorylated by a particular kinase. Some capping proteins may be sequestered by PIP_2, releasing G-actin or severing activity. There is an influx of calcium during chemotactic stimulation, which is important to the regulation of many of the proteins that participate in the rapid response of the cytoskeleton. Many of these events are described in more detail in Chapter 8.

The cytoskeleton is controlled to a great extent by small GTPases, of which *Dictyostelium* has the appropriate families. Some of these have functions other than during motility, such as in cytokinesis (Larochelle *et al.*, 1997). Overexpression of a Rho family GTPase causes cells to reorganize the actin cytoskeleton into blebs on the surface of amoebae (Seastone *et al.*, 1998). RasG mutants grow well, but have severe defects in motility and pseudopodial extension (Tuxworth *et al.*, 1997). These mutants seem to make actin-based structures, including excessively long filopodia.

A cell-free system in which actin polymerization can be followed would allow us to ask what role small GTPases (or other regulatory molecules) have on actin polymerization. Small GTPases have major effects on actin polymerization and the generation of ruffles and other membrane protrusions in animal cells. A cell-free system had been developed for *Dictyostelium* and it has been used to show that the Rho protein relative Cdc42 (but not Rho or Rac) causes the polymerization of actin in extracts of *Dictyostelium* or neutrophils in the presence of GTPγS (Zigmond *et al.*, 1997). This polymerization occurs at barbed ends; thus, we should be able to trace the events in a cell from reception of the chemotactic signal, to stimulation of conventional GTP-binding proteins, to induction of effector molecules such as guanylyl cyclase, and ultimately through a series of kinases and small GTPases to the complex molecules of the cytoskeleton.

6.8 Cytokinesis

The previous discussion focused on cell movement, but the cytoskeleton also participates in movement of organelles and chromosomes. *D. discoideum* is particularly useful in the study of cytokinesis because mutants that fail to carry out effective cytokinesis grow and become multinucleated, but can be rescued by attachment-mediated cytofission, as described in Section 6.2 for myosin II. This phenomenon was first observed with the myosin II mutants and has since been found in an increasing number of mutants with defective cytoskeletal elements or defective small GTPases. For a review, see Wolf *et al.* (1999).

Cytokinesis has four phases (Fig. 6.6). During the first phase, the position of the cleavage plane is established, and this depends on the orientation of the spindle, such that the plane is orthogonal to the spindle and at its center. What cues determine where actin should be assembled is not known. The second step is the assembly of actin at the plasma membrane where the furrow will form, and this is followed by the localization of myosin II, which generates the force of contraction. Contraction of the ring structure constitutes the third step, followed by the fourth phase, membrane fusion to separate the two daughter cells (Wolf *et al.*, 1999).

The REMI methods used by Larochelle *et al.* (1996, 1997) revealed a member of the Ras family of signaling molecules called RacE, which appears to play a role in cytokinesis. RacE does not localize to the contractile ring, which forms normally in its absence. Rather, RacE seems to function at a later stage in a way that does not require all of it to locate to the ring and may act indirectly by controlling cortical tension (Gerald *et al.*, 1998). Another REMI screen, reported by Adachi *et al.*, revealed a gene called *gapA*, which codes for a Ras-GTPase activating protein and acts very late in cytokinesis, after the contraction of the furrow, perhaps by activating a Rac or Rho family member during the membrane fusion step (Adachi *et al.*, 1994, 1997). A second GTPase activating protein controls a distinct step in cytokinesis. This gene is called *RgaA* or *DGAP1*, and its deletion causes cytokinesis defects distinct from *gapA* (Faix and Dittrich, 1996; Lee *et al.*, 1997a). *RasG* has been deleted,

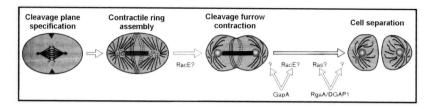

Figure 6.6 The stages of cytokinesis and the role of small GTPases. The circumferential lines represent actin, the asters are microtubules. Small GTPases and their regulatory molecules are required at several stages (Chisholm, 1997). (Reprinted with permission from Elsevier Science.)

which results in defective cytokinesis such that giant multinucleated cells are formed in suspension, again with a block that is late in the process. The *rasG* mutation causes other motility defects and influences the actin cytoskeleton.

Other proteins are known to participate in cytokinesis but many of these, like myosin II, have roles in other motile events and were not originally identified by the fact that they cause multinucleated cells to form in suspension. These molecules include the regulatory myosin light chain and the myosin essential light chain (Chen *et al.*, 1994, 1995), as well as coronin (de Hostos *et al.*, 1993). Disruption of the clathrin gene revealed its unexpected role in cytokinesis (Niswonger and O'Halloran, 1997).

6.9 The microtubule cytoskeleton

In comparison with the actin cytoskeleton, the microtubule (MT) cytoskeleton of *Dictyostelium* has been relatively unexplored. Nonetheless, *Dictyostelium* provides an attractive system in which to study MTs genetically and biochemically. *Dictyostelium* possesses only one α-tubulin gene and one for β-tubulin, in contrast to the multiple genes in most other eukaryotes (Trivinos-Lagos *et al.*, 1993). Diversity of function at the level of tubulin may be conferred by posttranslational modifications although to date, modifications known in complex eukaryotes such as detyrosination or acetylation, have not been described in *Dictyostelium*. The sequences of the tubulins in *Dictyostelium* are similar to those of other tubulins, but there are regions of sequence divergence that have been hypothesized to be responsible for some of the unusual properties of *Dictyostelium* tubulin (Trivinos-Lagos *et al.*, 1993). For example, *Dictyostelium* MTs are insensitive to cold or calcium treatments, which cause MTs from most other systems to disassemble (White *et al.*, 1983). A MT-associated protein (MAP), related to the *Drosophila* 205-kDa MAP has been identified, although its function has not been explored (Kimble *et al.*, 1992). *Dictyostelium*, as we have seen, has been a useful system to study the role of myosin motor proteins and is becoming one in the study of MT motor proteins, dynein and kinesin. Indeed, *Dictyostelium* dynein has been extensively studied and the protein has been purified and cloned (Habura et al., 1999; Koonce et al., 1994; Koonce and Samso, 1996). Overexpression studies point to a role for dynein in the maintenance of the interphase array of MTs in vegetative cells (Koonce and Samso, 1996), and the dynein heavy chain appears essential for viability (Koonce and Knecht, 1998). Active kinesin motor protein was purified from *Dictyostelium* a number of years ago (McCaffrey and Vale, 1989), and recent PCR cloning strategies have revealed six kinesin-related proteins, several of which are developmentally regulated (de Hostos et al., 1998). The complete sequence of one such developmentally regulated kinesin (termed K7) showed it to be a novel member of the kinesin superfamily, similar to conventional kinesin, but without an extensive coiled-coil region in its non-motor tail domain. Cells null for K7 showed no gross developmental abnormalities, but were mostly absent from the prestalk zone when cultured with wild-type cells, suggesting that K7 is involved in differentiation. The role of the other kinesin-related proteins has not been explored, but they are likely to be involved with the extensive cytoplasmic transport of vesicles and organelles during *Dictyostelium* migration and differentiation (Roos et al., 1987).

Microtubules are the predominant structural component of the mitotic spindle in *Dictyostelium* and other eukaryotes. Early studies of the mitotic spindle in *Dictyostelium* revealed that it has a fine structure that is similar to MTs described in other eukaryotes (Jensen et al., 1991; McIntosh et al., 1985; Moens, 1976). Spindle and interphase MTs appear to be nucleated by an unusual MT-organizing center (MTOC) or centrosome that is associated with the nucleus and has a layered structure, somewhat similar to the spindle pole bodies in yeast (Omura and Fukui, 1985; Ueda et al., 1999).

The *D. discoideum* centrosome has been purified and its components are being characterized by immunochemical and molecular means (Euteneuer et al., 1998; Gräf et al., 1998; Kalt and Schliwa, 1996). This difficult purification has opened the possibility that *D. discoideum* will become an exceptionally useful experimental tool in the study of spindle formation. One early indication of this comes from studies in which the γ-tubulin gene has been fused to the GFP-coding sequence. Using this construct, cells in which the behavior of the centrosome can be followed have been created. The centrosome cycle was shown to take about 20 minutes, starting in prophase with an increase in size of the centrosomal core structure and loss of the corona. The formation of the spindle poles has been examined in detail (Ueda et al., 1997).

The role of MTs in *Dictyostelium* motility has also been explored. Whereas the actin cytoskeleton, with associated myosin motors, is responsible for generating the force necessary to propel amoebae forward, the MT cytoskeleton may be involved in cell polarization during motility. In many crawling cells, the position of the MTOC is not random and usually lies between the nucleus and the leading edge of the cell. The MTOC in *Dictyostelium* is located between the nucleus and the leading edge in some, but not all conditions in which motility was stimulated (Sameshima et al., 1988). This suggests that the correlation in MTOC positioning does not hold in *Dictyostelium*. Nonetheless, other studies point to a role for MTs in stabilizing directional movement of *Dictyostelium*.

For example, if amoebae are separated into nucleated and anucleate cell fragments, only the nucleated fragment, bearing the MTOC, is capable of locomotion (Swanson and Taylor, 1982). Also, a study with GFP-γ-tubulin, which allowed the examination of centrosome position in living cells, revealed that the centrosome moved considerably during cell movement and that its repositioning adjacent to a previously formed protrusion appeared to stabilize that protrusion (Ueda *et al.*, 1997). Many of the earlier studies were based upon analysis of fixed samples, and this latter study highlights the dynamics of the system and underscores the importance of examining living cells.

As the genetics and genomics of *Dictyostelium* develop, many other aspects of the connection between the stimulation of the cells and the local recruitment of the cytoskeleton should be clarified. In addition to the molecular genetic capabilities that these cells offer, there is a capacity to carry out large-scale biochemical studies; moreover, with GFP as a marker, it will also be possible to perform an array of cytochemical experiments.

7

The Transition from Growth to Development: From Starvation to Self-Sustaining cAMP Signal Relay

Starvation is a crisis that the cells confront with immediate action. Polysomes are degraded; transcripts that are required for growth disappear; and the cell cycle halts. All of development is accomplished without the addition of new metabolic reserves, and so there is selective advantage in using the reserves that have been accumulated during the trophic phase as parsimoniously as possible, without wasting energy on constituents necessary only for growth. Slowly, after the onset of starvation, new protein synthesis begins – only after a few hours do many of the transcripts and proteins that will mediate aggregation appear. This earliest period of development leads to the induction of genes necessary for aggregation, and it is constructive to think of it as a separate and essential series of events. Chemotaxis and aggregation are not the earliest developmental events.

7.1 Cells can detect imminent starvation

In the laboratory, when development is induced by abruptly washing away nutrients, the onset of starvation is sudden. In the soil, depletion is more gradual and the cells have mechanisms to sense when hard times are approaching. There are two density-sensing mechanisms that function during the early stages of development. One mechanism is mediated by a molecule called prestarvation factor (PSF) and controls induction of certain very early genes (Rathi and Clarke, 1992). The other mechanism, mediated by a molecule called conditioned medium factor (CMF) (Gomer *et al.*, 1991), helps the cells to assess density at a slightly later period – during aggregation (see below). PSF is synthesized during growth and accumulates in the micro-environment according to the density of the cells. The gene inductive effect of PSF is inhibited by the presence of bacteria. As Clarke and her colleagues have pointed out, the level of PSF that is detected by the cells indicates the ratio of *Dictyostelium* amoebae to bacteria (Clarke *et al.*, 1992).

The role of PSF is difficult to study because it has not been purified to homogeneity, nor has its gene been cloned. PSF is a glycoprotein with a mass of 65,000 Da and is sensitive to proteases and to heat. Using partially purified PSF, a number of genes that were previously thought to be induced by complete starvation have been induced in growing cells. These genes include *pdsA*, which encodes the secreted cyclic nucleotide phosphodiesterase (Lacombe *et al.*, 1986), members of the discoidin gene family (Clarke *et al.*, 1987), and the α-mannosidase gene, *manA* (Schatzle *et al.*, 1992). *carA* encodes the major cAMP receptor during early development and is also regulated in this manner (Louis *et al.*, 1993; Rathi *et al.*, 1991; Sun and Devreotes, 1991). Like the discoidins, *carA* is part of a multi-gene family (Saxe *et al.*, 1991b). The α-mannosidase is representative of a group of glycosidases that are induced early in development (Ashworth and Quance, 1972; Dimond *et al.*, 1976). Cells growing on reduced amounts of bacteria, which do not express these genes, will express them if PSF is added (Clarke *et al.*, 1992). At least for these genes, there is no requirement that the cell cycle be arrested before they can be expressed. Since all of these studies were done on populations, it remains possible that the induction of the various transcripts occurs in a subset of the cells which have stopped dividing. The synthesis of PSF declines as development proceeds. PSF does not promote further development in the absence of starvation.

There is a difference in early enzyme induction of cells grown on bacteria and those grown in semi-defined axenic medium. A number of enzymes, particularly the glycosidases mentioned above, are induced in axenically growing cells. When true starvation is imposed, the cells make more of these enzymes. In cells growing on bacteria – a relatively more luxuriant condition – these genes are completely inactive, but are induced when starvation occurs. This result suggests that there are sets of genes that respond differently to the degree of starvation (Ashworth and Quance, 1972; Dimond *et al.*, 1976).

Among the genes that are induced differently when cells are grown on bacteria or on axenic medium are the discoidin genes. This small family of genes codes for a lectin-like activity that appears to be involved in cytoskeletal function, but the role of these proteins has been the source of some controversy (Alexander *et al.*, 1992). The discoidin lectin domain is conserved in higher organisms, where it is expressed as part of a receptor tyrosine kinase in carcinoma cells (Johnson *et al.*, 1993a). An analysis of the discoidin gene promoter has revealed a sequence called the dIE element, that is essential for the prestarvation response (Vauti *et al.*, 1990). A TTG sequence was found in the discoidin promoter and in other promoters which are inactive during growth on bacteria, but active when cells are grown axenically. The

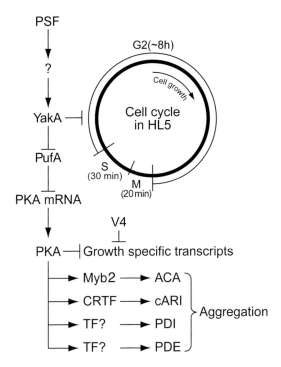

Figure 7.1 A model of the growth to development transition mechanism. The lengths of the periods of the cell cycle are shown for cells growing in axenic medium. The transcript of the catalytic subunit of PKA is induced by starvation, but is not translated because of the binding of PufA to nanos response elements. YakA represses the transcription of *pufA* and thus allows the translation of the PKA mRNA. How YakA inhibits the cell cycle is not known. A Myb transcription factor participates in the activation of the adenylyl cyclase gene (see text). CRTF regulates cAR1 induction and perhaps other genes. The arrows impinging on the effectors of chemotaxis are not meant to imply that this is the sole mode of regulation, only that there is a dependent sequence.

steps between binding of PSF to cells and induction of the various genes described above require protein synthesis and are likely to be complex. We can predict a PSF receptor and an associated signal transduction pathway, as shown in Fig. 7.1.

Other secreted proteins may play roles in the growth to development transition. When amoebae are given a small amount of nutrient, their development is blocked – unless conditioned medium from starving cells is added. Iijima *et al.* (1995) asked what protein in the conditioned medium promotes development, and have purified a trimeric protein to near homogeneity. How this protein functions to overcome the inhibition of development caused by low levels of nutrients is not known, but since the proteins making up the 450-kDa complex have been purified, we can look forward to the identification of the genes involved.

7.2   Growth-specific events cease during development

Much of the emphasis in development has been on the induction of new genes, but it is also important to rid the cell of the expenses of the trophic phase. When cells are washed free of their growth medium and starved, there is an immediate unloading of mRNA from polysomes and an increase in monomeric ribosomes (Margolskee *et al.*, 1980). The rapid collapse of polysomes may be an artifact of the sudden effects of centrifuging away the food source, but it is part of the reprogramming process, even if under other conditions it happens more gradually. On an individual gene basis, the transcripts of several biosynthetic genes disappear. These include transcripts for CprD, a growth stage cysteine proteinase (Souza *et al.*, 1995), UMP-synthetase and GMP-synthetase *(guaA)* (Jacquet *et al.*, 1988; van Lookeren Campagne *et al.*, 1991). The latter are pyrimidine and purine biosynthetic enzymes, respectively. Synthesis of rRNA is reduced, but does not stop completely during development (Kessin, 1973). Among the genes that are repressed during early development, Singleton has studied several ribosomal protein genes, whose transcription is rapidly reduced after starvation begins. Analysis of the V18 promoter suggests a model in which a positive regulator is lost during development (Ken and Singleton, 1994). The cyclic nucleotide phosphodiesterase gene, *pdsA*, plays a critical role during aggregation and other stages of development. One of the promoters of this gene is active during vegetative growth and then is repressed after a few hours of starvation, such that the 1.8-kb transcript disappears (Podgorski *et al.*, 1989). The disappearance of this transcript is prevented in a mutant that carries a defective gene that codes for the catalytic subunit of cAMP-dependent protein kinase (Wu *et al.*, 1995b).

The cell cycle is the major event that ceases during starvation, although there remains some controversy about this point. Originally, at least one round of DNA replication was thought to occur during development, but more recent results suggest that chromosomal DNA synthesis ceases within a few hours of starvation (Shaulsky and Loomis, 1995). Mitochondrial DNA synthesis continues (Shaulsky and Loomis, 1995). *Dictyostelium* lacks a G_1 period in its cell cycle – cells which have undergone mitosis immediately enter S phase and then spend a long period in G_2 (Weijer *et al.*, 1984b). The amount of DNA in mono-nucleate amoebae is therefore 2N. One might imagine that starving cells which had passed a point of commitment at the G_2–M boundary, would pass through M and then develop in G_1, but this does not appear to be the case. If it were, we would predict that an asynchronous population of cells would contain nuclei with different DNA contents. Fluorescence-activated sorting experiments with isolated nuclei show only one peak (Weijer *et al.*, 1984b; D. N. Dao and R. H. Kessin, unpublished results). Earlier studies were confounded by the fact that mitochondrial DNA synthesis continues. As we have seen (Section 4.4), mitochondrial DNA constitutes a large percentage of the total cellular DNA.

Although the cell cycle of *Dictyostelium* is regulated by many of the same components that regulate yeast or other cell cycles (Weeks and Weijer, 1994), we do not know how it is shut off during starvation, nor how it is reactivated during spore germination. Overexpression of the kinase YakA stops cell division, but how this is accomplished is not known (Souza *et al.*, 1998). *YakA* null cells are smaller than their isogenic parents, which means they enter mitosis earlier in G_2 and also divide more rapidly. The *yakA* mutants also grow to a higher cell density. Arrest of the cell cycle constitutes a critical puzzle for development of *Dictyostelium* and for other organisms that convert from a trophic feeding form to a differentiated one. It is the same problem faced by any stem cell, whether of an animal or of a more simple eukaryote.

We do not know how the cells sense starvation. The amoebae need only glucose and amino acids to grow, but we do not know how the levels of these nutrients are monitored (Franke and Kessin, 1977; Marin, 1976). One could guess from results of experiments with *Saccharomyces cerevisiae* which genes are involved, but so far this has not been productive. A second approach which may bear fruit is to isolate mutants which aggregate and develop in the midst of nutritional plenty, when their wild-type parents do not. Several such mutants, which form aggregates in the feeding zone of colonies where bacteria have not been completely consumed, have been isolated and their characterization is being undertaken (B. Wetterauer and W. Nellen, personal communication).

7.3   The first events after starvation

Early experiments used two-dimensional polyacrylamide gel electrophoresis (2-D PAGE) analysis to study the synthesis of individual proteins after the cells were starved. As noted above, when cells are suddenly removed from the medium, there is a degradation of polysomes, followed by a resynthesis. While most proteins are reduced in the hours after starvation, the synthesis of several proteins is transiently induced (Margolskee and Lodish, 1980). At the time this work was done, it was not possible to determine what these proteins were. However, more recent molecular studies have revealed a number of early transcripts. Singleton reported the cycloheximide-resistant induction of several genes (Singleton *et al.*, 1988), but it is too early to tell whether these play a critical role in early development because they have not been sequenced and their genes have not been disrupted. However, they remain attractive candidates for a role in early development. Another interesting gene is V4, also isolated by Singleton and colleagues (McPherson and Singleton, 1992). Antisense disruption of V4 expression leads to a failure to deactivate the transcription of vegetative stage genes, and also leads to a reduction in the transcription of genes that are involved in the events of chemotaxis. The V4 gene is not necessary for growth of the cells and is apparently not needed for later development because its transcript disappears shortly after development begins. V4 codes for a 17.8-kDa protein, with no known homologues. An unexplained feature of the gene is that overexpression also blocks aggregation. Preliminary experiments suggest that the V4 promoter is responsive to nutrient levels, and several nutrient responsive elements have been defined (McPherson and Singleton, 1992, 1993).

Protein kinase A (PKA) plays a critical role during the early stage of development and at all later ones. The enzyme was purified and studied by Veron and his colleagues (De Gunzburg *et al.*, 1984), and is expressed early in development. The PKA of *Dictyostelium* is a dimer of one regulatory and one catalytic subunit, rather than the tetramer of higher organisms, and the catalytic subunit has a long amino-terminal extension, the role of which remains unknown (Anjard *et al.*, 1993; Etchebehere *et al.*, 1997).

The catalytic subunit gene was recovered in the laboratory of Reymond, and also by Firtel and colleagues (Burki *et al.*, 1991; Mann and Firtel, 1991; Mann *et al.*, 1992). The regulatory subunit gene was recovered following the purification of the protein (Mutzel *et al.*, 1987). The mRNA for the regulatory subunit of the PKA is present at low levels during growth and the levels increase 10–20-fold during the first 3 hours of development (De Gunzburg *et al.*, 1986). Expression of catalytic and regulatory subunits is not coincident, and there may be a period (after starvation begins) when the catalytic subunit is uncontrolled by a regulatory subunit. Removing the regulatory subunit by deleting its gene releases the catalytic subunit from inhibition and causes development to proceed very rapidly; this was first observed in a rapid development mutant, *rdeC* (Abe and Yanagisawa, 1983; Simon *et al.*, 1992). Overexpression of the catalytic subunit also causes rapid development (Anjard *et al.*, 1992; Mann *et al.*, 1992).

Deletion of the catalytic subunit of PKA causes an aggregateless phenotype, without affecting growth. Several of the genes involved in chemotaxis – *acaA* (adenylyl cyclase), *pdiA*, the phosphodiesterase inhibitor, and *carA*, the major cAMP receptor in early development, are not transcribed at all in the absence of the PKA catalytic subunit (Mann *et al.*, 1997; Wu *et al.*, 1995b). The failure of PKA-null mutants to develop is not due to the loss of adenylate cyclase, because this function can be replaced by expressing the adenylate cyclase gene under the control of a constitutive actin promoter without restoring development (Mann *et al.*, 1997). We conclude that the PKA is responsible for the induction of a number of genes that are essential for chemotaxis or aggregation.

As we will see, activation of PKA is critical at a number of transitions during development. This presents a problem because it suggests that the PKA should be down-regulated between critical junctures in development. No one has succeeded in measuring PKA activity *in situ*, so it is not possible to know if this occurs. If we make a transformant that modestly overexpresses the catalytic activity of PKA, development can proceed in an orderly way (Wang and Kuspa, 1997). These facts suggest that some other form of constraint is important in regulating the major transitions during development, perhaps in gating the cAMP responses. Recent evidence suggests that this role can be fulfilled by a series of sensor histidine kinases that integrate with the cAMP signaling events (Loomis, 1998; Thomason *et al.*, 1998).

7.4   YakA kinase regulates the growth to development transition

The sequence leading from starvation to PKA induction can be extended thanks to the discovery of an aggregateless mutant that is defective in a gene called *yakA*. *YakA* was recovered in a mutant screen employing REMI (Souza *et al.*, 1998). The PKA catalytic subunit mRNA appears as normal in a *yakA*-null mutant, but PKA enzyme activity does not show the characteristic increase after 5 hours of starvation. *YakA* transcript levels increase during growth and reach a maximum at about the time of starvation. Genes dependent on PKA are not expressed in *yakA* mutants. Equally important, *yakA*-null cells, like those of V4 mutants, do not turn off genes that are expressed in growing cells. These are normally down-regulated during development. If YakA is overexpressed in a conditional manner, it arrests growth, as mentioned above. Thus, a single mechanism employing YakA appears to mediate much of the transition from growth to development. YakA shares many functions with the yeast protein Yak1p. The yeast protein also mediates growth to development transitions, and the *Dictyostelium* protein can substitute for the yeast one. There are homologues in mammalian cells to the yeast and *Dictyostelium* kinases – the minibrain kinases. These appear to manage cell fate decisions in neuroblast differentiation (Tejedor *et al.*, 1995). A mechanism for converting cells from a stem cell population to a differentiating one may have been conserved over at least 600 million years of eukaryotic evolution, back to Cambrian times or before.

To identify additional components of the YakA signaling pathway, Souza and her colleagues carried out a REMI suppressor screen on *yakA*-null cells and have identified a mutant gene, *pufA*, that reverses the aggregateless phenotype of *yakA*-null cells (Souza *et al.*, 1999). The suppressor has no effect on the growth associated phenotypes of *yakA* mutants. *PufA* encodes a member of the Puf (pumilio/FBF) family of proteins, which function in the translational control of key regulators of anterior–posterior patterning in *Caenorhabditis elegans* and *Drosophila* (Forbes and Lehmann, 1998; Wharton *et al.*, 1998; Zamore *et al.*, 1999; Zhang *et al.*, 1997). Puf proteins are sequence-specific RNA-binding proteins which bind to the 3′ ends of mRNAs that encode key developmental regulators. The *Drosophila* pumilio protein binds to the 3′ end of hunchback mRNA and, together with the nanos protein, inhibits the translation of hunchback protein in the posterior of oocytes. The RNA sequences that pumilio binds to, nanos response elements (NREs), have been defined. Since a likely candidate for regulation by PufA is PKA, Souza and Kuspa examined PKA mRNA for sequences related to the NRE control elements of the *Drosophila* hunchback gene. Several sequences that are similar to hunchback's NREs were found in a 200-bp segment in the 3′ end of the coding sequence of PKA – and PufA binds specifically to these elements. *PufA*-null mutants and *pufA/yakA* double mutants resemble YakA overexpressing cells in their rapid development, precocious expression of ACA and increased PKA activity. Inactivation of *pufA* restores PKA catalytic subunit mRNA expression in *yakA*-null cells, but to levels lower than those found in wild-type cells.

YakA is essential for the repression of vegetative gene expression at the onset of development. In *yakA* mutants, expression of some vegetative genes persists throughout development, confirming a role for this kinase in controlling the expression switch that occurs during the growth to development transition. *PufA* mRNA is present during growth and disappears by 8 hours of starvation. In *yakA*-null cells, the *pufA* mRNA persists well after 2 hours of starvation. Thus, YakA is required for the loss of *pufA* mRNA at the onset of development. The fact that the inactivation of *pufA* restores development to *yakA*-null cells indicates that the inhibition of *pufA* expression or protein function is required for the initiation of development. This probably occurs by regulating the translation of PKA catalytic subunit mRNA. Defects in such negative regulators of *pufA* mRNA translation are exactly what we would expect from suppressor analysis using REMI.

At the onset of development, as the cells acquire the ability to congregate by chemotaxis, a dependent sequence of gene expression has been established, in which expression of several genes depends on activity of the PKA catalytic subunit. Between the expression of these genes and the genes that code for the important elements in chemotaxis, there must be a series of transcription factors. The first of these are becoming known. The adenylyl cyclase gene is dependent for its activation on the transcription factor Myb2 (Otsuka and van Haastert, 1998). REMI mutants which lack Myb2 do not aggregate and do not activate the adenylyl cyclase gene. Myb2 seems to be specific for *acaA* because restoring the adenylyl cyclase gene under the control of another promoter restores aggregation. Myb2 may control the expression of other genes, but if so, they are not required for aggregation.

By careful dissection of the *carA* promoter, Kimmel and his colleagues have found a transcription factor, called CRTF (cAMP response transcription factor), which is essential for the induction of the critical cAMP receptor of early development (Mu *et al.*, 1998). This work defines a short sequence, that when put in the promoters of genes that are not responsive to cAMP pulses, causes them to be induced. The late promoter of *carA*, which functions after aggregation, does not require CRTF. CRTF is a 100-kDa zinc-binding protein that is itself constitutively expressed (A. Kimmel, personal communication). CRTF activity may be regulated by a signaling pathway that functions through cAR1, the product that it regulates. *Dictyostelium* cells that lack CRTF express *carA* at only a basal level and do not aggregate. It is not yet known whether CRTF induces other genes, or is relatively specific as Myb1 appears to be.

One of the virtues of *Dictyostelium* is that elegant genetic selection and screening systems can be combined with REMI and used as a form of mutagenesis (see Section 4.9). One of these schemes takes advantage of the well-studied discoidin genes, which are expressed in the growth to development transition. The discoidin genes are induced by PSF and are expressed very early in development, before visible changes in morphology take place. By using a screen that selects for premature expression of discoidin it has been possible to recover strains that express discoidin when it is normally repressed. These strains carry a mutation in a gene called *gdtA* (growth–development transition) which greatly overexpresses discoidin and accelerates early development. The active gene represses early development and discoidin induction, a marker for early development, potentially by monitoring the nutrient levels. GdtA appears to be a large trans-membrane protein that can be phosphorylated by PKA (Zeng *et al.*, 2000).

Another interesting early gene, *amiB* (aggregation minus), has recently been described (Kon *et al.*, 2000). This mutant fails to turn off a growth-specific gene, and, like *myb*2 mutants, fails to induce the ACA gene. Introduction of the Myb2 gene controlled by a constitutive promoter rescues the *amiB* mutant, putting *amiB* prior to *myb*2 in the induction pathway shown in Fig. 7.1. The *amiB* gene codes for a novel protein.

There are many unanswered questions about the growth to development transition. Some genes such as *yakA* and *pufA* have emerged and, like pioneers, will eventually accumulate families and partners through genetic and biochemical analysis. Some sequences which are involved in starvation sensing, vegetative gene shut-off, and initial induction of aggregation-specific events are known, but have been inadequately studied. We have already seen conservation of function with *yakA* and *pufA*. Such transition regulators may be important in lower eukaryotes – particularly in pathogenic organisms such as *Plasmodium falciparum* and *Entamoeba histolytica*.

Chemotaxis and Aggregation

Free-living *Dictyostelium* amoebae must be able to find their way toward prey or, in the face of starvation, toward each other. *Dictyostelium* amoebae share a chemotactic capacity with leukocytes and many other motile cells, and employ many of the same mechanisms during the detection of the chemotactic molecule, the activation of signal transduction pathways, and the mobilization of the cytoskeleton (Devreotes and Zigmond, 1988; Parent *et al.*, 1998; Parent and Devreotes, 1999). Chemotactic molecules bind to cell-surface receptors and stimulate G protein-mediated signal transduction pathways in amoebae and in mammalian cells. Agonists are degraded to steepen gradients and to overcome the effects of adaptation. Despite evolutionary distance, the cytoskeletons of leukocytes and *Dictyostelium* employ similar cytoskeletal rearrangements to move in the right direction. The advantage of *Dictyostelium* in the study of chemotaxis, motility and aggregation is that the gene products involved in each event can be eliminated by mutation, and the contribution of each element can be studied. The biochemical advantages that stem from synchrony of development and quantities of material have also been invaluable. This is not to say that we understand completely how a cell perceives that a gradient exists or how it moves toward higher concentrations of cAMP, but we are beginning to understand how this complex process works.

8.1 An overview

The outline of the chemotactic relay system employed by *Dictyostelium* was understood by Bonner, Shaffer, and their colleagues nearly 40 years ago, but it was the discovery that cAMP was the chemotactic molecule (acrasin) that put the field on a sound biochemical basis (Barkley, 1969; Bonner *et al.*, 1969; Konijn *et al.*, 1967). Once cAMP was identified as a chemotactic molecule, proteins to produce and detect cAMP became necessities, and the degrading enzyme described by Shaffer became a cAMP phosphodiesterase. Over the past years, other macromolecules that are required for aggregation have been described. Some modulate the response to cAMP outside the cell, such as the density-sensing molecule CMF, the secreted cAMP phosphodiesterase, and an inhibitor of the phosphodiesterase. At the plasma membrane, cAMP receptors detect the signal by binding cAMP, and trimeric GTP-binding proteins transfer information on the occupancy of the cAMP receptors to signal transduction pathways that engage the motile apparatus and command the synthesis and secretion of more cAMP. The interaction of the G protein subunits with adenylyl cyclase involves novel components. The activated G proteins also interact with a novel guanylyl cyclase, and the resulting cGMP is responsible for initiating events in motility. The number of components known to be involved in aggregation has increased dramatically since the discovery of cAMP as the mediator of chemotaxis.

The components that synthesize, detect, and destroy cAMP are induced in starving cells, although low levels of some of the components are expressed in growing cells. No crucial role for cAMP is known in growing cells; the amoebae grow normally without cAMP receptors, adenylyl cyclase, cAMP phosphodiesterase, or other molecules that are critical to chemotaxis and development. Perhaps there are specialized growth conditions that require cAMP as a second messenger, but none has been observed under laboratory conditions. How this extraordinary segregation of function evolved is not known.

There are a number of questions that await more complete answers. Following the binding of cAMP, the receptors and other components must adapt if chemotaxis is to be successful. The mechanism of adaptation is not understood, and remains one of the crucial questions about chemotactic aggregation. The synthesis of cAMP occurs about every 6 minutes, and the cells have a system to control this period. The mechanisms that control periodicity remain to be explained. The basis of secretion of cAMP is not known, except to say that it probably does not involve vesicles. The details of second messenger activation of the cytoskeleton remain to be elucidated in detail, although hints exist. The inducibility of the many genes involved in chemotaxis is an area that we do not understand thoroughly enough, though in all of these areas, individual investigators have made important advances that are now being exploited.

8.2   There are several ways to study the cellular response to cAMP binding

The response of living cells to cAMP occurs rapidly, and several systems have been devised to examine the properties of the response to agonists. Devreotes and colleagues developed ways to study the signaling properties of cells aggregating on solid substrata by passing streams of cAMP around the aggregating cells and collecting the effluent, which contained the cells' cAMP response (Devreotes *et al.*, 1979; Devreotes and Steck, 1979; Dinauer *et al.*, 1980a,b, c). Aggregating cells were shown to respond to cAMP with a pulse of their own that was made and secreted with a time course that is discussed below. Cells stimulated with cAMP underwent a period of adaptation during which they would not respond to a second application of cAMP unless the pulse was of a higher concentration. Thus, cells stimulated with a 1 nM application of cAMP do not make and secrete more cAMP if after a minute another pulse of 1 nM cAMP is added, but they will respond to 10 nM cAMP. The concentrations capable of eliciting a cAMP secretion by the cells span the range from 1 nM to 1 μM or more. By characterizing the adaptive properties of the amoebae and then examining the physical changes to the cAMP receptor that occur during adaptation, this series of studies led to the purification of the cAMP receptor, as described below.

To study the second messenger responses to cAMP, another experimental system was developed which takes advantage of the fact that starving cells, once they have elaborated all of the molecules that are important to chemotaxis, secrete and respond to cAMP in suspension (Gerisch and Hess, 1974). The light-scattering properties of the cells change as they detect a pulse of cAMP and mobilize their cytoskeletons. The change in shape occurs whether the cells are on a solid substratum (Fig. 8.1) or in suspension (Fig. 8.2). In cell suspensions, the contractile properties of the cells can be followed as a periodic change in light scattering. The phase of the oscillations of cells in suspension can be shifted with exogenous cAMP. The advantage of suspended cells signaling in synchrony is that large numbers of amoebae can be rapidly sampled in biochemical experiments (Gerisch and Hess, 1974; Gerisch *et al.*, 1975). This capacity is shown nicely in the intracellular cAMP measurements that correlate with optical density changes in Fig. 8.2. The changes in shape of cells in suspension mimic those of cells on a solid substratum. These also respond to increases in cAMP by mobilizing their cytoskeletons and rounding up (Fig. 8.3). This rounding process was originally called the cringe response by Futrelle *et al.* (1982), who showed that it is the same whether the cells are in suspension or on a substratum.

8.3 Second messengers and cytoskeletal events can be studied with suspended cells

Using synchronized suspensions of cells and a number of other techniques, several events have been observed in rapid succession during each oscillation. These have been arranged in Fig. 8.4 to show their relative time courses. As might be expected from the rapid change in shape of the cells, there is major rearrangement of cytoskeletal macromolecules following a cAMP signal. Actin, which exists in soluble and filamentous cytoskeletal forms, rapidly leaves the soluble pool and within a few seconds is bound to the cytoskeleton. This binding is transient and rapidly reversed. Actin binding rises again within a few seconds, as shown in Fig. 8.4, panel A. Myosin II, the other major

Figure 8.1 Moving and stationary cells have different optical properties. The light bands of amoebae are moving and are therefore elongated. Stationary cells are not moving and are rounded. They are seen as darker bands in dark-field microscopy. The outermost (sixth) band of moving cells is responding to a wave of cAMP that initiated at the center about 36 minutes earlier and was gradually propagated outward. Cells move as long as the gradient is positive. When the wave passes and the slope reverses, the cells stop moving and await the next wave, which appears in about 6 minutes under standard conditions. Note that where waves collide, they are extinguished. Waves cannot be propagated by cells that are adapted because of the recent passage of a colliding wave of cAMP. The wave pattern eventually gives way to streams of adhesive cells, as is already happening at the center of one of the aggregation territories. (Courtesy of Peter C. Newell, Department of Biochemistry, University of Oxford.)

cytoskeletal component, incorporates into the cytoskeleton on a slower time scale, after a cascade of phosphorylation by myosin heavy and light chain kinases (Moores *et al.*, 1996; Silveira *et al.*, 1998; Wilson *et al.*, 1992). Modifiers of the actin cytoskeleton, such as talin and coronin, appear in the leading edges of filopodia and pseudopodia, as does F-actin (Dharmawardhane *et al.*, 1989; Gerisch *et al.*, 1995; Hall *et al.*, 1988; Kreitmeier *et al.*, 1995).

Much of the mobilization of the cytoskeleton is accomplished through the cGMP signal transduction pathway (Silveira *et al.*, 1998). The rise in cGMP

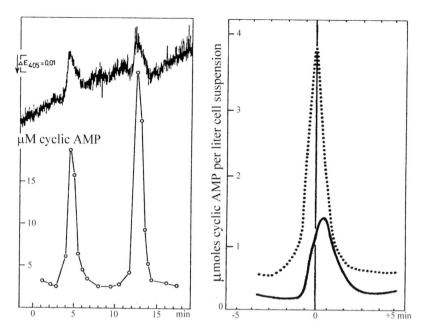

Figure 8.2 Periodic changes of light scattering can be observed in cells in suspension. These oscillations occur spontaneously after about 7 hours of starvation. For details of the method, see Gerisch and Malchow (1976). Samples removed at intervals of a few seconds and analyzed show the rise in intracellular cAMP (left panel). The right panel shows the rise in extracellular cAMP (solid line) superimposed on the intracellular cAMP concentration (dotted line). See also Fig. 8.5. (This experiment was done by U. Wick (Gerisch and Wick, 1975). Reprinted with permission.)

Figure 8.3 The cell-shape change can be seen in individual cells. The panel on the left shows an amoeba prior to receiving a sudden increase in cAMP. The cell is asymmetric, highly ruffled, and moving. The center panel shows a scanning electron micrograph of a cell 30 seconds after it has received a 1 μM increase in cAMP so that all receptors are occupied. Under these conditions the cells cringe, rounding up. At 30 seconds after the cringe the cells have returned to the motile state by adapting to the ambient cAMP (right panel). (Courtesy of A. Hall and J. Condeelis (Hall *et al.*, 1988). Reprinted by permission of Wiley-Liss, Inc, a division of John Wiley and Sons, Inc.)

peaks within a few seconds and subsides, except in mutants to be discussed below (Bumann et al., 1986; Mato et al., 1977a; Valkema and van Haastert, 1994). The reasons we believe that cGMP is important to mobilization of the cytoskeleton will be discussed after we have dealt with the intracellular rise in cAMP. Cell motility has been discussed more thoroughly in Chapter 6.

After a cAMP signal there is a transient influx in calcium (Bumann et al., 1986; Milne and Coukell, 1991; Milne and Devreotes, 1993). Protons and potassium rapidly flow out (Aeckerle and Malchow, 1989; Aeckerle et al., 1985; Aerts et al., 1987; Caterina and Devreotes, 1991). A slower intracellular increase in cAMP occurs about 60 seconds after activation of the cAMP receptors, and most of this is secreted as a chemotactic signal, as shown in Fig. 8.2 and Fig. 8.5 (Dinauer et al., 1980a). cAMP is not stored within the cells, nor is it known to be secreted in vesicles as a neurotransmitter would be. Adenylyl cyclase is activated and deactivated with each round of synthesis and secretion. The control of this effector is complex, involving subunits of a GTP-binding protein, several novel components, MAP kinase, Ras, and other gene products. One of the best characterized components of the adenylyl cyclase complex is a protein called CRAC, or cytosolic regulator of adenylyl cyclase. It is a molecule that responds to the differential concentrations of cAMP across a cell by binding to the membrane on the side that encounters the highest concentration of cAMP. Although CRAC mediates cAMP synthesis, it binds to the membrane rapidly, on a time scale of seconds, as shown in Fig. 8.4, panel C. We will return to CRAC after we have established the fundamentals of cAMP responses. For reviews of how the cells respond to cAMP, see Furukawa and Fechheimer (1997); Newell (1995a); Noegel and Luna (1995); Parent and Devreotes (1996b); and Parent and Devreotes (1999).

Second messengers are employed to relay the cAMP signal and to mobilize the cytoskeleton. They are also important to maintain the induction of necessary genes. As we will see, the elements of the chemotactic apparatus are auto-induced – the more pulses of cAMP a cell detects, the more induction of key genes that code for proteins that are important in chemotaxis. Weak responses can be seen within 1–2 hours of starvation (Futrelle et al., 1982). On a solid substratum at standard cell densities, about 20 rounds of signaling and movement bring the amoebae into an aggregate. Once the cells are in an aggregate, many of the genes that are important to aggregation are repressed, and a large number of new ones are induced. During the later stages of aggregation, the cumulative effects of multiple signals and responses prepares the cells for the next step of development – adhesiveness and formation of the mound in which the precursors of stalk and spore cells form.

8.4 Components of the cAMP signal transduction and relay pathway

Over the past 20 years, some of the important molecules of the chemotactic system of D. discoideum have been purified and their genes cloned. In some

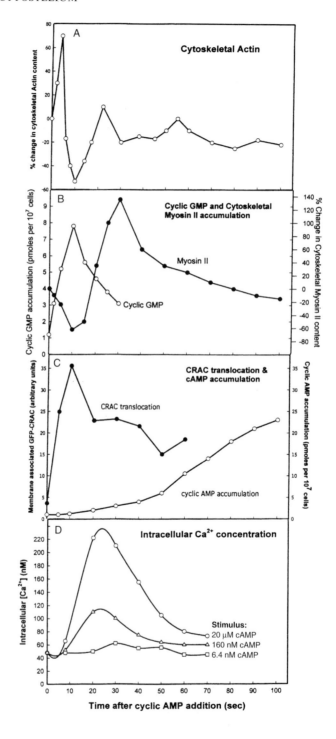

cases, this involved long struggles of protein purification, sequencing, and cloning. More recently, important elements have been recovered by genetic methods that avoid the moral rigors of biochemistry. Some of the molecules that are important for chemotaxis vary from one species to another. *Polyspondylium violaceum* uses the dipeptide glorin as a chemotactic molecule (Shimomura *et al.*, 1982), and *D. lacteum* uses pterin (de Wit *et al.*, 1988; van Haastert *et al.*, 1982). These species have receptors that remain unknown, but are probably no less complex than those we have come to know in *D. discoideum*. Those species that use cAMP as their chemotactic molecule during aggregation have evolved elaborate methods to synthesize and secrete it, as well as to detect and degrade it. We will concentrate on the components of the cAMP-mediated systems and then we will try to understand their regulation during a cAMP pulse and a cell movement step.

8.4.1 The cAR1 receptor is essential to aggregation

A central element in the chemotactic mechanism is the cAMP receptor, cAR1. Other molecules, while no less important to the aggregation process, regulate the responsiveness of the cAMP receptor, or in the case of effector molecules, are activated by it. The cAMP receptor is a seven trans-membrane domain glycoprotein that is distantly related to receptors in animals, plants and other simple eukaryotes and, like them, is coupled to trimeric GTP-binding proteins (Parent and Devreotes, 1996b). The cAR1 receptor and three related proteins, cAR2, cAR3, and cAR4, are the only receptors known for which the extracellular ligand is cAMP. The cAR2 and cAR4 receptors have low affinities for cAMP and control events during later times in development (Louis *et al.*, 1994; Saxe *et al.*, 1993, 1996). While they have much in common with cAR1, deletion of the genes coding for cAR2 and for cAR4 affect events after aggregation, and we will return to them. At this point it is sufficient to realize that the development of *Dictyostelium* is mediated in part by switches in the cAMP receptor subtypes. A second type of receptor, the histidine kinase class, also switches subtypes during development and may, in combination with the cAMP receptors, mediate an amoeba's interaction with its environment during development (Loomis *et al.*, 1998).

Figure 8.4 A summary of the immediate responses of a cell to cAMP. Mobilization of cytoskeletal elements and synthesis of second messenger molecules are shown. These results are compiled from a number of experiments, and it is not certain that the exact conditions were maintained in all cases, so the timing of events is approximate. See Fig. 8.10 for additional data. These data are from the following sources: Liu *et al.* (1993); Liu and Newell (1991); McRobbie and Newell (1983), (1985b); Menz *et al.* (1991); Nebl and Fisher (1997); Parent *et al.* (1998); van Haastert (1987). (The data for this figure were analyzed and formatted by P. C. Newell of the University of Oxford.)

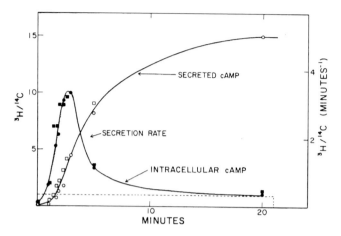

Figure 8.5 Externally applied cAMP elicits one round of cAMP synthesis and secretion. The graph shows the amount of ^{3}H-cAMP secreted by cells during a prolonged stimulation by 10^{-6} M cAMP (dotted line). To monitor recovery, cAMP is measured as the ratio of a ^{3}H/^{14}C cAMP (Dinauer *et al.*, 1980a). (Reproduced from the *Journal of Cell Biology*, by permission of The Rockefeller University Press.)

The regulation of receptors and the intracellular consequences of their activation employs a sometimes confusing vocabulary. We speak of *desensitization* as the decline of a response during persistent stimulation. This can happen by several mechanisms. If there is no loss of a receptor from the cell surface, the decline in responsiveness is called *adaptation*. Adaptation usually involves phosphorylation of the receptor or an associated protein. In response to long-term stimulation, the receptor may be removed from the cell surface, which is called *sequestration*, and finally, receptor molecules may be destroyed, which we call *down-regulation* (Parent and Devreotes, 1996b).

One of the consequences of cAMP binding to the receptor is cAMP synthesis and secretion. The time course of these events is shown in Fig. 8.5, which was derived from the perfusion experiments described above. Amoebae were labeled with ^{3}H-adenosine and then allowed to develop on a filter. A series of pulses of cAMP was applied with a special apparatus, and the cells become synchronized in their responses (Devreotes, 1983; Dinauer *et al.*, 1980a). After this period of preparation, a steady flow of 10^{-6} M cAMP was applied and both intracellular and extracellular cAMP were measured. The amount of cAMP made by adenylyl cyclase is proportional to the level of the extracellular stimulus, and the amount of extracellular cAMP depends on how much intracellular cAMP was made. The process is energy-dependent. If the flow of cAMP stops abruptly, there is degradation of the intracellular cAMP internally (Dinauer *et al.*, 1980a).

This is an adaptive process – a constant level of cAMP causes one burst of synthesis and secretion and then a halt – as shown in Fig. 8.5. There is no loss

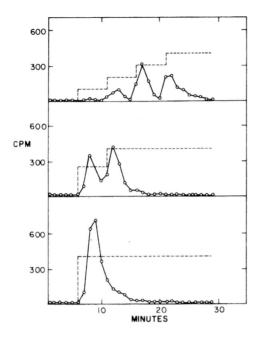

Figure 8.6 Adaptation of the cAMP response occurs at constant cAMP concentrations. Amoebae at the aggregation stage of development were perfused with a gentle flow of cAMP, as indicated by the dotted lines. They respond by making and secreting cAMP with the time course shown. The lower panel shows the response of developing amoebae to a sudden rise in cAMP to 10^{-6} M (saturating). The response adapts – it ceases in the presence of continuous 10^{-6} M cAMP, as shown in Fig. 8.5. A series of 10-fold increases in concentration from 10^{-9} to 10^{-6} M causes multiple rounds of synthesis, as shown in the upper panel. The middle panel shows the effect of two increases in cAMP concentration. De-adaptation occurs over a period of minutes when cAMP is removed (Devreotes, 1983; Theibert and Devreotes, 1983). (Courtesy of Peter N. Devreotes.)

of cAR1 from the cell surface. Figure 8.6 shows that increasing the cAMP concentration causes a greater receptor occupancy and overrides the adaptation. Saturating levels of ligand cannot be overridden with more cAMP.

The process of de-adaptation can take as much as 15 minutes after cAMP is removed by cAMP phosphodiesterase (PDE). Adaptation is an internal process that continues after all extracellular cAMP is removed by PDE. Adaptation does not depend on intracellular concentrations of cAMP (Theibert and Devreotes, 1983). The time of half-maximal adaptation of the cAMP response is about 110 seconds at 20°C. In contrast that for guanylyl cyclase is only a few seconds, suggesting that two mechanisms are involved (Kim *et al.*, 1996; van Haastert and Kuwayama, 1997).

Binding of cAMP causes receptor phosphorylation, and the affinity of the receptor for cAMP is reduced about 5-fold as a consequence. Phosphorylation is not the reason that many of the responses of the cells are suppressed by constant levels of cAMP (Kim *et al.*, 1997b). Altering the serines that are phosphorylated does not prevent the cells from adapting or aggregating well. Adenylyl cyclase activation, cGMP synthesis, myosin light and heavy chain phosphorylation, and actin polymerization, all cease normally after receptor stimulation in mutants without the serines in the carboxyl terminus of the receptor. The mechanism of transcriptional regulation is also adaptive and transcription of most aggregation-specific genes is severely curtailed in a population of cells that is exposed to constant level of cAMP. Many of the genes whose transcription is suppressed code for important elements of the chemotactic system. The exception to this rule is the induction of the PDE gene, which occurs whether cAMP is applied in pulses or in constant saturating amounts (Wu *et al.*, 1995b). If the PDE were not secreted under circumstances when cAMP is saturating, paralysis of the cells would continue.

8.4.2 The cAR1 protein has several important domains

The key to purification of cAR1 and eventually cloning *carA* was photoaffinity labeling with 8-azido-^{32}P-cAMP. Applied to cells, the label identified two bands on SDS–polyacrylamide gels, which showed a mobility shift on stimulation with cAMP (Juliani and Klein, 1981; Theibert *et al.*, 1984). The shift in mobility was caused by phosphorylation of a 40,000-kDa glycoprotein. This labeled protein was purified and an antibody was made against it, so clones could be recovered from cDNA libraries (Klein *et al.*, 1988). The proof that the recovered sequence coded for a cAMP receptor depended on showing that overexpression of the cDNA led to an increase in cAMP binding to growing cells (Johnson *et al.*, 1991; Klein *et al.*, 1988). The most compelling demonstration derives from knockout experiments in which *carA* was deleted. Deleting the gene removed almost all cAMP binding and completely blocked aggregation (Sun and Devreotes, 1991).

The cAMP receptor sequence and its presumed structure are shown in Fig. 8.7. The characteristic seven hydrophobic regions of G protein-coupled receptors occur in cAR1. A short N-terminal domain is located on the extracellular face of the membrane. The C terminus of this protein, as in other members of the class, is occupied by a serine-rich element that is the target of the phosphorylation discussed above (Hereld *et al.*, 1994; Klein *et al.*, 1987). There is about a 20-fold increase in phosphate incorporation on cAMP stimulation. This phosphorylation alters the affinity of the receptor for cAMP, reducing it about 5-fold. This may be important when cells confront a passing wave, but it is not the basis of adaptation. PKA affects receptors of this category in higher organisms, but the *Dictyostelium* receptor lacks a PKA phosphorylation site. The basis of adaptation for the cell's responses to cAMP remains unknown.

cAR1 is the major phosphoprotein in isolated membranes and seems to be contained within subdomains of the membrane (Xiao *et al.*, 1997). By immunocytochemistry and GFP labeling, these are known to be evenly distributed, rather than at the leading edge of a moving cell (Xiao *et al.*, 1997). The direc-

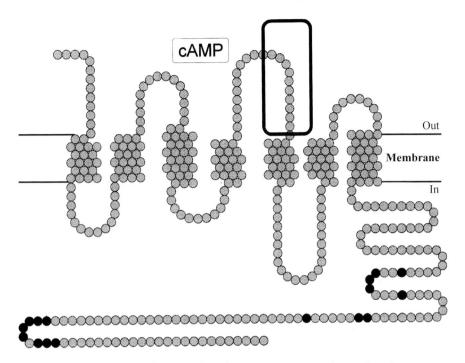

Figure 8.7 The structure of cAR1. cAR1 is a seven trans-membrane domain receptor and is one of a class that binds G proteins. The phosphorylated serines in the carboxy-terminal region are shaded in black. Phosphorylation reduces the affinity of the receptor for cAMP about 5-fold, but cells that carry a truncated receptor, lacking the serine-rich domain, still adapt their cAMP responses in the presence of continuous cAMP. Extracellular amino acids that form the major determinant of affinity for cAMP are boxed. (Courtesy of P. N. Devreotes and C. Parent.)

tional movement of chemotactic cells is not the result of polarized receptor distribution.

There are additional complexities to cAR1. Long before the protein was purified or the gene was known, binding studies detected a receptor on the cell surface and showed that it was both developmentally regulated and kinetically complex (Green and Newell, 1975; Henderson, 1975; Malchow and Gerisch, 1974; van Haastert, 1983; van Haastert and de Wit, 1984). cAR1 binds cAMP with affinities of 25 and 230 nM, and has several activation states. Adenosine has a direct effect on cAMP binding to cAR1. Newell and Ross determined that 5 mM concentrations of adenosine inhibit the formation of individual signaling centers, but do not affect chemotaxis to cAMP or relay of the signal (Newell and Ross, 1982b). In gradients of adenosine prepared in Petri dishes, this leads to huge aggregation territories. Adenosine binds to the cells, probably on the cAMP receptor, cAR1 itself, with a K_i of about 300 μM (Newell, 1982; van Lookeren Campagne *et al.*, 1986). Binding of adenosine inhibits cAMP relay and blocks adaptation in perfusion experiments (Theibert and

Devreotes, 1984). A role for adenosine in later development when pattern is formed has also been postulated, but remains unproven (Schaap and Wang, 1986). One question that arises in these studies of adenosine is whether the concentrations needed to affect the interaction of cAMP with cAR1 are ever reached during normal development.

The domains of cAR1 that are responsible for its various functions have been localized by random mutagenesis and retransformation of *carA*-null cells (Kim *et al.*, 1997a; Milne *et al.*, 1997). Amoebae which have no functional cAR1 provide an assay for functionally important residues. In this type of experiment, mutations are introduced randomly into the cAR1 cDNA and the resulting sequence is introduced into the mutant cells. If an altered amino acid leaves the function of the receptor intact and does not impede its transport through the rough endoplasmic reticulum (ER) and Golgi to the plasma membrane, the transformants aggregate. If essential amino acids are altered, the transforming cDNA will not rescue the phenotype, or the effect will be incomplete. Using this powerful technique to analyze structure–function relationships, domains that participate in the binding site or the interaction with GTP-binding proteins have been determined. Some of the mutations permit one cAR1-stimulated event – calcium influx – while blocking others. These experiments demonstrate that the receptor has multiple activation states and its connection to signal transduction pathways depends on discrete parts of the molecule (Kim *et al.*, 1997a).

Three serpentine receptor genes, all displaying the same topography and intron placement as cAR1, have been recovered (Johnson *et al.*, 1992; Rogers *et al.*, 1997; Saxe *et al.*, 1991a,b). Some of these receptors have affinities that differ from the major aggregation-specific receptor cAR1 (Hereld and Devreotes, 1992; Johnson *et al.*, 1992; Pupillo *et al.*, 1992). The most important of these for aggregation is *carC,* coding for cAR3. Although *carA* is expressed first, *carC* has an overlapping time of expression (Johnson *et al.*, 1993b). *CarA*-null mutants can be encouraged to aggregate if they are given high levels of cAMP, commensurate with the affinity of cAR3. Double mutants of *carA* and *carC* are completely insensitive to extracellular cAMP and will not aggregate under any circumstances. The *carB* and *carD* genes, coding for cAR2 and cAR4, are expressed later in development, have lower affinities for cAMP, and are cell type-specific. The dissociation constant of cAR2 for cAMP is more than 2 μM, compared to 25 nM and 230 nM for cAR1. Deletion of *carB* causes an arrest at the tight-aggregate stage during the formation of the tip (Saxe *et al.*, 1993). Deletions of *carD* result in an alteration of cell type-specific gene expression such that prestalk gene synthesis is reduced and pre-spore gene expression is increased (Louis *et al.*, 1994).

Chimeric genes have been created between *carA* and *carB* and have been used to determine which domains of the resulting proteins are responsible for the dramatic differences in cAMP affinity. The major determinant of affinity has been mapped to a five-residue domain in the third extracellular loop (Kim and Devreotes, 1994). The third intracellular loop is required for all signaling activity (Caterina *et al.*, 1994).

8.4.3   *cAR1 initiates G protein-dependent signal transduction pathways*

The cAMP receptor is a complex molecule because in its sequence there exist domains to bind cAMP, to segregate to subdomains of the plasma membrane, and to drive several signal transduction pathways. One molecule that we expect to interact with such a receptor is a trimeric GTP-binding protein composed of α, β, and γ subunits. Activation of this type of receptor causes a loss of GDP from the α subunit, which is replaced by GTP, and a dissociation of the $\beta\gamma$ subunits from the α subunit. β and γ subunits form a heterodimer. Both Gα-GTP and the $\beta\gamma$ heterodimer may transduce the signal to effector molecules. There is only one Gβ subunit and it is closely related to its mammalian ortho-logues (Lilly *et al.*, 1993). The fact that the Gβ subunit is encoded by a single gene is useful because its disruption blocks all G protein-mediated responses. The small γ subunit, which tethers the trimeric complex to the plasma mem-brane, has recently been described (Ning Zhang, personal communication).

Eight α subunit genes have been cloned, and sequences for several more can be found in a database of cDNAs. Four of these – Gα2, Gα3, Gα4, and Gα5 – are known to have a visible effect on morphogenesis if their genes are deleted (Parent and Devreotes, 1996b). Gα2 mutants do not aggregate, Gα3 mutants require applied cAMP pulses to aggregate (Brandon and Podgorski, 1997; Brandon *et al.*, 1997), Gα4 mutants make no spores (Hadwiger and Firtel, 1992), and Gα5 mutants have a defect in tip formation (Hadwiger *et al.*, 1996). Under certain conditions, there is a role for Gα1 (Dharmawardhane *et al.*, 1994; Rietdorf *et al.*, 1997). There are no obvious defects in Gα7 or Gα8 mutants (Wu *et al.*, 1994). The Gα subunits are 30–35% identical when com-pared among themselves, and have about the same relation to their mammalian counterparts. They cannot be correlated with G_s, G_i or G_q families of mam-malian cells (Parent and Devreotes, 1996b).

Gα2 controls the expression of many genes that are important to aggrega-tion and was originally discovered as a so-called *frigid* mutant by Coukell (Coukell *et al.*, 1983; Kesbeke *et al.*, 1988). The combination of Gα2/G$\beta\gamma$ and cAR1 is the most important one during aggregation. Deletion of Gα2 or Gβ blocks aggregation and development at an early stage. In such mutants there is no stimulation of adenylyl cyclase, guanylyl cyclase, phospholipase C, or polymerization of actin – all events that normally follow a bolus of cAMP. Gβ-null cells are even more impaired than the Gα2 nulls, which can respond to folic acid gradients (Kesbeke *et al.*, 1990). Gβ-null cells cannot do even this and therefore we assume that all chemotaxis, not just that toward cAMP, is mediated through G proteins.

In purified membranes, cAR1 and Gα2 interact with each other. GTP-bind-ing proteins reduce the affinity of their associated receptors for cAMP when stimulated by addition of GTPγS, a non-hydrolyzable analogue of GTP (Kumagai *et al.*, 1991; Snaar-Jagalska *et al.*, 1988). This effect does not occur in Gα2-null cells. It is possible to create strains with a dominantly active Gα2 subunit by altering residues that control the hydrolysis of GTP, and this creates an aggregateless phenotype (Okaichi *et al.*, 1992). The probable explanation is that the constitutively active subunit causes an unabated desensitization of the cAMP receptor or other components in cAMP secretion.

8.4.4   The $\beta\gamma$ subunit stimulates adenylyl cyclase (ACA) in conjunction with other proteins

Dictyostelium contains three adenylyl cyclase genes – ACA, the adenylyl cyclase of aggregation, and ACG, the adenylyl cyclase of spore germination (Pitt *et al.*, 1992). Another adenylyl cyclase, ACB, active in double mutants of ACA and ACG, has been described, but its role is not known (Kim *et al.*, 1998). ACA is the enzyme that is expressed during aggregation and, like the enzymes of mammalian cells and *Drosophila* with which it shares homology and topology, it is a 12 trans-membrane domain protein with two large intracellular loops. ACA depends on GTP-binding proteins for activity, but ACG does not, nor does ACB. The pathways for cAMP and cGMP production diverge after the activation of the GTP-binding protein. The activation of guanylyl cyclase and chemotactic functions by exogenously supplied cAMP are retained in *acaA*-null cells. The time course of cGMP production and cAMP production after a sudden pulse of cAMP, with cGMP appearing so much faster, makes it unlikely that cAMP is required for cGMP synthesis, as shown in Fig. 8.4, panels B and C.

ACG is expressed during spore germination and is structurally unlike ACA, passing the membrane only once and acting as an osmosensor during spore germination (van Es *et al.*, 1996). ACG, when expressed in the *acaA*-null mutants, rescues the phenotype partially – the cells now aggregate and make small fruiting bodies. The cAMP produced by ACG is transported to the extracellular space and is used for chemotaxis, albeit inefficiently (Pitt *et al.*, 1992). This is an informative result because it means that ACA does not mediate the secretion of cAMP. Many ATP-dependent transporters have the same topology as ACA, and so the idea that ACA could be both enzyme and transporter was reasonable, but the basis for cAMP export lies elsewhere. We can expect the eventual isolation of cAMP transporter genes from REMI screening or genome projects. There is no evidence for regulated exocytosis of vesicular cAMP, although there is one report of vesicle formation in response to a cAMP pulse (Maeda and Gerisch, 1997).

The adenylyl cyclase of aggregation may be activated by the $\beta\gamma$ subunit since GTPγS activates the enzyme in membranes that lack Gα2, but not in membranes from Gβ-null mutants (van Haastert, 1997; Wu *et al.*, 1995c). There is precedent for this in type II and type IV adenylyl cyclase from mammalian cells (see Parent and Devreotes (1996b) for references). The putative sites in the intracellular loop through which G proteins activate ACA have been identified in ACA loss of function mutants (Parent and Devreotes, 1995). Constitutively active adenylyl cyclase has also been made, and these proteins retain activity in the absence of G proteins or the other associated proteins described next (Parent and Devreotes, 1996a).

The *Dictyostelium* adenylyl cyclase is regulated by more than the $\beta\gamma$ subunit – something we would not have known had it not been for the isolation of a mutant called synag7, whose adenylyl cyclase cannot be activated by GTPγS or by receptor activation. GTPγS activates the adenylyl cyclase in the synag7 mutant – if a cytosolic extract from wild-type cells is added. This activation of the adenylyl cyclase became the basis for an assay for the CRAC, which was subsequently purified, leading to the cloning of the gene (Lilly and Devreotes, 1994). The sequence revealed that CRAC contains a PH (pleckstrin homology) domain at its amino terminus (Insall *et al.*, 1994). The phenotype of mutants that lack CRAC, as expected for a cell that cannot activate adenylyl cyclase, is a failure to aggregate (Insall *et al.*, 1994). No CRAC homologues have yet been recovered in mammalian cells.

CRAC is a large hydrophilic protein that binds to the inner face of the plasma membrane when the cells are stimulated with chemoattractants or when lysates are treated with GTPγS. In lysates in which Gβ is missing, there is no binding of CRAC to membranes (Lilly and Devreotes, 1995). This result suggests a model in which CRAC binds to $\beta\gamma$ and is essential for its function. CRAC binding sites are rapidly exposed after cAMP binds cAR1. The binding of CRAC is far more specific than initially thought. CRAC binds so swiftly (see Fig. 8.4, panel C) that in a situation in which a cell finds itself with more cAMP on one side than on the other, even transiently, CRAC binds to the membrane region that is exposed to the highest cAMP concentration (Parent *et al.*, 1998). The rapidity of CRAC binding translates the polarity of a cAMP wave into a polar response. The astute reader will note that there is a lag between the binding of CRAC and the activation of ACA, shown in Fig. 8.4, panel C. The gap suggests a need for other elements in cAMP activation.

REMI mutagenesis has led to the discovery of another gene, *pianissimo*, which codes for a 130-kDa cytosolic protein that, like CRAC, is necessary for adenylyl cyclase activation by cAMP *in vivo* or by GTPγS *in vitro* (Chen *et al.*, 1997). Reconstitution with pianissimo restores the activation of adenylyl cyclase, but not in the absence of CRAC, indicating that the functions of CRAC and pianissimo are not redundant. The adenylyl cyclase of a double mutant can be activated only if both molecules are added. Like CRAC, pianissimo is a previously undescribed protein, but has now been found in *S. cerevisiae*, where its deletion is lethal (Chen *et al.*, 1997). Although we are accumulating a repertoire of proteins that regulate ACA, there are a number of proteins that regulate mammalian G proteins that have not yet been identified in *Dictyostelium*, including arrestin and homologues of the β-adrenergic receptor kinases.

The activation of ACA also requires a guanine nucleotide exchange factor called aimless (Insall *et al.*, 1996). Mutants without an aimless gene do not activate ACA, even when GTPγS is added to extracts to promote the release of the $\beta\gamma$ subunit. This suggests that aimless and an associated Ras gene product are involved in activation of the adenylyl cyclase, possibly by generating

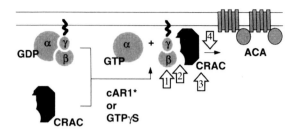

Figure 8.8 The activation of adenylyl cyclase, including CRAC and pianissimo. GDP is initially bound to the Gα subunit. On activation of cAR1 or by binding of GTPγS, Gα and the $\beta\gamma$ subunit dissociate. Note that β and γ form a stable dimer and that the γ subunit is attached to the inner face of the membrane by a lipid moiety. CRAC is transiently recruited to the plasma membrane and activates adenylyl cyclase. The arrows with numbers indicate where pianissimo could function (Chen *et al.*, 1997). (Reprinted by permission of Cold Spring Harbor Laboratory Press.)

CRAC binding sites or by affecting the binding of $\beta\gamma$ subunits associated with CRAC (Parent and Devreotes, 1996b). How aimless and Ras participate in the activation or adaptation of adenylyl cyclase is not known. *Aimless* mutants are not chemotactic, even though a failure to activate ACA should not by itself prevent chemotaxis toward cAMP. The whole process of adenylyl cyclase interaction and activation is inhibited by caffeine in a relatively specific manner, which results in an increase in the average chemotactic de-adaptation time (Siegert and Weijer, 1989; Brenner and Thoms, 1984). Caffeine also has effects later in development.

8.4.5   The signal transduction components beyond ACA

There are other proteins that function downstream of ACA to control cAMP production and responses. REMI analysis has revealed mutants with defective proteins that are incapable of undergoing chemotaxis. One of the mutants that fails to aggregate is defective in ErkB, a MAP kinase that is phosphorylated rapidly and transiently when cAMP binds cAR1 (Gaskins *et al.*, 1996; Segall *et al.*, 1995; Wang *et al.*, 1998). Activation of ErkB by phosphorylation requires cAR1, but not Gα2 or Gβ. The activity of ErkB increases about 40-fold in activated cells over the course of nearly 1 minute, and then declines during the following 5–8 minutes, which is more or less the time course of de-adaptation. The kinases and phosphatases involved in this process are not known.

In the absence of ErkB, activation of guanylyl cyclase is normal, but there is no accumulation of cAMP in response to extracellular cAMP. The failure to accumulate the normal amounts of cAMP may be because ErkB is necessary for activation of adenylyl cyclase. Deletion of *erkB* may cause activation of an intracellular PDE , called RegA (Shaulsky *et al.*, 1996, 1998; Thomason *et al.*, 1998). Genetic evidence suggests that ErkB represses RegA activity and this occurs by phosphorylation (A. Kuspa, personal communication). Deletion of *regA* causes the normally aggregateless *erkB* mutant cells to aggregate. The genetic evidence suggests that the problem in *erkB* mutants is that RegA is not inhibited and cAMP is destroyed as fast as it is made. Deleting *regA* allows cAMP to accumulate and to be used for activation of PKA and for relaying the chemotactic signal.

8.4.6 *RegA is an unusual phosphodiesterase (PDE) and affects the activity of cAMP-dependent protein kinase (PKA)*

The *regA* gene product is a type IV cAMP phosphodiesterase with no ability to hydrolyze cGMP (Shaulsky *et al.*, 1998; Thomason *et al.*, 1998). This PDE is cytosolic, placing it in a position to control the activity of PKA. In addition to being inhibited after phosphorylation by ErkB, RegA can be controlled in another way. In its amino terminus, RegA has sequences that are response regulator (RR) components and the receivers of two-component signal transduction systems. Two-component signal transduction pathways are initiated by sensor histidine kinases that catalyze the phosphorylation of histidine residues in their own intracellular domains. The phosphate is then transferred to aspartate residues in RR regions that control an enzymatic activity. In the case of RegA, this is a phosphodiesterase, although an intermediate phosphate shuttle protein called RdeA is also implicated (Chang *et al.*, 1998). RdeA is the homologue of the yeast phosphotransfer protein Ypd1p, which can complement the *Dictyostelium rdeA* mutation. The phosphodiesterase activity of RegA is increased by phosphorylation of the RR region (Chang *et al.*, 1998; Thomason *et al.*, 1998). Two-component histidine kinases are active signal transducers in bacterial chemotaxis, in budding yeast, and in plants (Loomis *et al.*, 1998). *Dictyostelium* has a number of these proteins (11 at the last count) and communication with the environment or other developing cells may be mediated by histidine kinases as much as it is by seven-trans-membrane domain receptors like cAR1. We will return to the details of two-component signal transduction pathways in Chapter 11, where they play a better characterized role. These signal transducing pathways and the roles that they play in *Dictyostelium* development have been reviewed (Loomis *et al.*, 1998; Thomason *et al.*, 1999).

8.4.7 *The pKA is essential for gene induction and regulates events in the cytoplasm*

During the growth to development transition, the catalytic subunit of PKA is regulated on a translational level, as discussed in Chapter 7. In its absence, a number of important elements of the chemotactic apparatus, including ACA and the phosphodiesterase inhibitor (PDI), are not made because their genes are not transcribed. The PKA is crucial at all stages of *Dictyostelium* development, not just during the growth to development transition. The PKA is a heterodimer of regulatory and catalytic subunits, as discussed in Section 7.3. The regulatory subunit of the PKA binds cAMP with high affinity (10 nM) and dissociates from the catalytic subunit, which is active upon release. In order to re-engage the regulatory subunit, intracellular cAMP must be degraded.

Elimination of the regulatory subunit leaves the active catalytic subunit uncontrolled and accelerates development, while deletion of the catalytic subunit blocks development (Mann *et al.*, 1992). It is possible to block PKA activity by expressing excess amounts of the regulatory subunit – especially a form that does not bind cAMP (Harwood *et al.*, 1992a; Simon *et al.*, 1989). This type of experiment, in which the modified regulatory subunit is expressed from promoters that function at particular times or in different cell types, helped to establish that active PKA is necessary for gene expression during the initial stages of starvation, during aggregation, during the transition to post-aggregative development, and during culmination.

8.5   The developmental regulation of chemotactic components

Many of the components of the chemotactic system are developmentally regulated, as summarized in Fig. 8.9, which was compiled by Peter Devreotes and Carole Parent from Northern blot analysis of many genes. Note that the genes coding for cARs 1–4 are expressed at different times, and that while there are multiple Gα subunits, Gα2 is the most critical for aggregation. Gβ gene expression is constitutive and is also involved in chemotaxis to folate (Blusch and Nellen, 1994; Kesbeke *et al.*, 1990; Mato *et al.*, 1977b). Chemotaxis to folate or derivatives is mediated by Gα4 during development (Hadwiger *et al.*, 1994). A temperature-sensitive Gβ is now available and has been used to show that Gβ is essential for the induction of aggregative gene expression mediated by the serpentine receptors (cARs). The temperature-sensitive allele was also used to show that Gβ is not necessary for later induction of prespore or prestalk genes (Jin *et al.*, 1998).

8.6   cGMP and calcium mobilize the cytoskeleton

Having discussed the elements of cAMP signaling and relay, let us return to the question of calcium and the cGMP pathway. Both of these inductions are mediated by cAR1 in a way that is at least partially independent of cAMP synthesis and relay. The timing of calcium influx in response to cAMP is shown in Fig. 8.4, panel D. This influx is affected in Gβ-null cells. Mutation of Gβ reduces the amount of calcium that enters by about 50% (Milne *et al.*, 1995). Therefore the flow of calcium, which may be essential for the induction of certain genes as well as for activation of cytoskeletal components, is a cAR1-dependent, but not a totally G protein-dependent event (Wu *et al.*, 1995b). The

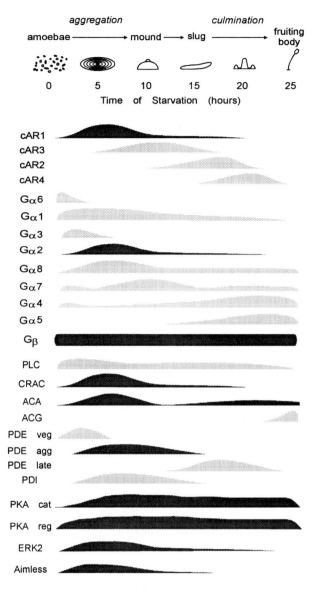

Figure 8.9 A summary of developmentally induced transcripts. The top panel indicates the stage of development, and the times of expression of the various transcripts are shown below. The catalytic subunit of PKA is translationally regulated (see Chapter 7). Those components that are known to be essential for early development (aggregation) are shown in black (Parent and Devreotes, 1996b). Most of the components are discussed in the text. (Courtesy of C. A. Parent and P. N. Devreotes. Reprinted with permission from the *Annual Review of Biochemistry*, Volume 65, © 1996 by Annual Reviews.www.AnnualReviews.org.)

receptors cAR1, cAR2, and cAR3 all mediate the uptake of calcium (Milne and Devreotes, 1993).

The difficulty in assessing the importance of calcium in regulatory events is that there are no mutants that block its uptake specifically. The channel responsible for calcium uptake in the plasma membrane is not characterized. It is possible to wash away calcium from the extracellular space and in this case cells aggregate well – potentially by release of calcium from intracellular stores. *Dictyostelium* contains the normal complement of calcium-binding proteins, including calcineurin (Horn and Gross, 1996), calmodulin (Marshak *et al.*, 1984) and other EF-hand proteins (André *et al.*, 1996). There has been a report of a calmodulin kinase, which could be an important element in calcium-mediated signal transduction (Dunbar and Wheldrake, 1994). We do not know the details of the signal transduction pathways that calcium stimulates. The cGMP response stimulates calcium uptake and calcium inhibits the guanylyl cyclase, but as yet there is no reliable information that calcium must come from a particular source. Calcium is essential for a number of enzymes that take part in motility and chemotaxis, including guanylyl cyclase, phospholipase C, calcineurin, α-actinin, and severin. Its increase may serve a facilitating function, rather than being involved in signal transduction (Malchow *et al.*, 1996a,b; Menz *et al.*, 1991; Newell, 1995b; Newell *et al.*, 1995; Schaloske and Malchow, 1997). There is evidence that later in development, when prestalk cells are induced by differentiation inducing factor (DIF), increased calcium levels are necessary for DIF induction (Schaap *et al.*, 1996a; Verkerke-van Wijk *et al.*, 1998). Recent results from the laboratory of D. Traynor (personal communication) have shown that an inositol tris-phosphate (IP_3)-like receptor is essential for calcium influx. Deletion of the gene for this receptor blocks the increase in intracellular calcium that occurs after stimulation with cAMP. The mutant develops normally.

8.7 cGMP controls motility during chemotaxis

There is a sharp rise in cGMP when cAMP binds to cAR1, as shown in Fig. 8.4, panel B. cGMP is essential for chemotaxis, but the events leading to its synthesis are less understood than those leading to synthesis of cAMP because most of the macromolecules that mediate the effects of cGMP, including the cGMP phosphodiesterase and a putative cGMP-dependent protein kinase, have not been purified, nor have their genes been cloned (Newell, 1995a,b). The expectations for a role of cGMP rest on its rapid but transient synthesis and degradation after stimulation with cAMP, as shown in Figs 8.4 and 8.10. The intracellular concentration of cGMP reaches a peak after 10 seconds and declines by 20 seconds. This precedes the phosphorylation of the heavy chain of myosin II and incorporation of myosin and actin into the cytoskeleton (Liu *et al.*, 1993; Liu and Newell, 1991, 1994). A non-chemotactic mutant, KI-10, makes no cGMP in response to cAMP and does not phosphorylate myosin II or incorporate it into the cytoskeleton, as wild-type cells do. The amount of cGMP in the cell can be elevated for a long period in the streamer F (*stmF*) mutant, which contains a defective cGMP phosphodiesterase, as shown in Fig. 8.10, panel A (Coukell and Cameron, 1985, 1986; Ross and Newell, 1981). If cGMP controls the cytoskeleton, then we would expect altered motility in the *stmF* mutants, and this appears to be the case (Chandrasekhar *et al.*, 1995). Cells that carry the *stmF* mutation move for long periods and form long streams during aggregation, although their initial movement response is inhibited and the speed of their movement is severely reduced. These results suggest a role for cGMP in modulation of the motile apparatus, probably through effects on myosin light chain kinase (Silveira *et al.*, 1998).

A guanylyl cyclase gene has recently been cloned by Roelofs *et al.* (P. van Haastert, personal communication). Surprisingly, this gene does not code for an enzyme with the topology of known guanylyl cyclases. These enzymes are usually either cytosolic or transit the membrane once. Instead, the *Dictyostelium* enzyme has the same trans-membrane topology as a standard adenylyl cyclase, with 12 putative membrane-spanning domains. The gene, when overexpressed in cells, causes an excess of cGMP, but not of cAMP. The cloning of the guanylyl cyclase gene and its structure make it likely that there is a direct interaction with $G\alpha2$ or with $G\beta\gamma$ subunits. Unfortunately, deletion of this gene does not affect the production of cGMP that occurs in response to a cAMP pulse, and therefore some other molecule is responsible for cGMP synthesis during chemotaxis (P. van Haastert, personal communication). Nevertheless, this result gives us the first component of the cGMP signal transduction pathway and could lead to greater understanding of how cGMP mobilizes the cytoskeleton. The other major enzyme that is needed to metabolize cGMP, the cGMP-specific phosphodiesterase, has neither been purified, nor has its gene been cloned. Attempts to purify the enzyme have been extensive but unsuccessful. Nothing is known about a cGMP-dependent kinase activity in *Dictyostelium*, but these enzymes mediate many of the effects of cGMP in mammalian cells.

KI-10 and other chemically induced mutations affect the synthesis or binding of cGMP to cellular components (Kuwayama *et al.*, 1993, 1995). These mutants are incapable of chemotaxis toward folate or toward cAMP. Unfortunately, these are not insertional mutations and we cannot recover the affected genes. One of these mutants blocks all response to folate. The cells are still chemotactic toward bacteria, which suggests that there is an undiscovered chemotactic system and that amoebae do not depend exclusively on folate to locate bacteria (Pan *et al.*, 1972; van Haastert and Kuwayama, 1997). One identified gene has been shown to be necessary for the cGMP response to cAMP stimuleton, this being a MAP kinase called DdMEK1 (Ma *et al.*, 1997). In *ddmek*1 mutant cells, cAMP responses are normal, but the cells of the mutant are incapable of chemotaxis and succeed only in making small aggregates and fruiting bodies.

A critical event in cell movement is the mobilization of myosin. This mobilization, as we saw in Figs 8.2 and 8.10, follows a discrete time course after cAMP stimulation and requires that the myosin heavy chain be phosphory-

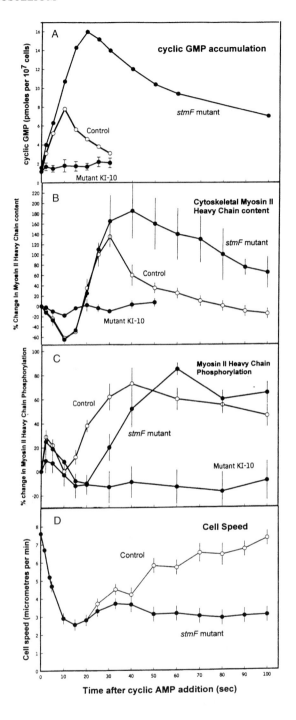

lated. This does not occur in the absence of cGMP, which is the best evidence for a direct role of cGMP in cytoskeletal mobilization and motility (Fig. 8.10, panel C). The evidence for a cGMP-dependent kinase is still weak, although there is a cytosolic cGMP-binding activity (Kuwayama and van Haastert, 1996; Parissenti and Coukell, 1990; Wanner and Wurster, 1990).

Cyclic GMP is also involved in resistance to hypertonic osmotic shock, the importance of which was discussed in Chapter 4 (Kuwayama *et al.*, 1996; Kuwayama and van Haastert, 1998b). The signal transduction pathway leading from cAR1 to the guanylyl cyclase could involve a GTP-binding protein, but at least for the osmotic shock response, $G\beta$-null mutants are not affected, which suggests that activation of guanylyl cyclase does not necessarily require G proteins (Parent and Devreotes, 1996b). Resistance to hypertonic shock probably requires mobilization of the cytoskeleton.

One of the problems with the analysis of guanylyl cyclase activation is that there is no adequate system for reconstitution *in vitro*. Such a reconstituted system led to the discovery of CRAC and pianissimo, but has not yet been applied with equal success to guanylyl cyclase. Despite this failure with broken cells, GTPγS does stimulate guanylyl cyclase activity in electro-permeabilized cells (Schoen *et al.*, 1996). Guanylyl cyclase cannot be activated in Gα2- or Gβ-null cells. These are suggestive results and, because of the structure of guanylyl cyclase, it is likely that there is a direct interaction of guanylyl cyclase and a GTP-binding protein. The cloning of a presumptive guanylyl cyclase gene is an important step in establishing the details of the cGMP pathway.

8.8 The amoebae control extracellular cAMP by secreting a PDE and a PDE inhibitor (PDI)

The genes coding for PDE and PDI are highly regulated. A sensory system must undergo adaptation, and as long as cAMP levels remain elevated in the extracellular space, the receptor will remain desensitized. All of the responses that depend on periodic stimulation of the receptor will cease after one round.

Figure 8.10 The cGMP-mediated events following a pulse of cAMP. Panel A shows the normal rapid synthesis and degradation of cGMP. The mutant KI-10 makes no cGMP and is not chemotactic. The *stmF* mutant makes an elevated level of cGMP and is defective in chemotaxis and regulation of cytoskeletal content of myosin II. Panel B shows that myosin heavy chains integration into the cytoskeleton depends on cGMP levels. Panel C shows that myosin heavy chain phosphorylation depends on cGMP. The *stmF* mutants also show a prolonged calcium uptake, as if the uptake or the secretion of calcium is regulated by cGMP. Panel D shows that increases of cGMP cause a movement defect of the cells. (The data were compiled by Peter C. Newell, University of Oxford, from the following references: Liu *et al.* (1993); Liu and Newell (1994); Newell and Liu (1992); Ross and Newell (1981); Segall (1992); and van Haastert *et al.* (1983).)

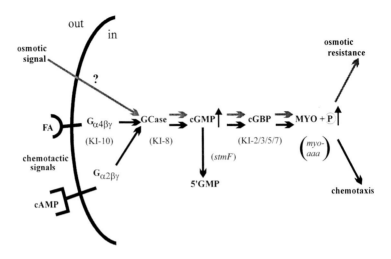

Figure 8.11 Pathways of guanylyl cyclase activation. cAMP, folic acid and osmotic stress all activate guanylyl cyclase (GCase). Folate and cAMP activate different Gα subunits. Osmotic stress does not require a G protein to activate guanylyl cyclase. The KI-10 mutation blocks the rise in cGMP and acts at or before guanylyl cyclase. The *stmF* gene codes for a phosphodiesterase that inactivates cGMP. Cyclic GMP-binding proteins have been detected, and kinases have been detected, but are not well characterized. Myo-aaa is a modified form of myosin heavy chain which cannot be phosphorylated. For reviews see: Newell (1995a); Newell (1995b); van Haastert and Kuwayama (1997). (Reprinted from van Haastert and Kuwayama (1997), with permission from Elsevier Science.)

That a cAMP PDE was needed as an essential part of the chemotactic response was first recognized by Shaffer, although at the time, before cAMP was known, the chemotactic molecule was called acrasin and PDE was called acrasinase (Konijn *et al.*, 1968; Shaffer, 1956). The secreted cyclic nucleotide PDE was first assayed by Chang (Chang, 1968) and then by Malchow and Gerisch, who found the PDE on the cell membrane and in the extracellular space (Gerisch *et al.*, 1972; Malchow *et al.*, 1972, 1975). A second molecule, an inhibitor of the PDE, was also discovered and is now called the PDI (Gerisch *et al.*, 1972). Our own laboratory undertook the purification of both proteins, and ultimately also the cloning and characterization of their genes (Franke *et al.*, 1991; Franke and Kessin, 1981; Orlow *et al.*, 1981).

The PDE is an unusual phosphodiesterase in that it is secreted. The homologues of the *Dictyostelium* PDE in *Schizosaccharomyces pombe* and in *S. cerevisiae* are not secreted, and no similar proteins have been found in higher organisms. The PDE is developmentally regulated and its expression is driven by three promoters – a weak promoter is active during vegetative growth, a strong promoter controls expression during aggregation, and a late promoter is expressed only in stalk cells (Faure *et al.*, 1990; Hall *et al.*, 1993). The three promoters and the times of their expression are shown in Fig. 8.12.

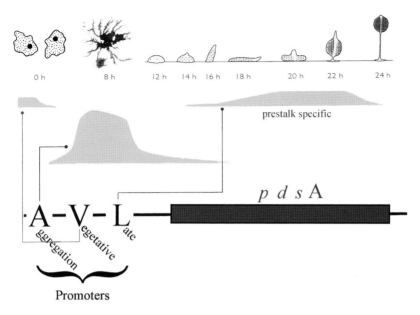

Figure 8.12 The structure of the PDE gene. Three promoters control a single structural gene and determine when and where it is expressed. A weak vegetative promoter (V) functions during growth, and a stronger promoter (A) controls transcription during the aggregation period when more PDE is made. The late promoter (L) limits expression to prestalk cells after aggregation. Deletion or overexpression of the gene blocks aggregation by causing unremitting adaptation or by destroying too much cAMP. Like other genes involved in chemotaxis and cell adhesion, the PDE is induced by cAMP. (Courtesy of Richard Sucgang.)

The PDE is regulated in a way that makes sense when the adaptive properties of cells are considered. Elevated levels of cAMP induce the PDE aggregation-specific promoter and more phosphodiesterase accumulates in the extracellular space, reducing cAMP, terminating adaptation, and preparing the cells for a new encounter with cAMP. Overexpression of PDE leads to rapid development, presumably because the cells spend less time in an adapted state (Alcantara and Bazill, 1976; Kessin, 1988; Wier, 1977). The cells are also sophisticated in their deployment of PDE. The PDE is retained on the cell surface in micro-domains where cAR1 also resides (Xiao and Devreotes, 1997), or it can be released into the extracellular space (Faure *et al.*, 1990; Hall *et al.*, 1993). Constant levels of cAMP promote secretion of PDE into the extracellular space and repetitive pulses promote an increase in membrane-bound PDE (Yeh *et al.*, 1978). Synthesis of the protein does not adapt to constant cAMP levels.

There is only one gene coding for secreted PDE. and deleting it leads to a completely aggregateless phenotype, devoid of membrane-bound or extracellular cAMP PDE activity (Barra *et al.*, 1980; Sucgang *et al.*, 1997). The ability to degrade very local concentrations of cAMP is important to the fate of the cells. If cells with a defective PDE are mixed with isogenic parental cells, the PDE nulls almost always end up as stalk cells and do not survive to the next

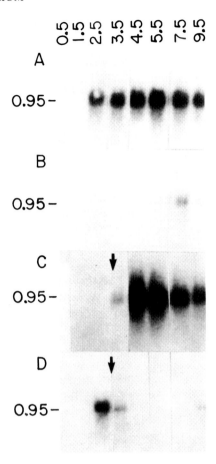

Figure 8.13 Transcription of the PDI gene is tightly regulated. These Northern blots show that the PDI transcript is normally made after several hours of starvation (Panel A). In the presence of extracellular cAMP, no transcript is made (panel B), but if cAMP is removed at the time indicated by the arrow, transcript is made rapidly (panel C). Cyclic AMP added to cells (arrow) that are transcribing the gene causes immediate repression, as shown in panel D (Wu and Franke, 1990). (Reprinted with permission of Elsevier Science.)

generation (Sucgang *et al.*, 1997). There is a strong selection to control extracellular cAMP levels in a very localized way.

One reflection of the selective advantage of controlling extracellular cAMP levels is the PDI. The PDI binds to the PDE, and acting as a competitive inhibitor for cAMP, effectively increases the K_m from about 5 µm to 2 mM (Kessin *et al.*, 1979). The PDI is a heat-stable glycoprotein with a large number of cysteine residues, and like the PDE it is secreted by the cells during development (Franke and Kessin, 1981). Its regulation is precisely opposite that of the PDE, as it is repressed by high levels of extracellular cAMP. The sensitivity of the PDI transcript to cAMP is shown in Fig. 8.13. Removing the cAMP from a population of developing cells leads to a dramatic synthesis and secretion of the PDI (Wu and Franke, 1990). PDE is inhibited in a teleologically satisfying way – apparently the cells never lose control over their extracellular cAMP concentrations.

Our laboratory has deleted the PDI gene, and the result for development is that aggregation territories are slow to get started, waves of cAMP are not propagated in spirals, and the fruiting bodies that form are small (Palsson *et al.*, 1997). The effect can be seen in Fig. 8.14. Palsson and colleagues postulate from theoretical considerations that a sudden release of PDI by a small number of cells is what causes a sudden rise in cAMP and initiates an aggregation territory (Palsson and Cox, 1996).

The extracellular PDE of *Dictyostelium* can have two roles: it degrades the background cAMP between the pulses that come about every 6 minutes, yet it is also likely that the enzyme plays a role in modulating the gradient as it passes the cells. Both Nanjundiah, for the membrane-bound enzyme, and Gomer, for the released form of the enzyme, have calculated that in addition to background degradation, PDE serves to steepen the passing wave and thus improve the chemotactic sensitivity of moving cells (Nanjundiah and Malchow, 1976; R. Gomer, personal communication). Simulations, in which a point source of cAMP is allowed to diffuse with and without continuous breakdown of cAMP, demonstrate that the gradient is steeper when very localized cAMP degradation occurs. The disadvantage of PDE-null cells in chimeras, despite the presence of PDE on surrounding cells, is consistent with the idea that localized degradation is important to the chemotactic sensitivity of the cells.

8.9   The cells can sense density and aggregate size

Cells will not aggregate below a certain density, though why this should be, is not clear. Populations of a few hundred cells are capable of aggregation, and very small fruiting bodies produce viable spores. Perhaps at low densities the energy expended in migration becomes too large, and cells in most species differentiate into microcysts. Alternatively, large fruiting bodies may be much more adaptive than small ones. A molecule called conditioned medium factor (CMF) is secreted by starving cells (Clarke and Gomer, 1995; Jain *et al.*, 1997; Mehdy and Firtel, 1985). CMF, an 80-kDa glycoprotein, has been purified and its gene cloned (Gomer *et al.*, 1991; Jain *et al.*, 1992). Using antisense techniques, strains were prepared which make no CMF, and these strains will not aggregate unless CMF is added (Yuen *et al.*, 1995). Cells without CMF will not respond to cAMP, as if the cAMP receptor is unavailable. One might imagine that CMF could bind to cAR1, but this does not appear to be the case (van Haastert *et al.*, 1996). Rather, CMF binds to its own receptor, which indirectly permits the binding of cAMP to cAR1.

A CMF receptor has been purified by affinity chromatography with CMF bound to columns. Sequencing the 50-kDa retained protein led to a probe and then to a CMF-receptor gene (Deery and Gomer, 1999). The disruption of this

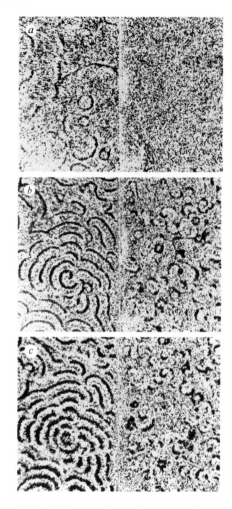

Figure 8.14 Deletion of the PDI gene affects the timing of aggregation and the size of the territories. The territories of the PDI-null cells form late and weakly compared to controls. In the wild-type (left panels), most territories are spirals, but the *pdi*-null (right panels) forms circular patterns. During the formation of spirals, waves of cAMP from adjacent territories collide and the signaling ceases because the cells at the collision point are all adapted and cannot propagate signals to produce overlapping arcs. The behavior of waves depends on the cell density and how far into development the cells have proceeded. The figure shows that the period of the waves decreases with time, reflecting more rapid signaling toward the end of aggregation. The upper two panels (a) show cells at 271 minutes of starvation; the cells in the middle panels (b) have been developing for 344 minutes; and those in the bottom panels (c) for 422 minutes (Palsson *et al.*, 1997). (Reprinted with permission of the National Academy of Sciences, U.S.A.)

receptor's gene causes a 50% reduction in binding of CMF to the cells and a loss of G protein-independent, CMF-induced gene expression. CMF also seems to signal through a G protein-dependent pathway and therefore there is probably another receptor, which would account for the remaining CMF-binding activity. The basis for the conclusion that CMF also works through a G protein mediated pathway is that GTPγS affects binding of CMF to its receptor in membrane preparations, just as it affects the binding of cAMP to cAR1. In cells that lack Gα1, GTPγS does not affect binding of CMF to its receptor(s) (Brazill *et al.*, 1998).

At low densities, wild-type cells do not make enough CMF to aggregate, but they will aggregate at low densities if recombinant CMF is added. Gα1-null cells are capable of aggregating at densities that are much lower than wild-type, without added CMF, and the suggestion is that they no longer require CMF because the signal transduction pathway that is normally inhibited by Gα1 becomes constitutive (Brazill *et al.*, 1997, 1998).

Dictyostelium employs a cell-counting mechanism to regulate the size of its aggregates (Brown and Firtel, 2000). The first gene involved in regulating size (*smlA*) was isolated using a mutagenesis technique that employed antisense cDNA to identify genes that are required for normal development (Spann *et al.*, 1996). The absence of the *smlA* gene product causes abnormally small aggregates to form. Recently, the *countin* gene was cloned. Disruption of the *countin* gene results in unusually large fruiting bodies (Brock and Gomer, 1999).

8.10   Auto-induction of the chemotactic response

In the early hours of starvation, a field of cells shows flickers of responsiveness as weak signals diffuse from a few cells, and other cells respond with half-hearted movements. The ability to respond appears before the ability of a few cells to secrete cAMP autonomously (Raman, 1976; Raman *et al.*, 1976). The early responses die out, but as time goes on, the rhythmic response becomes stronger and stronger. This can be appreciated from the increasingly powerful response of the wild-type cells shown in Fig. 8.14. Some aggregation centers signal more frequently than their neighbors and gain dominance, entraining other cells. This progression from a weak, barely visible response to robust signaling and migration is due to an increase in the elements of the chemotactic system on all of the cells as they are induced by cAMP over a period of hours. cAMP induces the cAR1 gene, the PDE gene, the ACA gene and also the contact site A gene, the product of which (gp80) mediates cell-to-cell adhesion. The transcription of these genes depends on conditions that permit de-adaptation of the receptor – with the exception of the PDE gene, which is transcribed even when the cells are adapted. When there is adequate PDE, adaptation is transient and cAMP impressively induces its own sensory apparatus.

Which signal transduction pathways carry out the induction of genes that mediate chemotactic aggregation? The first step in induction is binding to cAR1. After this, the data are more difficult to interpret. As we have seen, the flow of information from cAR1 can follow G protein-dependent and independent pathways. The PDE is inducible by a G protein-independent pathway (Jin *et al.*, 1998; Wu *et al.*, 1995b). None of the transcription factors that bind the PDE aggregation-specific promoter is known. Other genes require a standard G protein-signaling pathway. The cAR1 and gp80 proteins are not induced in a Gβ-null mutant, and in a mutant that carries an allele that codes for a temperature-sensitive Gβ, there is no auto-induction effect at restrictive temperature. Therefore these two proteins (and probably others) are induced through a G protein-dependent mechanism (Jin *et al.*, 1998). A cAR1 transcription factor, CRTF (cAMP receptor transcription factor) has been purified and cloned. This zinc-finger domain protein acts in a developmentally regulated manner. Deleting it prevents accumulation of cAR1 above basal levels (A. R. Kimmel, personal communication). The adenylyl cyclase gene requires the transcription factor Ddmyb2 for activity, and this in turn requires PKA (Otsuka and van Haastert, 1998). Whether Ddmyb2 and CRTF act to coordinate other genes that are expressed during the aggregation period is not known (see Fig. 7.1).

8.11 Mobilizing the chemotactic machinery

How do the amoebae sense direction? The foregoing rather lengthy descriptions detail the components that the cells employ to synthesize, detect, and destroy signals. In addition, we have an idea of the complexity of the networks that cause secretion of cAMP to be periodic. We have not yet confronted the problem of how cells move up a gradient of cAMP. Amoebae measure cAMP across their cell diameters. They will do this in a static gradient or in an oncoming wave. Unlike bacteria which sample, move, and then sample again to determine whether an attractant or repellant is increasing or decreasing, *Dictyostelium* amoebae can be placed in a gradient and will immediately extend pseudopods toward the highest concentration of cAMP or folate. The cells can measure differences across their cell bodies that are very small – a differential of receptor occupancy of perhaps 2%. For reviews see van Haastert (1995); Caterina and Devreotes (1991); Chen *et al.* (1996); Devreotes and Zigmond (1988); Parent and Devreotes (1996b, 1999); Pitt *et al.* (1990); and van Haastert and Devreotes (1993).

The desensitization of the cAMP response is essential for proper directional movement. Aggregating amoebae can orient when the number of occupied receptors in the front is 500 and the number in the rear is 400, or when the numbers are 5,000 and 4,900. Adaptation functions like the tare on a balance, the absolute difference in receptor occupancy being the important number. This ability depends on adaptation, which removes the effect of the first 400 or 4,900 receptors, leaving the cells able to register the effect of 100 activated receptors. We do not understand how adaptation works, but we do know that it is rapid, does not depend on intracellular cAMP production or phosphorylation of cAR1, and survives in cell lysates. If a population of cells is stimulated with cAMP such that *in vivo* the adenylyl cyclase adapts, we cannot then break the cells, stimulate with GTPγS and activate adenylyl cyclase (Theibert and Devreotes, 1986). The adenylyl cyclase can be activated by GTPγS in broken cell preparations if cells were not previously stimulated with cAMP and have not adapted.

In stable gradients, the amoebae are capable of detecting a 2% difference in receptor occupancy between the fronts and the backs of the cells. Moving *Dictyostelium* amoebae are highly polarized with a leading pseudopod and a trailing structure called the uropod. The distance from leading edge to uropod is about 10 μm. This translates to a very small difference in receptor occupancy across the cells – on the order of 250 receptors out of 40,000–200,000 (Mato *et al.*, 1975). Van Haastert calculates that the difference in the number of occupied receptors between front and back is even less and approaches the theoretical minimum (van Haastert, 1995). Recall that the cAR1 receptors are found in microdomains spread evenly over the plasma membrane (Xiao *et al.*, 1997).

The fact that cells can move in stable gradients has been useful for measuring their sensory capability, especially with the advent of analytical microscopy capable of recording and analyzing the responses of many cells (Fisher *et al.*, 1989). These experiments demonstrated the developmental regulation of the response to cAMP, and showed that the maximum efficiency of the response, measured by the accuracy of chemotaxis, occurs in a gradient of 25 nM/mm with a midpoint of 25 nM, which is not far from the K_d of cAR1. Orientation in less steep gradients is less efficient, with cells extending pseudopods more frequently in a direction away from the cAMP source. When the cAMP concentration is far above or below the K_d, a steeper gradient is necessary for orientation. It has been suggested that cells are incapable of migrating in a stable gradient, but this is thought to be incorrect (Vicker, 1994).

There is a difference in the response of a cell placed in a stable gradient and a cell with waves of cAMP sweeping over it. A wave has a steeper differential, and this is not constant. During aggregation, cells confront waves of cAMP and these can be visualized and quantified by an isotope dilution technique (Tomchik and Devreotes, 1981). In the experiments shown in Fig. 8.15, we see the actual cAMP waves detected by competition of labeled cAMP bound to a cAMP-binding protein. This method was a conclusive demonstration that chemotaxis is mediated by cAMP, and allowed Tomchik and Devreotes to determine the amplitude of the cAMP wave. Under standard conditions, the amounts of cAMP can vary between 10^{-6} M and 10^{-8} M at the peak and the trough of a wave, respectively. The time between peaks, when directly measured, varies between 6 and 8 minutes. The behavior of the cells is consistent with movement up the gradient as long as it is positive, and then a cessation of movement as the peak passes. The cells do not reverse direction and head back up the receding wave when the sign reverses. A cell migrating in a gradient can be made to reverse direction by application of cAMP at its rear, but this is due to very large amounts of cAMP. In the case of the cells shown in Fig. 8.16, the reverse gradient is less than the peak of cAMP they have just experienced and adaptation of movement responses will prevent reverse migra-

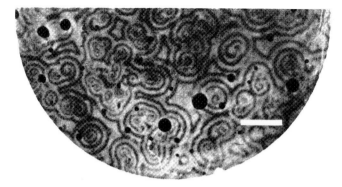

Figure 8.15 Isotope dilution methods can be used to detect cAMP waves. The figure shows the localized concentrations of cAMP (light areas) derived from fields of aggregating cells. The bar represents 1 mm. Waves of unlabeled cAMP produced by the cells displaced ^{3}H-cAMP bound to a cAMP-binding protein. The results confirmed the periodic nature of cAMP signaling on a substratum. See Tomchik and Devreotes (1981) for experimental details. (Reprinted with permission from the American Association for the Advancement of Science.)

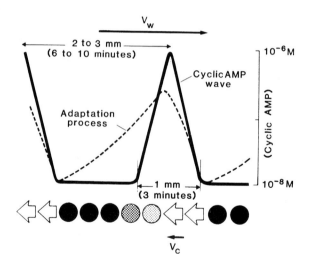

Figure 8.16 The movement of cells depends on the dynamics of a passing wave. This schematic, created by Tomchik and Devreotes from data like that presented in Fig. 8.15, shows that the concentration of cAMP in the wave varies over two orders of magnitude. The wave moves at about 300 μm/min, toward the right. A single radial line of cells is represented along the bottom of the figure. Under the experimental conditions used, each symbol represents a band of about 50 cells. Arrows show cells that perceive the rising gradient and move at 10–20 μm/min. The two shaded circles show cells that are in a falling gradient. The dark circles represent cells that are no longer in a gradient and move randomly. Note that the adaptation process initiates as soon as the amoebae perceive the first rise in cAMP. As long as the cAMP concentration is going up, adaptation does not stop signaling or directional movement. As soon as the peak passes, adaptation blocks signaling and begins to decay. (Courtesy of Peter Devreotes. From Tomchik and Devreotes (1981). Reprinted with permission from the American Association for the Advancement of Science.)

tion. Without adaptation there would be no net inward movement, and the exercise would cease to have any selective advantage. Although phosphorylation of the serine residues in the carboxyl-domain of cAR1 is not responsible for adaptation, this phosphorylation does cause a reduction in affinity of the cAMP receptor, and this reduction may have adaptive value in preventing the cell from reversing its direction.

In a study by Wessels *et al.* (1992) the movement of cells was followed with sophisticated optics and recording. The results are consistent with those of Tomchik and Devreotes, but instead of concentrating on the gradients, Wessels *et al.* observed the behavior of individual cells. In these experiments, the period between waves was calculated to be about 7 minutes. During the first 150 seconds, the slope of the wave was positive, the cells moved toward the aggregation center with a high velocity and relatively little lateral pseudopod extension, as shown in Fig. 8.17. As the wave moves, there is a period in which the slope becomes zero at the crest and there is no change in concentration. During this phase, which lasted about 1 minute, the cells stop moving, make no pseudopods, and even suppress the movement of particles within the cells, as if they are frozen. The phase is analogous to the cringe response, which occurs when cells perceive an instantaneous rise in cAMP on all sides. As the wave passes, the slope becomes negative, the cells stop their cringe response, and start to extend pseudopods randomly for about 180 seconds. Thus the amoebae behave according to the temporal dynamic of the passing wave, as described by Tomchik and Devreotes. The speed of amoebae migrating up a gradient is about 10 μm/min, or about one cell diameter per minute. With each passing wave they move two or three cell diameters toward the aggregation center.

8.12 Polarity of movement and the polarity of G-protein activation

What provides the basis for the directional sensing of the cell? The cells have polarity of movement, but cAR1 and Gβ are evenly distributed around the cell, including the thinnest filopods. Amoebae maintain chemotactic sensitivity at all points on their surface. The same is true of neutrophils, whose chemotactic receptors are also evenly distributed and do not concentrate at one place on the cell as chemotaxis is induced. As has been pointed out by Parent and Devreotes (1999), this arrangement of receptors is ideally suited for directional sensing through a spatial mechanism. In contrast, the elements of the cytoskeleton that mediate movement are localized. Actin and actin-binding proteins such as filopodin, cofilin, and coronin localize to newly extended pseudopods (Aizawa *et al.*, 1997; Gerisch *et al.*, 1995; Kreitmeier *et al.*, 1995; Westphal *et al.*, 1997). Myosin and myosin heavy chain kinase are stimulated to move to the cell cortex by cGMP (Abu-Elneel *et al.*, 1996; Dembinsky *et al.*, 1996).

Parent and colleagues reasoned that the basis of the polarity, and the connection between signal reception and mobilization of the cytoskeleton, must lie downstream of the cAMP receptors (Parent *et al.*, 1998; Parent and Devreotes,

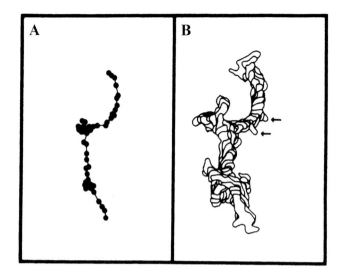

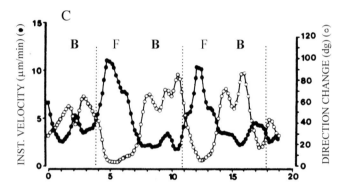

Figure 8.17 The behavior of individual cells in a gradient. In panel A, a wave of cAMP came from the bottom, and the dots mark the movement of the centroids of an individual cell. Panel B shows the perimeter images of a cell and indicates that there is suppression of lateral pseudopod extension (arrows). Panel C shows that when a cell is in the front of a wave, its velocity increases (black dots) and its random extension of pseudopods decreases (open circles). F and B indicate the deduced fronts and backs of several successive waves. (Reprinted from Wessels *et al.* (1992) by permission of Wiley-Liss, Inc., a division of John Wiley & Sons, Inc.)

1999). To visualize G-protein activation in live chemotaxing cells, the green fluorescent protein (GFP) gene was fused to the CRAC gene. This construct, which produces a functional CRAC protein, was used because it is known that CRAC is recruited to the plasma membrane following chemoattractant receptor stimulation (Parent *et al.*, 1998).

If cells are given a saturating dose of cAMP, the CRAC–GFP protein leaves the cytoplasm and binds within seconds around the entire periphery of the cell.

Under these circumstances, the cells cringe – they contract and round up and do not move. Binding of CRAC–GFP to the intracellular face of the membrane can be induced with as little as 1 nM cAMP. If cAMP is applied from one side of the cell – either in a stream or in a pulse – CRAC–GFP moves within 1–2 seconds to the membrane at the side that receives the highest dose of cAMP (Fig. 8.18). For a more complete view of these events consult the Web pages listed in Chapter 13. The binding to the membrane depends on the pleckstrin homology (PH) domain. The protein or lipid in the membrane to which the PH domain of CRAC and other proteins binds is not yet known. The binding of CRAC to one side of a cell in a gradient implies that adenylyl cyclase is activated on the upstream side of the cell, and not in the rear of the cells. The delay in the activation of adenylyl cyclase means that other components, including pianissimo, must also be recruited. Among the PH proteins that behave in a similar way is protein kinase B (PKB) (Meili *et al.*, 1999; Firtel and Chung, 2000). Stimulus-induced recruitment of cytosolic proteins to the plasma membrane is an emerging theme in chemotaxis and signal transduction.

Recruitment to the membrane does not depend on actin or cell movement, because inhibitors of actin polymerization do not affect movement of CRAC. One might imagine that the extra width that filopodia bestow on a cell would be essential for detecting a spatial gradient and while this may occur, it is not essential. Binding of CRAC to the uphill side to the cells does not depend on filopodia, which collapse in the presence of actin inhibitors.

Recruitment to one side of the cell and adaptation is a way of amplifying an initial difference in concentration across the cells. Receptor occupancy is not coincident with CRAC–GFP localization. The difference in receptor occupancy across the cell is about 10%, but there is a much more dramatic localization of CRAC at the leading edge of chemotactic cells. This localization tends to be persistent. Only if cAMP is removed or applied uniformly across the cells does the localization of CRAC–GFP disappear. In the biochemical experiments described earlier, stimulus increments which are applied uniformly, induce biochemical events that adapt and subside. Yet cells in a static gradient respond by moving up the gradient, and they are persistent. Therefore the presentation of agonist at the leading edge overcomes the tendency of the cells to adapt.

One model that accounts for these results has previously been presented (Caterina and Devreotes, 1991; Devreotes and Zigmond, 1988; Fisher, 1990). Parent and Devreotes (1999) speculate that the sites on the inner face of the membrane that recruit CRAC and PKB could also recruit other PH-domain proteins, including exchange factors that induce actin polymerization. These PH-binding sites (as yet uncharacterized) would be generated by a membrane-bound enzyme that is rapidly activated in direct proportion to the number of locally occupied cAMP receptors. The sites would be removed by another enzyme that is soluble and is slowly activated by the average number of occupied receptors of the entire cell. Even in a chemotactic cell, many activated receptors will not be at the leading edge and will contribute to the activation

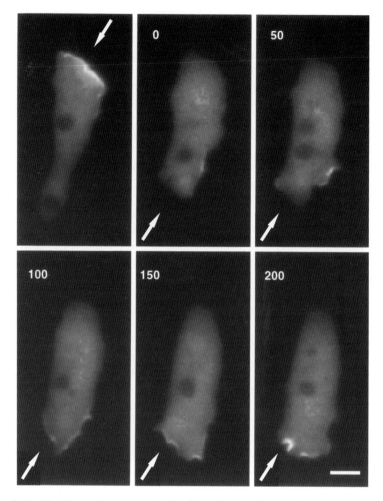

Figure 8.18 CRAC translocates to the leading edge of chemotaxing cells. A micropipette filled with 1 μM cAMP was positioned to the upper right corner of the frames. The micropipette was then moved to the lower left and CRAC accumulated there. Numbers in the upper left corner are seconds after the observations began. The scale bar represents 6 μm. Protein kinase B behaves in a similar way (Parent *et al.*, 1988; Meili *et al.*, 1999). (Courtesy of Carole Parent and Peter Devreotes. Reprinted with permission from Cell Press.)

of the diffusible enzyme that removes PH-binding sites. In the case of the biochemical experiments described earlier in this chapter, a sudden increase of cAMP would cause a rapid increase in the PH-binding sites and then a slow decrease as the degrading enzyme is activated. The downstream pathways would also subside, as observed in Figs 8.4 and 8.10. If a naïve cell were placed in a gradient, it would initially respond by making PH-binding sites at all points, but slightly more at the higher concentration. If the rate of degradation of PH-binding sites is high enough to remove them all, except at the point where the

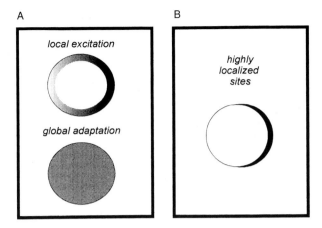

Figure 8.19 A model for directional sensing. Imagine a naïve cell placed in a gradient of chemoattractant, with the highest concentration on the right side and the lowest on the left side. Panel A shows that excitation occurs locally near the membrane and is only slightly graded in proportion to receptor occupancy. Inhibition (of PH-binding sites) occurs everywhere at an intermediate level determined by the average fraction of occupied receptors (lower circle of the panel A). Panel B shows that with time, global inhibition has created a sharp difference. Essentially, the lower figure in panel A has been subtracted from the upper figure, creating a persistent region of activated signal transduction pathways and a locally dynamic cytoskeleton. (Courtesy of Carole Parent and Peter Devreotes. Reprinted from Parent and Devreotes (1999), with permission from The American Association for the Advancement of Science.)

gradient is highest, the response will become refined and localized to the leading edge, as shown in Fig. 8.19. This would mean that the binding sites for PH domains and the associated downstream responses could persist indefinitely in the front of the cell and the cell would continue to move up the gradient.

A great deal remains to be learned about the signal transduction cascades and the cytoskeletal mobilization that controls chemotactic movement. Among the open questions are: do other signaling events, in addition to CRAC and PKB activation, occur at the leading edge of the cell? What is the nature of the PH-binding site, and what are the enzymes that generate and remove these sites? Is there more than one type of site? What are the molecules that connect signal perception with mechanical movement? There is good reason to believe that the directional sensing mechanisms used in *Dictyostelium*, both for signal transduction and for cytoskeletal mobilization, are also used in leukocytes and other chemotactic cells (Devreotes and Zigmond, 1988; Parent and Devreotes, 1999).

8.13 Mathematical simulations to explain signaling in *Dictyostelium*

The important thing about modeling is that a model uses applied mathematical techniques to test a series of assumptions about rate constants, substrate concentrations, K_ms, and receptor dynamics, and to ask whether the modeling assumptions yield a system with the properties observed by experiment. If it does, then this implies we have an understanding of the system that is more than the simple sum of the experimental observations. Signaling in *Dictyostelium* is one of the most thoroughly modeled of all biological phenomena, in the sense that there is now a mathematical model, based on the biochemical properties of the system, that accurately reproduces cAMP wave initiation and propagation. The model also suggests how events at the molecular level can be used to explain the evolution of territory size, and a possible function for spiral rather than circular waves of cAMP. It integrates information on time and space scales to explain the macroscopic behavior of developing *Dictyostelium*. A variety of experimental tests that might not otherwise have occurred to us can be inferred from a close examination of the mathematical models.

We have seen earlier how cAMP synthesis, secretion, and degradation, when properly coupled, can lead to cAMP waves that propagate over many millimeters as circular or spiral wave forms. Much of our progress in understanding wave propagation comes from the early work of Martiel and Goldbeter, who asked if the known enzymatic properties of the cAMP signaling system could be used in a mathematical model to simulate sustained cAMP oscillations in a well-stirred reaction vessel (Martiel and Goldbeter, 1984). Using a realistic set of values for cAMP synthesis, degradation, transport, and receptor occupancy, Martiel and Goldbeter were able to show that cells (actually virtual cells) in suspension could be perturbed by a small cAMP pulse in such a way that the entire cell population would continue to oscillate spontaneously (Goldbeter, 1996). Their model has been widely used by others and has the following essential features. First, it is an excitable system, which is to say that prior to the addition of a small cAMP pulse, it is in a state that can be easily perturbed by small changes in cAMP concentration. Discovering the parameters of the system that can give rise to an excitable state and successive responding and adapting states is no mean feat, and is one of the main goals of mathematical modeling in this system. The addition of a small pulse of cAMP causes a large non-linear increase in the rate of cAMP synthesis and secretion, and this (as we have seen earlier in this chapter) leads to an adaptive response where cAMP synthesis decreases and cells become non-responsive. However, in the simulations, cAR1 occupancy begins to decrease as cAMP levels drop, and this leads to another round of cAMP synthesis and secretion. This process continues, the cells oscillate spontaneously, and cAMP levels rise and fall, in good agreement with the observations presented earlier.

When the Martiel and Goldbeter model was first formulated, many components of the cAMP system had not been characterized. It is thus to be expected that many of the details of the model would have to be reformulated. However, the insights concerning the non-linearities built into CAR responses are not likely to change, since they are at the heart of how the oscillator that controls aggregation works.

The models led to an oscillator that can send out a regular cAMP pulse. What will happen when a few spontaneously oscillating *Dictyostelium* cells on a damp surface secrete cAMP? The cAMP diffuses, and under certain conditions, a coherent wave can propagate across a field of starving, i.e., excitable cells, as first observed by Cohen and Robertson (1971a,b) and studied by Alcantara and Monk (1974) and Gross *et al.* (1976). To describe wave propagation, a spatial term must be introduced to the kinetic model of Martiel and Goldbeter, and this spatial information is contained in an expression describing the diffusion of cAMP. The same kinetics describing cAMP synthesis, binding, and degradation apply, but gradients of cAMP can now build up locally and diffuse away as the cells secreting the molecule down-regulate their receptors. Meanwhile, cells at the low edge of the cAMP gradient are stimulated to produce cAMP, and a wave is propagated.

Several groups have modeled wave propagation and the subsequent break-up of circular waves (Dallon and Othmer, 1997; Lauzeral *et al.*, 1997; Levine *et al.*, 1996; Palsson and Cox, 1996; Palsson *et al.*, 1997; Segel *et al.*, 1986; Tang and Othmer, 1995; Tyson *et al.*, 1989; Vasieva *et al.*, 1994). This theoretical work has led to deeper insights in two areas. First, it suggests that our understanding of wave propagation is essentially complete; and second, it has stimulated our thinking about the possible significance of the geometry of waves, circular or spiral. An analytical model for spiral wave formation was first described by Tyson and Murray, who explored how spirals persist and propagate (Tyson *et al.*, 1989). Unlike circular cAMP waves that expand from an oscillating center, spirals rotate about a small core of non-excitable cells. They do not need an oscillatory center and can thus persist in a weakly excitable field of starving amoebae once a few waves centered on oscillating cells have been established (see Palsson *et al.* (1997) and Lauzeral *et al.* (1997) for brief discussions). Spiral waves propagate at a higher frequency than circular waves because of the higher curvature of spirals versus circles propagating in an excitable system. Consequently, when circular and spiral waves collide and extinguish each other, spirals dominate with time because the next crest of a given spiral arrives at the previous extinction point before the next circle arrives. Because of their higher frequency, circles also extinguish oscillators (Lee *et al.*, 1996).

There are several competing models for spiral formation (Dallon and Othmer, 1997; Lauzeral *et al.*, 1997; Levine *et al.*, 1996; Palsson and Cox, 1996). One of them predicts that PDI, the inhibitor of the cAMP PDE, alters excitability locally, favoring spiral formation (if PDI is secreted by randomly placed cells, the cAMP concentration will quickly rise near the secreting cell, and this will cause a wave to propagate and break the symmetry of a nearby circle). This theoretical result has been tested by studying strains that produce no PDI (see Section 8.9 and Fig. 8.14). In these strains, spirals rarely form (Palsson *et al.*, 1997). The aggregation territories and fruiting body sizes of the PDI mutants are much smaller than those of the wild-type parent. The regulatory circuits governing PDI and PDE synthesis may have evolved to regulate territory and fruiting body size. In wild-type *Dictyostelium,* the larger territory swept out by a spiral cAMP wave may lead to a larger and a more more easily dispersed fruiting body. One of the attractions of this area of *Dictyostelium* research is the opportunity to tie molecular function at the level of biochemical action to the various intermediate levels of organization through to the behavior and fitness of the whole organism.

Othmer and Schaap (1998) have recently presented a comprehensive model that examines the subjects discussed above, and is particularly strong in its discussion of the diffusion of cAMP from a signaling cell. It is also a good study of theory of signal transduction, but like all such studies, it underestimates biochemical complexity. The mathematical simulation of developmental events is not limited to aggregation. Refined mathematical models have been produced which define the behavior of cells during mound formation and at later stages of development (Bretschneider *et al.*, 1997; Schaap *et al.*, 1996b; Vasiev *et al.*, 1997; Vasiev and Weijer, 1999).

There are other areas where modeling has been applied in *Dictyostelium*. Complex kinetic models that simulate carbohydrate metabolism have been constructed by Barbara Wright and colleagues (Albe and Wright, 1994; Wright, 1991; Wright and Albe, 1994). While this has been controversial, it led to the purification of a number of important enzymes. These include trehalase (Killick, 1983), trehalose-6-phosphate synthetase (Killick, 1979), and UDPG pyrophosphorylase (Franke and Sussman, 1973; Gustafson *et al.*, 1973).

9

Differentiation and Adhesion in the Aggregate

As they collect in the mound, cells begin the transcription of cell type-specific genes, develop adhesion mechanisms, and secrete an acellular covering called the sheath. The loose mound goes on to form a compact hemispherical aggregate and prepares the anterior–posterior pattern of the future slug. The mound has been called the crucible of cellular differentiation and morphogenesis. In earlier days of *Dictyostelium* research, it was difficult to detect the complex gene induction events and cellular movements that occur within the mound, but as markers of cell-type differentiation have improved, the nature of the differentiation events has begun to unfold. Great efforts in microscopy, notably by the laboratories of Weijer, Siegert, McNally, and Soll, have detected elaborate coordinated movements of cells, not just during aggregation, but within the confines of the aggregate. In this chapter we will examine some of the early events that occur after the amoebae aggregate – how they secrete a covering and how they induce new genes in the loose aggregate. The aggregate is where most prestalk- and prespore-specific genes are first transcribed. This subject encompasses a large and complex literature, and an attempt will be made to put this in context. Following an examination of how the pattern of prestalk and prespore cells forms, the cellular adhesion mechanisms that *Dictyostelium* employs to hold cells together will be analyzed. Normally, the formation of the mound and its organizing tip are considered together with the slug, which topologically is an elongated tipped aggregate that has developed the ability to move. To simplify the subject, most of the discussion of the organizing tip and slug behavior has been deferred until Chapter 10.

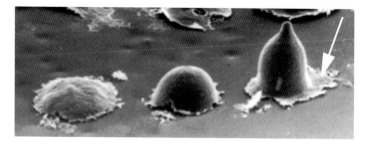

Figure 9.1 An acellular material forms over the aggregated cells. In this scanning electron micrograph, the sheath can be seen covering the aggregated cells, so that no individual cells are visible. The arrow points to sheath material at the base of a tipped aggregate. Note that the loose mound gradually compacts and becomes hemispherical, prior to forming a tip. (Courtesy of Mark Grimson and R. Lawrence Blanton, Texas Tech University.)

9.1 Amoebae form a sheath during aggregation

One of the events that occurs in the aggregate, before the induction of cell type-specific genes, is the formation of an acellular sheath that covers the aggregate (Hohl and Jehli, 1973). Amoebae form a coating on their exterior and this persists during development (Fig. 9.1). For a review, see Wilkins and Williams (1995). The sheath forms over the mound and encases the slug. As the slug moves, it leaves behind a collapsed casing of sheath material with the imprint of cells on it (Early *et al.*, 1988a; Vardy *et al.*, 1986). This material does not seem to be extensive between the cells in the aggregate, but can be seen as a glistening layer on the surface of the aggregate and is visible as the trail behind a slug. Cells invest considerable energy in the sheath. The primary structural component is cellulose, but there is a large glycoprotein component comprising about 50% of the dry weight of the sheath material isolated from migrating slugs (Freeze and Loomis, 1977; Hohl and Jehli, 1973; Wilkins and Williams, 1995). These glycoproteins, which appear from antibody studies to be developmentally regulated, can be released by denaturing reagents or cellulase digestion (Grant and Williams, 1983).

A number of roles have been proposed for the sheath. A simple role of protection from marauding nematodes, which cannot penetrate the sheath, was suggested (Kessin *et al.*, 1996). As has been pointed out in an earlier chapter, nematodes and *Dictyostelium* occupy the same ecological niche and nematodes feed on individual amoebae, at least in the laboratory. Alternatively, Loomis and Farnsworth have suggested that the sheath, which varies in thickness from the front to the back of the slug, has a more developmentally important role (Farnsworth and Loomis, 1975). Perhaps the sheath constrains the diffusion of critical signaling molecules. A mutant of *D. mucoroides* that lacks a sheath nevertheless manages to develop, although with weak and curly stalks, so there may be no other essential requirement for the sheath in development (Larson *et al.*, 1994). A mutant that cannot make cellulose because of a defect in cellulose synthase is not blocked in the formation of the tight aggregate. Therefore synthesis of the cellulose component of the sheath cannot be responsible for the compaction that occurs during the transition from loose aggregate to tight aggregate (R. Blanton, personal communication). These mutants fail at a later stage when cellulose forms a structural component of stalk and of spores, as well as of the extracellular sheath.

9.2 The transition from aggregation to loose mound is mediated by a transcription factor named GBF

How is the transition from aggregation to true multicellular development managed? Part of the answer has come from examining genes whose deletion causes an arrest just after aggregation, in the loose mound stage, when the aggregate is still flat and has not begun to assume the hemispherical shape of the tight aggregate, and has not formed a tip. One protein that acts at this stage is a transcription factor called G box-binding factor or GBF (Haberstroh *et al.*, 1991; Hjorth *et al.*, 1989, 1990; Powell-Coffman *et al.*, 1994; Schnitzler *et al.*, 1994). GBF has two zinc-finger domains and binds to the G-rich regions that are present in the promoters of many genes that are expressed after aggregation (Firtel, 1995). Mutation of these G-rich sequences reduces the binding of GBF *in vitro* and limits the expression of fused reporter genes *in vivo*. The G-box sequences to which GBF binds do not confer cell-type specificity, but rather the ability to be expressed during multicellular development – they can be exchanged between prestalk and prespore genes where they work efficiently and confer gene expression at the correct time, but do not alter cell type-specific expression (Ceccarelli *et al.*, 1992; Hjorth *et al.*, 1990; Powell-Coffman *et al.*, 1994).

Mutant cells without GBF do not induce the expression of genes that are specific to prestalk cells or to prespore cells. Aggregation is normal in these mutants, which indicates a precise time of action of GBF. GBF must also be essential for the proper induction of adhesion systems in the mound, since null mutants make only loose aggregates (Firtel, 1995; Schnitzler *et al.*, 1994). Whether GBF is required to make the components of the sheath is not known.

How is expression of GBF induced to appear just before the cells need it? If cells in suspension are given nanomolar pulses of cAMP every few minutes, they mimic aggregation and induce a number of the aggregation-specific genes that were discussed in Chapter 8. These have become known as pulse-induced genes. There are also several cAMP pulse-repressed genes (Mann and Firtel, 1989; Vauti *et al.*, 1990; Wu and Franke, 1990). A sustained increase in extracellular cAMP is thought to occur as the cells crowd into an aggregate. The idea of high constant cAMP levels after aggregation derives in part from the high levels of cAMP required to induce other post-aggregative genes in cell suspensions (Chisholm *et al.*, 1984; Kay, 1979; Landfear and Lodish, 1980; Town and Gross, 1978). If cells that have gone through a cAMP-pulse regime in suspension are treated with high levels of cAMP in the micromolar range, they induce a new set of genes. One of these is GBF, whose induction by high levels of extracellular cAMP requires the cAMP receptor cAR1, but does not require other signal transduction elements such as GTP-binding proteins or the catalytic subunit of PKA (Schnitzler *et al.*, 1995). Its induction is therefore different from that of the genes that mediate chemotaxis (Jin *et al.*, 1998).

The transcription of *carA*, upon which GBF depends, would normally adapt – high constant levels of cAMP would cause the gene to cease transcription. However, when the cells have aggregated, the induction of *carA* switches to a mechanism that responds to high constant levels of agonist. This switch is mediated by a second promoter that is only activated after aggregation (Louis *et al.*, 1993). Thus, continuous induction of GBF can be explained even under adaptive conditions.

These experiments on the induction of GBF and other post-aggregation genes are based on addition of cAMP to cells in suspension. However, is there evidence that in the normal aggregate, cAMP levels actually rise to micromolar levels? In one study, Traynor *et al.* (1992) measured the total amount of cAMP in cells aggregated on a filter. By making assumptions about the total intercellular volume, it is possible to calculate that in the aggregate and later stages, cAMP levels may approach micromolar levels. Abe and Yanigasawa (1983) also measured total cAMP during aggregation, and similar numbers were found. The most compelling study was done by Brenner (1978), who measured intracellular and intercellular cAMP through development. There is a peak of cAMP at the aggregate stage, but substantial levels exist throughout development. Brenner's calculations are also consistent with a micromolar intercellular cAMP level. This level is sustained in the face of a substantial extracellular phosphodiesterase activity and thus requires a continuous synthesis of cAMP by the cells (Wu *et al.*, 1995a).

It is reasonable to assume that *in vivo* high extracellular levels of cAMP acting through cAR1 induce GBF, and that its induction is critical for development beyond the loose aggregate stage. Are there are other genes that function during this period? A mutation in *lagC* causes a phenotype that is similar to that of the GBF-null cells – no prestalk- or prespore-specific markers are induced and there is a block at the loose aggregate stage (Sukumaran *et al.*, 1998). Overexpression of GBF in the *lagC* mutant strain suppresses the *lagC* defect, which is evidence that these two elements form part of a pathway that is essential for the transition from the aggregation to the early multicellular phase. LagC has been suggested as an adhesion molecule (see below), but the fact that the mutant phenotype can be suppressed by overexpression of GBF argues against such a role.

Gollop and Kimmel (1997) have shown by analysis of the *carC* gene that it is prestalk cell-specific and can be induced by GBF. There are two GBF-binding sites in the *carC* promoter, and both are required for prestalk-specific expression – deleting one causes cAR3 to be produced only in prespore cells. GBF is necessary, but not sufficient for induction of prestalk or prespore cells. Overexpression of GBF during vegetative growth does not cause prestalk- or prespore-specific genes to be expressed early. However, genes that are normally induced in the mound can be induced in vegetative cells if vegetative cells that constitutively

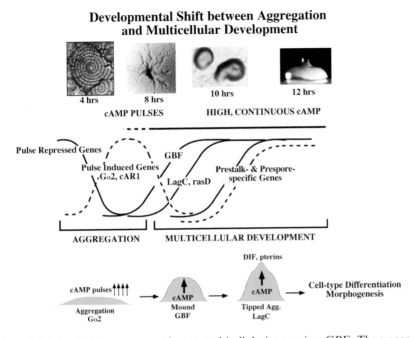

Figure 9.2 The shift from aggregation to multicellularity requires GBF. The upper panel shows the cells at the various stages of development. The cartoon in the middle shows that certain genes are repressed by cAMP pulses and others are induced. GBF is induced as the cells enter into aggregates. *LagC* and *rasD* are post-aggregative genes that are induced after GBF, but not in mutants that lack GBF. The lower panel shows the presumed levels of cAMP during aggregation and in the mound. Other signaling molecules, including DIF (see below) and pterins (Hadwiger *et al.*, 1994), may be involved in further morphogenesis. (Courtesy of Richard Firtel, University of California at San Diego.)

express GBF are given high levels of cAMP (Schnitzler *et al.*, 1995). This result indicates that there is a mechanism that interacts with the GBF induction pathway to induce cell type-specific genes and that it is present in vegetative cells that express GBF. The result also infers that vegetative amoebae must have low levels of cAMP receptors. All of the data indicate that GBF is employed in the transition to multicellularity, and while not itself cell type-specific, it is essential for the induction of genes that are expressed in only one of the two cell types. Several reviews give a more detailed account of GBF and its role (Brown *et al.*, 1997; Firtel, 1995), and a possible sequence of events is illustrated in Fig. 9.2.

9.3   The discovery of cell type-specific genes provided important tools to study pattern formation

As the cells undergo the transition from loose aggregate to tight mound, they begin to synthesize products that will be necessary for the differentiation of the stalk and the spore cells. These follow GBF by several hours and include a variety of structural proteins, some of which are listed in Table 9.1. *EcmA* and *ecmB* are expressed exclusively in the stalk lineage. Other gene products also qualify as prestalk-specific – including the low-affinity cAMP receptor cAR2, which is expressed on a subset of prestalk cells (Saxe *et al.*, 1996). *TagB* is a very early marker of the prestalk lineage (Shaulsky *et al.*, 1995). The late promoter of the phosphodiesterase gene is expressed exclusively in prestalk cells (Hall *et al.*, 1993). All of the promoters of these genes have potential sites for binding of GBF and presumably other factors. The *rasD* gene and the cysteine proteinase 2 gene are expressed in prestalk cells, but some workers believe that their expression may not be exclusive (Esch *et al.*, 1992; Jermyn and Williams, 1995). This has been a matter of some controversy. Several genes, such as *pspA*, *cotA*, and *cotB*, were found to be excellent markers of the spore lineage (see Table 9.1 for references).

The cell type-specific genes and their promoters were isolated and examined in detail (see below). One of the most useful approaches is to fuse the promoter of a prestalk or prespore gene to the β-galactosidase structural gene and transform the amoebae. Transformants produce a pattern of β-galactosidase expression that is exclusive to prespore or to prestalk cells, depending on the promoter (Dingermann *et al.*, 1989). Within a lineage, a gene, such as *ecmA* (extracellular matrix) may be expressed before another prestalk gene, for example, *ecmB*. The details of some of these promoters and their expression patterns will be delayed until Chapter 10, when the slug and its distinct zones will be discussed.

9.4 Pattern formation begins in the mound

Classical experiments by Raper (1940b) determined that cells in the anterior of the slug gave rise to stalk, and those in the posterior gave rise to spores. Cells can convert from one to the other and are not irreversibly determined. What these experiments did not establish was when the first cells of each lineage appear, and under what conditions. From experiments with vital dyes and antibodies it has long been known that products that are specific to prestalk or to prespore cells appear before the slugs form.

There are two general models that attempt to explain how patterns of differentiated cells develop. The first model contemplates a gradient of inducer such that cells at the highest concentration are driven to one fate, while those at a lower concentration have another outcome. A series of gradients containing positional information is the means by which the body plan of *Drosophila* is established. One of the predictions of such a positional information model is that differentiation of cells occurs in a block – stalk cells, for example, should develop in an area of high stalk cell inducer. At a particular concentration of inducer, all cells should form stalk. The initial differentiation of spore cells from multipotent amoebae should also occur within a group of cells. The pattern of the slugs, with stalk cells in the front and spores in the rear is

Table 9.1. *Prestalk- and prespore-specific markers*

Expression Site	Locus	Product	Likely function	Null phenotype	References
	pspA (D19, *psA*)	SP29	Extracellular matrix	None known	Early *et al.* (1988b)
	cotA	SP96	Spore coat	Permeable spore coat?	Orlowski and Loomis (1979)
Prespore	*cotB*	SP70	Spore coat	Permeable spore coat?	Fosnaugh *et al.*
	cotC	SP60	Spore coat	Permeable spore coat?	(1994)
	spiA	Dd31	Spore coat	Spore instability	Richardson and Loomis (1992)
	ecmA (*pDd63*)	EcmA (ST430)	Extracellular matrix	Slug morphology	Morrison *et al.* (1994)
Prestalk	*cprB*	CP2	Cysteine proteinase	None known	Datta *et al.* (1986)
	ecmB (*pDd56*)	EcmB (ST310)	Extracellular matrix	None observed	Williams and Morrison (1994)

Note: The genes and their products listed in this table are some of the most important in the resolution of specific problems in *Dictyostelium*. The list is not complete – a number of other genes are known to be prestalk- or prespore-specific (Firtel, 1995). Additional prespore- and spore-specific genes are described in the text and in Section 12.2. Names in parentheses are from the older literature, but are still occasionally used.

consistent with a gradient from front to rear or rear to front. In this model, the initial differentiation and the anterior–posterior pattern are established at the time that the cells interpret the gradient.

The cell-sorting model of pattern formation makes an opposite prediction – that the cells do not differentiate in blocks, but rather that differentiation depends on something else – perhaps with regard to when the cells last divided. Expression of cell type-specific genes would initially occur in a salt and pepper fashion with both primordial cell types mixed. Subsequently, they would sort out, in response to a gradient of some signal, perhaps emanating from the tip and dissipating in the substratum. In this case, initial differentiation does not form the pattern, which depends on an additional cell movement event. By this model, the striped pattern of prestalk cells in the front and prespore cells in the rear of a slug does not imply that the expression of prestalk or prespore markers takes place in the slug – it may have occurred much earlier. For reviews on these problems, see Bonner (1952a, 1992); Bonner and Cox (1995); Gross (1994); Inouye (1992); Kay (1992); MacWilliams (1991); MacWilliams and Bonner (1979); Nanjundiah (1997); Schaap *et al.* (1996b); and Takeuchi *et al.* (1982).

One way to distinguish such models is to find genes or proteins that identify one cell type and are never expressed in the other, such as those shown in Table 9.1. With cell type-specific genes in hand, it becomes possible to ask when they are expressed and if they are expressed in blocks or in a salt and pepper fashion and then sort out. Transplantation experiments also become possible. If we only consider prestalk and prespore cells, marked so that they can be distinguished as soon as possible after differentiation begins, it appears that differentiation occurs in an interspersed way in the loose aggregate (Datta *et al.*, 1986; Williams *et al.*, 1989). The cells then sort out to form the pattern that we observe – with most prestalk cells at the apex and the prespore cells at the base. Thus, a positional information model is ruled out.

We have to explain the fact that neighboring cells could have such different fates. The support for intrinsic differences is relatively strong, with many groups reporting that position in the cell cycle influences developmental choice between prestalk and prespore cells (Araki *et al.*, 1994; Gomer and Firtel, 1987; Leach *et al.*, 1973; McDonald and Durston, 1984; Ohmori and Maeda, 1987; Weijer *et al.*, 1984a). The nature of these cell cycle effects will be discussed below.

9.5 The position of PstA, PstO, and PstAB cells

The cells that will come to occupy the front of the slug are known as pstA cells because they express the gene *ecmA* most strongly, as described in Chapter 10 on slug morphology. Behind the pstA cells are a group of cells called pstO cells, which express the *ecmA* gene at a lower level. A third cell type that is relevant to the current discussion is the pstAB cell, which expresses both *ecmA* and *ecmB*. These occupy a core in the front of the slug (Jermyn *et al.*, 1989). PstA cells are first detectable in low numbers in loose aggregates (Traynor *et al.*, 1992). At this time, they are randomly scattered among prespore cells and are not clustered at the apex (Williams *et al.*, 1989). When sections were analyzed, the mixture of prespore and prestalk cells was confirmed (Traynor *et al.*, 1992). During the course of development, a morphologically distinct tip appears and the pstA cells become predominantly localized to it (Fig. 9.3). As Williams and Jermyn (1991) have pointed out, the pstA cells are not induced to differentiate in the tip – rather this happened elsewhere, before the tip formed, and then they migrated to the tip.

There may be a role for positional gradients of information in the initial differentiation of two types of prestalk cell. If we consider the subtypes of prestalk cells, the situation becomes more complex and we have to raise the possibility that within one lineage – the prestalk cells – a gradient of inducer may force the cells to differentiate in a position dependent manner (Early *et al.*, 1995). The promoter of *ecmA* has been analyzed and portions of it have been found that direct transcription in pstO cells but not in pstA cells, or in pstA cells but not in pstO cells (Early *et al.*, 1993). Using these sophisticated promoters fused to β-galactosidase it has become possible to identify two prestalk

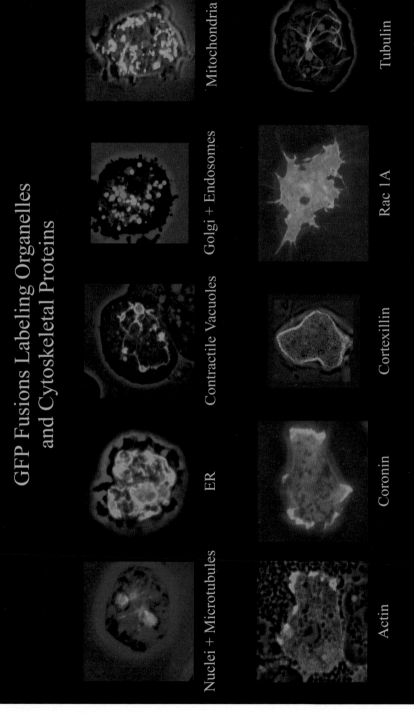

GFP Fusions Labeling Organelles
and Cytoskeletal Proteins

Nuclei + Microtubules   ER   Contractile Vacuoles   Golgi + Endosomes   Mitochondria

Actin   Coronin   Cortexillin   Rac 1A   Tubulin

10 μm

Plate 1 The compartments of a living *Dictyostelium* cell can be visualized. The upper panels show organelles in *D. discoideum* cells visualized by GFP fusion proteins. From left to right: nuclei labeled with histone 2 (and anti-tubulin antibody showing microtubules in red); membranes of the endoplasmic reticulum labeled with calnexin; the contractile vacuole network labeled with dajumin; the Golgi apparatus labeled with golvesin, and endosomes filled with a fluid-phase marker in red; mitochondria loaded with a construct that carries a targeted signal. In all panels except for the Golgi apparatus, fluorescence images are superimposed to phase-contrast images. Lower panels, GFP constructs representing dynamic cytoskeletal structures: actin and the actin-binding protein coronin accumulating at motile protrusions of the cells; cortexillin decorating the entire cell cortex; overexpressed Rac1 inducing filopods; microtubules and the centrosome visualized by α-tubulin. (Courtesy of Professor Günther Gerisch, with contributions by Mary Ecke, Jan Faix, Daniela Gabriel, Annette Huettig, Jana Koehler, Natalie Schneider, and Monika Westphal.)

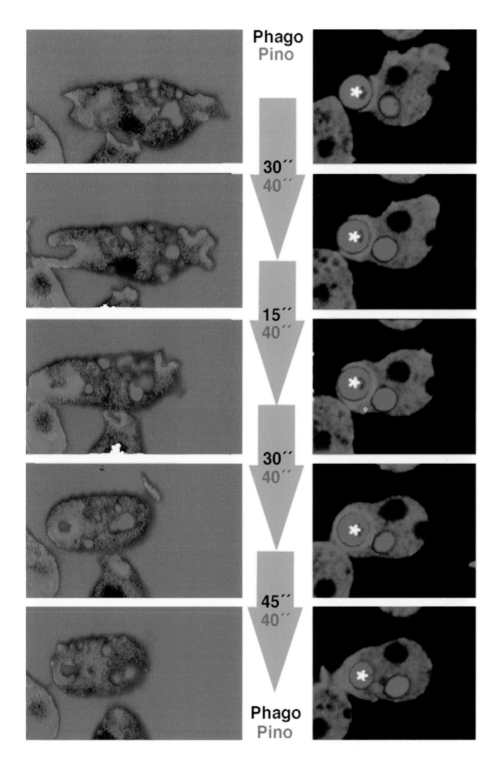

Plate 2 Comparison of cytoskeletal dynamics during macropinocytosis and phagocytosis *in vivo*. Fluid-phase uptake (left) is visualized with a red-fluorescent tracer added to the medium. For particle uptake (right) covalently labeled yeast cells (asterisk) were used. The actin cytoskeleton is revealed by a fusion of coronin to the green fluorescent protein (GFP). Single optical sections were taken at intervals using a confocal laser scanning microscope. The times between the consecutive frames are given in seconds. For each endocytic process, the following stages are shown (from top to bottom): (1) Accumulation of the cytoskeleton at the periphery of the cell. (2) Membrane protrusion toward the left of the cell. (3) Late stage just prior to engulfment of the liquid droplet (left) or yeast particle (right). (4) Endocytic vesicle surrounded by cytoskeletal coat. (5) Dissociation of the coat and release of the endosome into the cytoplasm. (Courtesy of Markus Maniak, University of Kassel.)

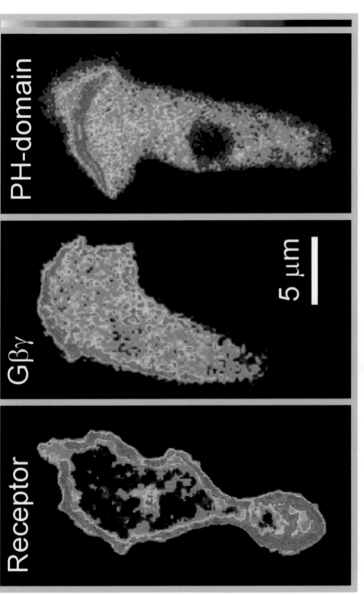

Plate 3 Orientation of proteins during chemotaxis. Left: Confocal image of GFP-tagged chemoattractant receptors in living amoebae undergoing chemotaxis. The chemoattractant gradient is established using a micropipet located just below the frame. (From Z. Xiao and P.N. Devreotes, Johns Hopkins Medical Institutions.) Center: Confocal image of GFP-tagged Gb-subunits in living amoebae undergoing chemotaxis. The chemoattractant gradient is established using a micropipet located just above the frames. (From T. Jin, N. Zhang, and P.N. Devreotes, Johns Hopkins Medical Institutions.) Right: Confocal image of GFP-tagged CRAC in living amoebae undergoing chemotaxis. CRAC-GFP specifically translocates to the leading edge of newly elicited pseudopods. The chemoattractant gradient is established using a micropipet located near the upper right corner of the frame. (From C.A. Parent and P.N. Devreotes, Johns Hopkins Medical Institutions.) See Tian Jin, Ning Zhang, Yu Long, Carole A. Parent, and Peter Devreotes, *Science* **287**, 1034–1036, 2000.

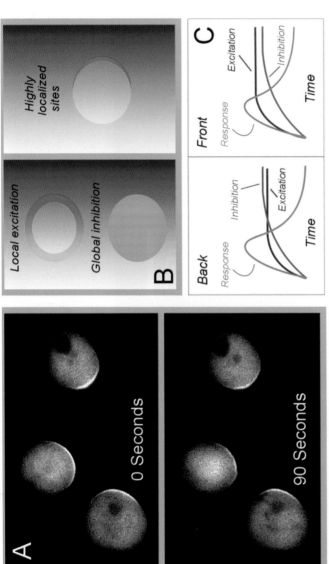

Plate 4 A model to explain directional sensing. Panel A shows that directional sensing does not require cell movement or the actin cytoskeleton. Amoebae expressing CRAC-GFP were treated with latrunculin for 20 minutes to disrupt the actin cytoskeleton. Note the round shape of the cells. A pipet containing cAMP was placed to the lower right center of the frames. The CRAC-GFP label accumulates on the side of the cells facing the source of cAMP and persists. Panels B and C show an illustration of a hybrid spatial-temporal model for directional sensing. The gradient of chemoattractant is depicted as a gradation in the gray background crossing the field from right to left. (Left) Excitation occurs locally near the membrane and is slightly graded (from yellow to red). Inhibition occurs throughout the cell and is uniform at an intermediate level (orange). The response, the difference between excitation and inhibition, is sharply localized on the right side of the cell (red). The steady-state situation, several minutes after the cells have been placed in a stable spatial gradient, is illustrated. When the cell is initially placed into the gradient, both sides will experience an increment in stimulus and respond. However, as steady state is reached, inhibition exceeds excitation at the back of the cell and excitation exceeds inhibition at the front, as shown in panel C. See text for further explanation and a website for on-line dynamic illustrations. (From Parent and Devreotes (1999), reprinted with permission. Courtesy of Carole A. Parent and Peter N. Devreotes.)

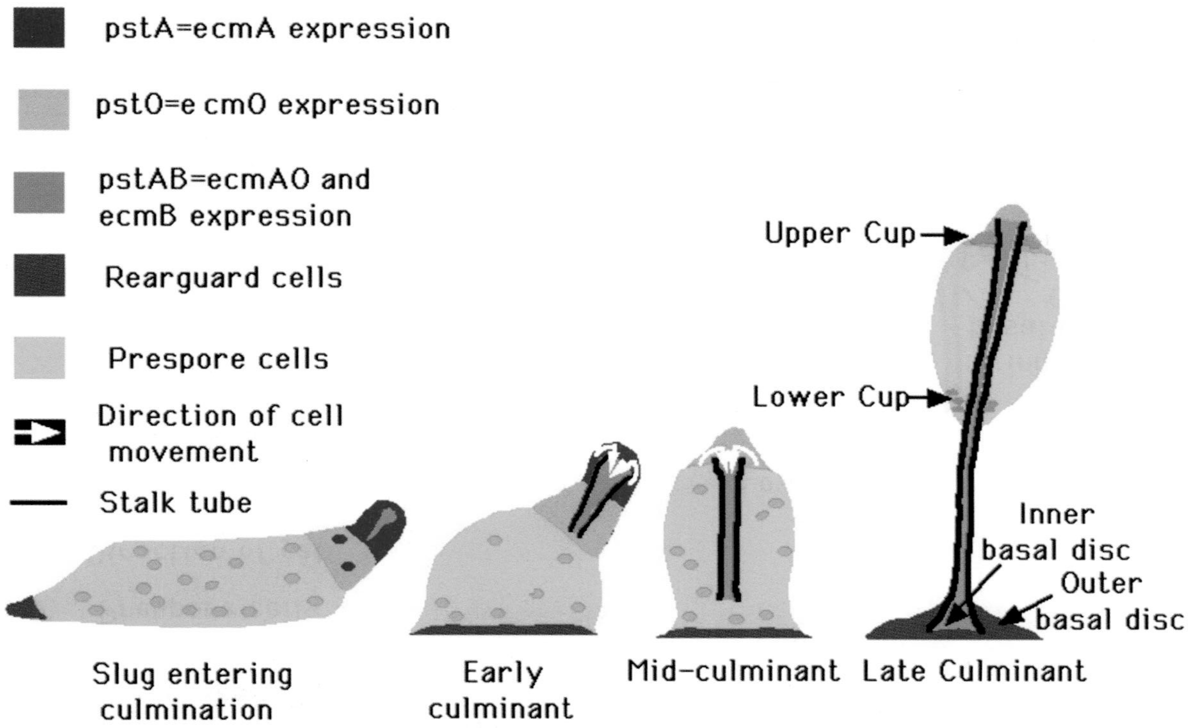

Plate 5 The movement of cells during culmination. This is a diagrammatic representation of culmination where, for the sake of clarity, the band of pstB cells that will form the outer basal disc (see Plate 6) is not shown. The *ecmA* promoter can be divided into two parts, a proximal part (the *ecmA* region) that directs expression predominantly in cells within the tip (i.e., in the pstA cells) and a distal part (the *ecmO* region) that directs expression in cells in the back of the prestalk region (the pstO cells) and in a subset of the anterior-like cells. The latter population has been termed the pstO:ALC (Abe *et al.*, 1995). The whole *ecmA* promoter (the *ecmAO* promoter) directs expression in all these cell subtypes, which collectively are called the pstAO population. (Courtesy of Jeffrey Williams, University of Dundee. Reprinted with permission.)

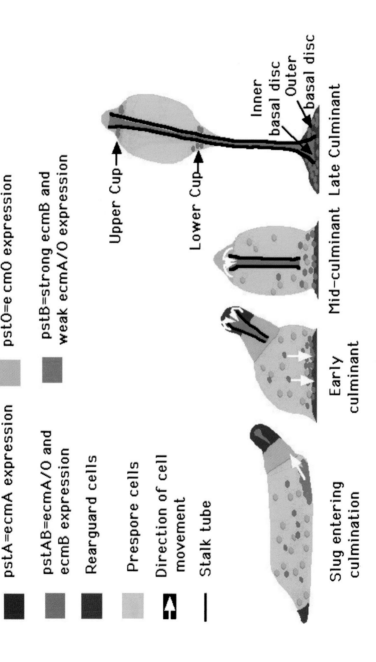

pstA=ecmA expression

pstO=e cmO expression

pstAB=ecmA/O and ecmB expression

pstB=strong ecmB and weak ecmA/O expression

Rearguard cells

Prespore cells

Direction of cell movement

Stalk tube

Slug entering culmination

Early culminant

Mid-culminant

Late Culminant

Upper Cup

Lower Cup

Inner basal disc

Outer basal disc

Plate 6 The movement of pstB cells at culmination. The pstB cells are defined by selective staining with neutral red and because they express the *ecmB* gene at a high level relative to the *ecmA* gene (Jermyn *et al.*, 1996; Dormann *et al.*, 1996). They have a complex movement pattern during slug migration (Dormann *et al.*, 1996). For the sake of simplicity, separate pstA and pstO populations are not shown, rather the behavior of the entire pstAO population is represented. (Courtesy of Jeffrey Williams, University of Dundee. Reprinted with permission.)

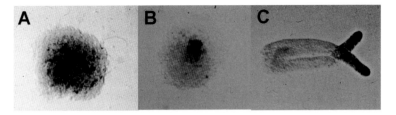

Figure 9.3 The pstA cells form in a dispersed fashion among the prespore cells. Panel A shows the prestalk cells marked with β-galactosidase in an aggregate before the formation of the tip. Panel B shows that the marked cells have migrated toward the tip. Panel C shows their eventual location in two intertwined slugs. For other experiments confirming these conclusions, see Williams *et al.* (1989); Traynor *et al.* (1992). (Courtesy of R. Insall, University of Birmingham, and R. Kay, Medical Research Council, Cambridge.)

cell types as the aggregate forms and develops (Early *et al.*, 1995). These experiments showed that pstA and pstO cells appear in spatially separate regions of the aggregate, with the pstA cells forming an outer ring surrounding the pstO cells. The pstA cells migrate through the pstO and prespore cells and eventually arrive at the tip. Below them lie the pstO cells and then the prespore cells. The authors suggest that, because there is a differentiation of the two subclasses of prestalk cells and this occurs in a block, a gradient of an inducing signal is responsible. The presumed inducer, which would qualify as a morphogen because it induces cell differentiation as a function of concentration, is DIF, a chlorinated hexanephenone, to be described below.

The conclusion from the pattern formation studies is that excellent cell type-specific markers have become available and they have been used to detect the earliest differentiation of the several classes of prestalk cells and of prespore cells. The two competing models of differentiation – the random differentiation-cell sorting model and the positional information model – may both be employed, although the major differences between the prespore and the prestalk progenitors appear to occur by a mechanism that does not depend on a cell's position. The positional information mechanism – differentiation *en bloc* in response to a gradient – may also occur, but additionally has the feature that the differentiated cells change position. Only the separation of the prestalk cell population into subclasses allowed the pattern to be observed. Evolution has been eclectic in the mechanisms it has forced on *Dictyostelium* development.

9.6   Differentiation inducing factor (DIF) and the origin of prestalk cells

The most thoroughly characterized inducer in *Dictyostelium*, other than cAMP, is differentiation inducing factor (DIF). The assay for this molecule was first developed by Kay and Town, who used a particular strain of

Dictyostelium plated in Petri dishes at a low density, such that the cells never touched and the complication of cell-to-cell interactions was removed. Under these conditions, isolated cells could be made to turn into stalk cells, complete with a large vacuole and a cellulose coat, if supernatants from developing cells were added to the culture (Kay and Jermyn, 1983; Town and Stanford, 1979; Town et al., 1976). The results of such an assay are shown in Fig. 9.4. Using this bioassay, Kay and his colleagues purified DIF and determined its structure in a *tour de force* of purification and structural analysis (Morris et al., 1987). The structure of DIF-1 and its presumed route of synthesis from a polyketide precursor are shown in Fig. 9.5 (Kay, 1998). Other DIFs, presumably breakdown products, exist (Kay, 1997). A DIF-1 binding protein has been detected and the K_d of the binding protein is about 2 nM (Insall and Kay, 1990). A dechlorinase, which removes the aromatic hexyl-chloride, has been assayed and may control the activity of the molecule. The dechlorinase is localized to the prestalk cells of the slug (Kay et al., 1993). DIF-like molecules have not been detected as natural products in other organisms, but DIF has been detected in other Dictyostelid species (Kay et al., 1992; van Es et al., 1994).

What does DIF do to cells? There is no detectable DIF in growing cells, but a strong rise occurs during aggregation with a peak at the slug stage (Brookman et al., 1982). The increase, initially detected by bioassay, has been confirmed by biochemical methods that employ labeling of DIF with radioactive choride (Kay et al., 1992). The concentration of DIF in a slug has been calculated to be between 50 and 200 nM (Kay, 1997). Using the bioassay, the relative concentrations of DIF have been determined along the axis of the slug. Slugs were dissected into prespore and prestalk regions, and pools of these fragments were extracted to purify DIF. The purified extracts were used to show that, on a normalized basis, there was approximately twice as much DIF in the prespore region than in the prestalk region. The prestalk region was later shown to be enriched in dechlorinase, so the lower levels in the prestalk region can be explained (Kay, 1997). There are obstacles in working with DIF and these result primarily from the difficulty of the assay. The absence of a means of detecting DIF immunologically has precluded efforts to determine whether it is present in a gradient in the mound, or in the slug.

For DIF to induce genes the cells must be given a prior exposure to cAMP, which activates protein kinase A and allows the cells to arrive at a point in development at which DIF can work. Whether sensitivity to DIF requires GBF activity has not been tested directly. If GBF is required for cell type-differentiation, the prediction is that without GBF, DIF would have no effect. DIF induces prestalk markers, such as *ecmA,* and represses prepore gene transcription. This is true for a number of prespore- or prestalk-specific markers, as shown in Table 9.2. The inductive power of DIF led to the isolation of the very useful *ecmA* and *ecmB* genes. When applied to migrating slugs, DIF increases the percentage of cells in the prestalk zone, shifting the cell-type ratio. DIF reduces the expression of some prespore genes, as the results in Fig. 9.6 demonstrate. An initial model stated that DIF in the front of the slug caused the differentiation of prestalk cells. This idea had to be modified because

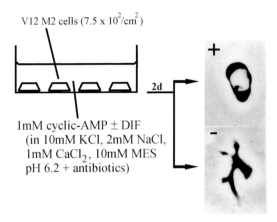

Figure 9.4 The bioassay for DIF. In the absence of DIF, cells incubated in conditioned medium and cAMP remain amoeboid. Addition of DIF causes the cells to convert to stalk cells after 48 hours. The upper cell is scored as a stalk cell, the lower one is not. Note the large vacuole of the stalk cell. A cellulose coat on stalk cells can also be detected with fluorescence microscopy. (Courtesy of R. Kay, Medical Research Council, Cambridge, UK.)

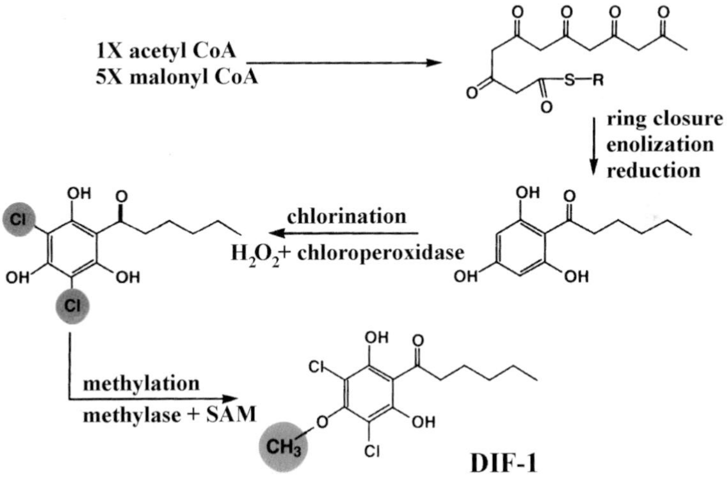

Figure 9.5 The structure and biosynthesis of DIF. A number of the enzyme activities that are required for the proposed route of synthesis have been detected (Kay, 1998). The degradation pathway is also known (Kay et al., 1993; van Es et al., 1994). (Courtesy of R. Kay, Medical Research Council, Cambridge, UK.)

of the discovery of lower levels of DIF in the anterior segments of slugs than in the posterior segments (Kay et al., 1993; Nayler et al., 1992). The initial induction of prestalk cells, whether by DIF or other compounds, occurs long before they have segregated into the anterior compartment of the slug (see Fig. 9.3). The DIF hypothesis predicts that all prestalk gene induction depends on DIF (Kay, 1997).

Table 9.2. *DIF responsive genes. A number of proteins or genes are induced or repressed by DIF*

Marker	Half-maximal effect (nM)	Effects	References
Stalk cells	0.18	Induced	Masento *et al.* (1988); Town *et al.* (1976)
Spore cells and prespore vesicles	0.8	Repressed	Kay (1989); Kay *et al.* (1983)
Acid phosphatase	ND	Induced (prestalk specific)	Kopachik *et al.* (1983)
2D gel spots	ND	18 stalk induced 4 spore repressed	Kopachik *et al.* (1985)
ecmA mRNA	30	Induced (prestalk specific)	Berks and Kay (1990); Williams *et al.* (1987)
ecmB mRNA	70	Induced (prestalk specific)	Berks and Kay (1990); Jermyn *et al.* (1987)
pspA (D19) mRNA	10	Repressed (prespore specific)	Berks and Kay (1990); Early and Williams (1988)
cotA, B, C	5–10 nm	Repressed (prespore specific)	Fosnaugh and Loomis (1991)
pdsA late transcript	ND	Induced, through specific promoter	Franke *et al.* (1991)
DIF-1 dechlorinase	30	Induced (prestalk specific)	Insall *et al.* (1992)
Cellulose synthesis	ND	Induced	Blanton (1993)
Glycogen phosphorylase 2 mRNA	ND	Induced in prestalk Repressed in prespore	Yin *et al.* (1994)

This table includes most of the known examples, but is not necessarily complete. The data in this table support the idea of DIF as an inducer of prestalk and a repressor of prespore gene activity. ND, not determined. (Compiled by R. Kay, MRC, Cambridge.)

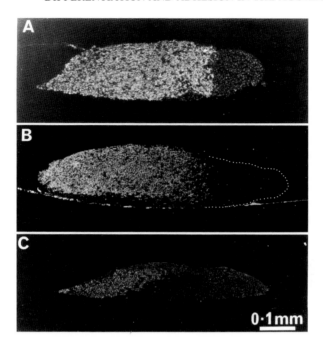

Figure 9.6 DIF-1 affects the prestalk/prespore pattern in slugs. Prespore cells were detected with an antibody. Panel A shows a control, without DIF. In panels B and C, reduced prespore zones can be seen in slugs that have developed and migrated on agar containing 0.2 μM DIF-1. For clarity, the slug in panel B has been outlined. The change in the ratio of cell types may occur before the slug forms (Kay *et al.*, 1988). (Courtesy of R. Kay, MRC, Cambridge. Reprinted by permission of Wiley-Liss, Inc., a subsidiary of John Wiley & Sons, Inc.)

9.7 Are any prestalk genes expressed without DIF?

If DIF causes undifferentiated cells to become prestalk cells, all stalk-specific genes that are induced during prestalk cell specification in aggregated cells, should depend on DIF. There are two stalk-specific genes, *carB* (Saxe *et al.*, 1996) and *tagB* (Shaulsky and Loomis, 1996), which are expressed at the tight aggregate stage. Neither is induced by DIF. Furthermore, *tagB* mutants make almost no detectable DIF, but still express the *tagB* promoter fused to β-galactosidase in prestalk cells in the absence of DIF. Controversy remains concerning the reliability of the DIF measurements that were done in the *tagB* experiments (Williams, 1997). If no DIF is made in the mutants, this would mean that, while DIF may have a role in stabilizing the prestalk cell phenotype, the earliest steps in prestalk cell specification may not require DIF. Because DIF is difficult to assay, it is hard to say that none is synthesized in the *carB* and *tagB* mutants.

The role of DIF could be clarified if a mutant existed that eliminates a step in DIF biosynthesis. A mutant called HM44 produces very low amounts of

DIF and fails to induce prestalk markers like *ecmA* and *ecmB* unless DIF is added (Kopachik *et al.*, 1983). The expression of *carB* or *tagB* in HM44 has not been studied, but if they are independent of DIF, we might expect them to be synthesized in HM44. It will take a defined mutant that affects only DIF biosynthesis to finalize the role of DIF in models of cell-type proportioning or pattern formation. Recently the gene that codes for the methylase that catalyzes the last step in DIF biosynthesis has been cloned and shown to be expressed in prespore cells (R. Kay, personal communication). The DIF hypothesis predicts that elimination of this gene should block DIF synthesis and prestalk cell differentiation at an early stage – at around the time of the loose aggregate – and prior to expression of prestalk-specific genes (Kay, 1997). Preliminary studies show that the methylase mutant lacks DIF – yet it develops relatively normally (C. Thompson and R. Kay, personal communication). It is always possible to say that DIF is, in fact, the normal inducer, but its absence invokes another normally quiescent system. Recently, a role for DIF-1 in the induction of pstO cells has been suggested (R. Kay, personal communication).

9.8 The history of a cell affects its fate

There is good evidence that the initial choice of cell types is biased by the conditions of the cells when starvation is imposed. If this is true, it should be possible to bias the fate of cells by growing them under rich or poor conditions and mixing them prior to development. Well-nourished cells may preferentially end up as spores. A number of workers have shown that it is possible to do this (Blaschke *et al.*, 1986; Inouye and Takeuchi, 1982; Leach *et al.*, 1973). Cells grown in axenic medium with extra glucose accumulate more glycogen reserves and are larger than cells grown with less glucose or cells grown on bacteria. If cells of the two populations are mixed, they compete to form spores, and the well-nourished population forms more than its fair share of spores, leaving the poorly fed population to die in the production of stalk. When starved alone, the well-nourished cells aggregate several hours faster than cells grown in limiting glucose (Forman and Garrod, 1977). Even within a well-nourished population, not all cells are equal. Cells that have just divided and have not had time to accumulate reserves when starvation is imposed, are different from those which have accumulated reserves for hours. In general, larger, better nourished cells (i.e., those in late G_2) form the spores. There is always heterogeneity in the population, whether genetic, temporal, or nutritional, and this heterogeneity affects the fate of a cell.

The effects of growth history on early prestalk or prespore differentiation may be traceable to the cell cycle and could be cell-autonomous events that do not depend on signaling among starving cells (Araki *et al.*, 1997; Gomer and Firtel, 1987; Maeda *et al.*, 1989; McDonald and Durston, 1984; Weijer *et al.*, 1984a; Zimmerman and Weijer, 1993). Cells that have been grown with extra glucose might be distributed differently in the G_2 portion of the cycle than cells grown with limiting glucose. The *Dictyostelium* cell cycle lacks a G_1 and therefore events that impinge on later development must occur during S or G_2. The effect of the cell cycle on cell differentiation can be seen when synchronized populations are allowed to starve at high density. Populations with a majority of cells in S or early G_2 form slugs with a higher percentage of prestalk cells, while populations with a majority of cells in late G_2 form slugs with a higher percentage of prespore cells (Wang et al., 1988a; Weijer et al., 1984a). The regulatory mechanisms first described by Raper are not able to compensate fully.

Normally, *Dictyostelium* amoebae do not differentiate to form prestalk and prespore cells if they are incubated at such a low density that the cells have no contact. This reluctance can be overcome by mutagenesis or by applying conditioned medium and cAMP. The cAMP acts through cAR1 to activate PKA, which is a requirement for aggregation. To examine the effect of cell cycle phase on the fate of cells, Gomer developed a video-microscopy technique in which cells plated at low density were observed under growth conditions and then development was induced by replacing the growth medium with phosphate buffer (Gomer and Firtel, 1987; Wood et al., 1996). The position of the cell in the cycle was measured as a function of the time since the previous cytokinesis. Mitosis and S phase constitute less than 15% of the cell cycle, with G_2 occupying the remaining time, which is normally about 8 hours in axenic medium. A cell that had just divided at the time buffer replaced the medium and was in early G_2 when starved had a strong tendency to turn into a prestalk cell and express prestalk-specific markers (see below). Curiously, the two daughter cells of a division had different fates. One cell differentiated and the other did not, remaining what is called a null cell. Null cells differentiate later. Cells that had grown for more than 2 hours after M and S had a tendency to become prespore cells and express the SP70 protein, a prespore-specific marker. The results in Fig. 9.7 are a graphical demonstration of this effect for a small number of individual cells. Cells that were in M, S, or early G_2 at the time of starvation, expressed a prestalk gene that codes for a cysteine proteinase. A certain percentage of the cells in either case did not differentiate at all (the null cells) and the expected 80:20 ratio was not established, indicating the need for additional mechanisms to control cell-type proportion. These biases toward prestalk or prespore formation were expressed in the absence of cell contact and were not due to intercellular communication, but rather to a cell-autonomous effect derived from the position in the cell cycle. One always makes the worrying assumption in these single cell experiments that the dilution of the cells has been enough to preclude the effects of secreted factors.

A mutant with an abnormally high percentage of stalk cells called *rtoA* has been isolated. It codes for a novel protein expressed during vegetative growth and then again after aggregation. The effect of the mutation, shown in Fig. 9.7, is to remove the dependency of prestalk or prespore differentiation on cell cycle position. In *rtoA* mutants, cells at any position in the cell cycle can differentiate into either cell type (Wood et al., 1996). This increases the frequency of prestalk cells. RtoA is part of an unexplored mechanism that recognizes the cell cycle phase at the time of starvation. With the gene and the protein in hand it

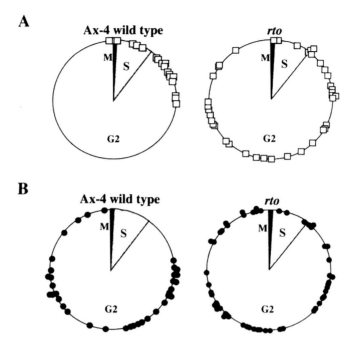

Figure 9.7 The *rtoA* gene and position in the cell cycle. Panel A: Normal cells in the first quarter of the cell cycle express prestalk rather than prespore markers when starved at low density. In a *rtoA* mutant, cells express a prestalk gene no matter where they are in the cell cycle (Wood *et al.*, 1996). Each open square represents a single cell expressing a prestalk marker (CP2). Panel B shows the same experiment with a prespore marker, the spore coat protein SP70. In wild-type cells, SP70 is induced only by cells that were in the last three-quarters of the cell cycle when they were starved. Mutant cells express SP70 with no regard to cell cycle position. Each closed circle represents a single cell expressing a prespore marker (SP70). (Courtesy of Richard Gomer, Rice University. Reprinted with permission from The Company of Biologists Ltd.)

will be possible to ask what proteins it binds, and eventually how it influences cell type decisions.

9.9 How are the constant ratios of prestalk and prespore cells to be explained?

Any explanation of the initial differentiation of prestalk and prespore cells must explain how contiguous cells become prestalk or prespore, and how a ratio of 80:20 is established. This ratio is maintained no matter how many cells are in the aggregate – several hundred or a hundred thousand. The initial differentiation of prestalk or prespore cells does not necessarily give rise to the precise ratio observed in later development – other mechanisms can be superimposed on the cell-autonomous events just described. Cell-autonomous primary differentiation explains the interspersed nature of the cell distribution. The problem of explaining the strict proportionality of cell types remains. One mechanism to explain the proportionality depends on lateral inhibition mechanisms such as have been described for vulval cell differentiation in *Caenorhabditis elegans* and in sea urchin mesenchyme development among many other cases (Ettensohn and McClay, 1988; Horvitz and Sternberg, 1991; Loomis, 1993). Lateral inhibition models predict that a cell that differentiates inhibits similar changes in neighboring cells, which are consigned to another fate.

In the case of *Dictyostelium*, Loomis has suggested that certain cells that are distinguished from their peers by their history – those in late G_2 when removed from their food – start to secrete an inhibitor of prespore cell differentiation (Loomis, 1993). Once they secrete the inhibitor, they become insensitive to it. As more and more cells starve long enough to begin to secrete the inhibitor its level rises and, once a certain concentration is reached, further formation of prespore cells is blocked. The default is to form a prestalk cell, which the inhibitor permits. In the initial differentiation that depends on the position of the cell in the cell cycle at the time of starvation, there are a large number of so-called null cells, which express neither prespore nor prestalk markers. Their conversion later in development may make up the stable 80:20 ratio. Such a mechanism could explain the salt and pepper distribution of precursor cells and could function in aggregates of any size. If there is a balance of synthesis and degradation of the inducer, a lateral inhibition model could account for the 80:20 ratio. This ratio is established gradually, beginning about 8 hours after the amoebae begin to starve, during the transition from loose aggregate to tight aggregate.

An inhibition model can also partially accommodate the results of surgical experiments in which the prestalk or the prespore regions are severed in a slug. Removal of the prespore region of the slug allows dedifferentiation of the prestalk cells that were in the prestalk (anterior) portion. The source of inhibitor would also be removed, allowing the cells in the front to form prespore cells, until the levels of inhibitor increased. Inouye has shown that isolated prespore cells (but not prestalk cells) produce a dialyzable inhibitor that prevents the conversion of prestalk cells to prespore cells (Inouye, 1989). The identity of the inhibitor remains unknown, although one candidate is DIF, which also prevents the conversion of prestalk cells to prespore cells (see Table 9.2). A weakness of the lateral inhibition model is that it does not explain how an isolated prespore segment regulates to produce prestalk cells.

A model with a different emphasis is based on DIF expression and sensing (Kay, 1997). This model suggests that DIF is made in the aggregate and then in greater amounts by prespore cells. The levels of DIF are determined by two mechanisms: it is destroyed by a dechlorinase in prestalk cells and made by the prespore cells. At early stages of development, which concern us here, cells moving in the mound (Fig. 9.8) are exposed to similar levels of DIF. The model predicts that cells have different thresholds for their responses to DIF, partially determined by the vegetative biases that were described in Section 9.8. The mechanism that determines the thresholds is not known. The difference in thresholds results in an 80:20 ratio of prespore and prestalk cells – thus at a given DIF concentration 20% cells will respond and the other 80% will not. DIF induces its own destruction so that the levels are internally controlled by negative feedback – the more DIF, the more dechlorinase. The model would give intermingled prespore and prestalk differentiation, would account for the vegetative biases, and would be regulative. Removing prespore cells from the mound would result in less DIF production but would also cause reduced synthesis of dechlorinase, resulting in subsequent increases in DIF, and the response of the cells. The model differs from the lateral inhibition mechanism in that it specifies an agent (DIF) and it also specifies that a general level of inducer works on a pre-existing bias in the population of responding cells. The lateral inhibition model states that a cell induces a neighboring cell to a fate that is different from the inducer. In the lateral inhibition model, effects are local in that they are confined to a cell and its immediate neighbors. There is considerable evidence, some of which has been presented above, for a role of DIF and feedback regulation in establishing the prestalk/prespore pattern (Kay, 1997). However, recent mutant studies cast some doubts on the role of DIF, as described in Section 9.7.

9.10   What is the basis of cell sorting?

We have established that the initial pattern that appears as the aggregate changes from a loose mound to a tipped aggregate, and then forms a slug, is due to cell sorting of previously mixed prestalk and prespore cells. As the mound develops it forms a tip, which is known to be a source of signals because it entrains other cells if grafted on the flank of a slug or deposited in a field of cells that are ready to aggregate. Therefore the question arises: toward what do the cells sort? cAMP is the first candidate, and several kinds of experiments lend credence to the idea that cAMP continues to fulfill a chemotactic role after aggregation when cells are moving within the mound. Using videomicroscopy, Weijer and his colleagues showed that cells move dramatically in the mound (Fig. 9.8). Signals diverge from the tip and sweep around the aggregate, with cells responding with shape changes and an inward spiral movement (Bretschneider et al., 1997; Rietdorf et al., 1996; Siegert and Weijer, 1995, 1997). These signals and movements are very closely related to the behavior of cells during aggregation. In chimeras in which a few cells lack cAMP phosphodiesterase, the mutant cells are at a severe disadvantage, indicating that cAMP gradients in aggregates are modified at the single cell level (Sucgang et al., 1997).

Experiments in which the extracellular cAMP was severely reduced by directing the secretion of a phosphodiesterase to the intercellular spaces of the mound had several effects (Traynor et al., 1992). The number of prespore cells in the mound was greatly reduced, and the migration of the prestalk

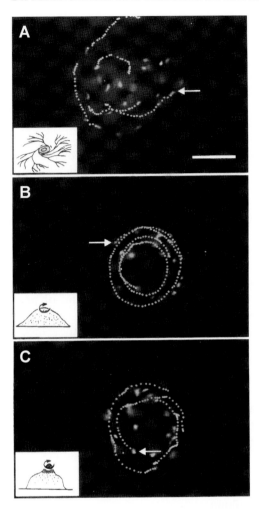

Figure 9.8 Cells in mounds migrate radially. Cells that express Green Fluorescent Protein were mixed with a 1,000-fold excess of unlabeled cells. The movement of a single cell is indicated by the white dots. The start position is at the arrow. Inserts indicate the stage of development and the direction of movement (Rietdorf *et al.*, 1996). (Courtesy of C. J. Weijer, University of Dundee. Reprinted with permission from Academic Press, Inc.)

cells to the apex of the mound was slowed dramatically. If cAMP was included in the substratum, the prestalk cells migrated toward the base. The difficulty of this experiment is that reducing cAMP outside the cells eventually leads to a failure to stimulate the PKA, and development is affected. While the cells in the mound were chemotactic toward cAMP, it is not necessarily true that this is the cause of their migration toward the apex as it forms the tip.

9.11   Overexpression of PKA can compensate for a lack of cAMP

It is clear that cAMP works in the extracellular space by binding to cAMP receptors, and in the intracellular space by activating PKA. What is in some dispute is whether extracellular cAMP, by virtue of its non-uniform distribution, provides positional cues that provoke the differentiation of cells. Wang and Kuspa (1997) asked whether cAMP serves a purely stimulatory function outside the cells, ultimately leading to activation of PKA, or whether it additionally provides directional cues. This was done by eliminating the major source of cAMP in developing cells – the adenylyl cyclase discussed in Section 8.7. Cells without ACA do not aggregate, nor do they make detectable levels of cAMP. If the endogenous PKA is activated by overexpression of the catalytic subunit, cells without measurable cAMP aggregate and form relatively normal fruiting bodies with spores. This only occurs if the cells are plated at high density – at lower densities, ACA and cAMP relay are essential for aggregation. These results suggest that cAMP does not necessarily form a gradient to attract cells to the apex of the mound or provide a differential gradient later in development.

There are two more known adenylyl cyclases; these include the ACG gene, which is expressed during spore germination (Pitt *et al.*, 1992). A third adenylyl cyclase activity, ACB, has been found and it is possible that this enzyme provides enough cAMP to assist in the formation of pattern (Kim *et al.*, 1998). ACB may provide enough cAMP for the cell sorting driven by the very sensitive cAMP receptor. The level of cAMP produced by ACB could be below the limits of detection used by Wang and Kuspa and still be recognized by cAMP receptors.

9.12   The cells induce several adhesion systems during formation of the mound

Dictyostelium has a number of induced adhesion systems, and has been a model organism for the study of adhesion. Work on cell–cell adhesion was pioneered by Günther Gerisch, who employed Fab fragments to detect and inhibit specific adhesion mechanisms. Gerisch's methods were later widely used by Edelman and others in mammalian cells. *Dictyostelium* cells, when they grow on a lawn of bacteria, are not adhesive, yet a short time after the bacteria are consumed the cells begin to adhere to one another and form clumps in liquid or aggregates on a surface (Gerisch, 1961, 1968). Several molecules that mediate this and other forms of adhesion during development have been cloned, and two – gp24 and gp80 – have been extensively studied (Bozzaro and Ponte, 1995; Fontana, 1993; Gerisch, 1980; Siu *et al.*, 1997; Siu and Kamboj, 1990). Despite this intensive effort, many questions remain unanswered about how cells adhere during aggregation, how they compact to form the tight aggregate, and how they change adhesion systems as they form slugs. Whether there is differential adhesion of prestalk and prespore cells is another question that has been investigated, but not settled. The conclusion from these studies is that adhesion systems are redundant, and gross observation of cells developing on a laboratory substrate is often inadequate to detect differences between mutant and parent. There are phenotypes when mutant cells are forced to develop in more challenging environments (Ponte *et al.*, 1998).

9.12.1 Gp24

Shortly after the cells start to starve, they become adhesive in a way that is sensitive to EDTA and EGTA, indicating a need for calcium. For many years this form of adhesion was referred to as "contact sites B"-mediated adhesion. The molecule that mediates this early form of adhesion, gp24, is not expressed at all in cells growing on bacteria, but is expressed in cells growing in axenic media (Beug *et al.*, 1973; Yang *et al.*, 1997). The gp24 molecule was studied by Loomis, who produced blocking antibodies (Knecht *et al.*, 1987; Loomis, 1988). Gp24 was purified by Brar and Siu, who showed that it binds to cell surfaces in an EDTA- and EGTA-dependent fashion, confirming calcium as an essential ion (Brar and Siu, 1993). Binding of labeled gp24 to the cells can be prevented by the antibody, as is re-association of cells that express gp24. The gene coding for gp24 was cloned by Siu and his colleagues, who named the protein DdCAD-1 because of its resemblance to well-defined mammalian calcium-dependent adhesion molecules, the cadherins (Sesaki *et al.*, 1997; Wong *et al.*, 1996). This homology is so distant as to be doubtful.

9.12.2 Gp80

The second type of adhesion mechanism originally discovered by Gerisch was mediated by gp80. Gp80- or "contact sites A"-mediated adhesion could be distinguished in the early experiments by several criteria. First, it did not appear until well into development; and second, it was calcium-independent because it is stable in the presence of EDTA or EGTA. As in the case of gp24, the cell adhesive mechanisms were blocked by Fab fragments (Beug *et al.*, 1970, 1973; Müller and Gerisch, 1978). The analysis of gp80 was the model by which mammalian cell adhesion molecules were studied. Gp80 is encoded by *csaA*, which produces a heavily glycosylated 53,703-Da peptide (Noegel *et al.*, 1986; Wong and Siu, 1986). Mature gp80 is bound to the cell surface and passes by the standard route through the rough endoplasmic reticulum and Golgi. The attachment to the plasma membrane is by a lipid glycan (Sadeghi *et al.*, 1988; Stadler *et al.*, 1989). The amino-terminal domain of gp80 is composed of three globular domains, followed by a sequence that is more rigid and has been called a stalk region, and then the membrane anchor, as shown in Fig. 9.9. Gp80 mediates adhesion by homophilic binding of a gp80 on one cell with another gp80 on a neighboring cell. A short amino acid stretch (YKLNVNDS)

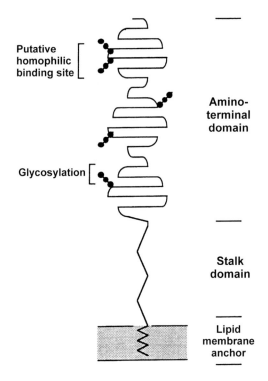

Figure 9.9 The structure of the gp80 cell adhesion molecule. An amino-terminal globular domain contains a single homophilic binding site (Kamboj *et al.*, 1989; Siu *et al.*, 1997). The globular domain is held away from the cell by a stalk region, and attachment to the membrane occurs by a lipid (Barth *et al.*, 1994; Stadler *et al.*, 1989). (Reprinted with permission from Wiley-Liss, a division of John Wiley & Sons, Inc.)

in the most amino-terminal globular domain mediates homophilic binding. This peptide, included in the medium, disrupts cell adhesion, although not completely. The peptide competes with the binding of labeled gp80 to the cells and also competes with binding to immobilized gp80 (Kamboj *et al.*, 1989).

Transcription of *csaA* is induced by cAMP pulses and appears before the cells enter the mound. Like the genes described in Chapter 8, it will not appear in the presence of constant levels of cAMP, which indicates that its induction mechanism is adaptive. The *csaA* transcript disappears after aggregation. If the gene is super-induced during development, aggregation is disrupted – aggregations streams fall apart and make small aggregates and slugs (Faix *et al.*, 1992). Transcription of *csaA* is stringently regulated, but when *csaA* is attached to a promoter that is active during cellular growth, the vegetative cells adhere, which is one of the strongest pieces of evidence in favor of its role as an adhesion protein (Faix *et al.*, 1990). This overexpression seems to induce

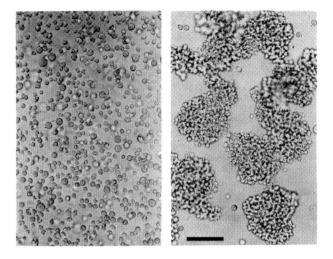

Figure 9.10 Overexpression of *csaA* in vegetative cells causes them to adhere. Deletion of *csaA* in wild-type cells does not cause a failure of aggregation, due to overlapping adhesion systems. The best evidence for a role in adhesion derives from the effects of overexpression in growing cells where there are no other adhesion systems. Control cells are shown in the left panel and cells that overexpress gp80 are shown in the right panel. In both cases, the cells have been treated with 10 mM EDTA to remove the effects of gp24 (Faix *et al.*, 1990). The bar represents 100 μm. (Reprinted with permission of the Oxford University Press.)

other developmental pathways – even in the presence of nutrients, perhaps because the adhering cells are isolated from the medium and can no longer undertake macropinocytosis to feed.

The difficulty arises when *csaA* is disrupted. This has been done, and strains carrying such a defective gene were incapable of EDTA stable adhesion when shaking in liquid (Harloff *et al.*, 1989). However, if the cells are allowed to develop on agar, as they normally do, aggregation is not impeded. Double mutants that lack both activities display a visually normal aggregation on agar (W. F. Loomis, personal communication). This result implies that there is more to the initial adhesion that creates the mounds than we currently know. One way to view the activity of gp80 is to assume that the cells evolved several systems of cell adhesion such that deleting one or two creates no obvious effects under normal laboratory conditions. The trick is to use more natural conditions, which has been done recently for strains carrying defects in *csaA* by Ponte, Bozzaro, and their colleagues. These authors showed that if *csaA* null cells are made to develop on the three-dimensional matrix of autoclaved soil, they are at a severe disadvantage (Ponte *et al.*, 1998). This phenotype is not obvious when the same cells develop on agar.

Gp24 and gp80 function during aggregation and in the mound, but other molecules may function at later times to hold the cells of the slug together. One additional molecule, gp150, has been suggested as a late stage adhesion molecule (Gao *et al.*, 1992). Gp150 is the product of *lagC* (Siu *et al.*, 1997), whose phenotype we have seen can be suppressed by overproduction of GBF. In the absence of LagC, but with excess GBF, development and adhesion of cells in the slug is apparently normal (Sukumaran *et al.*, 1998). This result means that gp150 cannot be the only cell adhesion molecule in late development. It may also be a signal-transducing molecule whose activities are interrupted when antibodies are bound (Dynes *et al.*, 1994).

9.12.3 Gp64

A substantial body of work exists on the adhesion systems of the related species *Polysphondylium pallidum* (Bozzaro and Gerisch, 1978). This species initially adheres to *D. discoideum* cells in a chimeric mixture, but then sorts out, probably on the basis of differential adhesion. The same immunological approach used with *D. discoideum* was taken and a molecule called gp64 was recovered and extensively characterized (Funamoto and Ochiai, 1996; Manabe *et al.*, 1994; Ochiai *et al.*, 1996, 1997). A monoclonal antibody detects a number of proteins on a Western blot, and therefore the role of this molecule in cellular adhesion is not certain. Like gp80, to which it bears no sequence homology, it is attached to the cell surface by a lipid moiety. Knockout mutants have not yet been prepared as this is not routine in *P. pallidum*, but antisense experiments showed an inhibition of adhesion (Ochiai *et al.*, 1997).

9.13 Substrate adhesion

Most adhesion studies focus on adhesion between cells, since this is one of the most extraordinary developmental properties of the organism. Yet the adhesion of cells to the substratum is also impressive and perhaps related to cell–cell adhesion. Even the debilitated axenic strains we use in the laboratory can move impressively over dirt, leaves of various species, and even the occasional insect carcass from which fruiting bodies sometimes spring. How do the cells grip such strange substrates? In fibroblasts, integrins in the plasma membrane interact with fibronectin the extracellular matrix and intracellularly with a variety of cytoskeletal proteins to form focal contacts. One of the proteins that links actin stress fibers to the membrane is talin, which also nucleates actin polymerization and cross-links actin to membranes. *Dictyostelium* is not known to have integrins or the focal adhesion complexes of fibroblasts, but may have some more primitive structure. The best evidence for this comes from mutants which lack talin (Niewohner *et al.*, 1997). These cells adhere poorly to substrates. Surprisingly, the mutant cells move rapidly, despite reduced cell substrate adhesion. The cells are defective in what may be related processes – cell-to-cell adhesion during growth and phagocytosis, but the separate adhesion systems that regulates cell-to-cell contact during development are not affected.

Discoidin I is a developmentally regulated lectin that is expressed at relatively high levels during development (Cooper and Barondes, 1984; Ma and Firtel, 1978). Discoidin was long thought to have a role in cell adhesion (Barondes *et al.*, 1982), but this does not appear to be the case. Antisense and other experiments suggest however, that in strains with much reduced discoidin levels, cell-to-substrate adhesion is altered (Barondes *et al.*, 1987; Crowley *et al.*, 1985; Springer *et al.*, 1984). A homology to fibronectin domains has been postulated, but never proved (Poole *et al.*, 1981; Springer *et al.*, 1984). There is close homology of discoidin with two mammalian receptor tyrosine kinases (Vogel *et al.*, 1997).

9.14   The genetic complexity of mound formation

The events of mound formation must include the events of cell-type differentiation, motility, adhesion, and chemotaxis. A fraction of the genes involved will, if mutated, disrupt development and hold development at the mound stage. In one study, employing chemical mutagenesis, Carrin, Kay and their colleagues isolated a number of chemically induced mutants that were arrested at the loose mound to the tipped aggregate stage (Carrin *et al.*, 1996). The mutants were analyzed for developmental arrest and fell into discrete groups. The first class, called post-aggregative mutants, did not express cell type-specific markers. The second class, pathway mutants, were blocked in the induction of prestalk or prespore differentiation. The third class, morphogenesis mutants, expressed differentiation markers of both classes, but were blocked at the tight aggregate to tipped aggregate stage. Most mutants were not cell-autonomous and developed in the presence of the wild-type. Nine complementation groups (*mndA-mndI*) contained 46 of the mutants, as defined by parasexual genetics. Statistical analysis suggests that this frequency of non-complementation is consistent with a requirement for about 118 genes to mediate the loose aggregate to tipped aggregate transition. This number does not include genes which influence the transition from loose aggregate to tipped aggregate, but which do not block it if mutated. This collection of mutants delineates important steps in mound formation, but until complementation analysis is available in *Dictyostelium*, it offers no way to know the nature of the affected genes. The analysis does allow us to gauge how many of the mound-specific genes we have so far recovered. Direct genetic complementation with libraries of cDNA carried in extra-chromosomal vectors is now being developed in several laboratories and it may soon be possible to recover genes affected by chemical mutations.

REMI methods have produced a number of mutations in defined genes that cause a block in morphogenesis at the aggregate stage. Some of these are shown in Table 9.3.

If we compare the number of complementation groups predicted by the parasexual genetic analysis of Carrin *et al.* with the number of specific genes that act during the period between the loose aggregate and the tight aggregate

Table 9.3. *Genes that function during the transition from loose aggregate to tight aggregate*

Locus	Gene product	Affected process	Terminal phenotype	Reference
gbfA	Transcription factor	Cell-type differentiation	Loose aggregate	Schnitzler *et al.* (1994)
lagC	Gp150, unknown function	Cell-type differentiation	Loose aggregate	Dynes *et al.* (1994)
gpaD	G protein, Gα4	Coupling with a potential pterin receptor	Tip elongation	Hadwiger *et al.* (1994)
gpaE	G protein, Gα5	Tip formation	Delay at aggregate stage	Hadwiger *et al.* (1996)
carB	Low-affinity cAMP receptor, cAR2	Cell-type differentiation	Tight aggregate, no tips	Saxe *et al.* (1993, 1996)
scar	Suppressor of cAR2	Negative regulator of cAR2 function	Multiple tips	Bear *et al.* (1998)
rasD	Small GTPase, developmentally regulated	Cell-type differentiation	Multiple tips during overexpression	Louis *et al.* (1997)
tagB	Putative serine protease, transporter	Prestalk signaling	Tight aggregate, no prestalk cells	Shaulsky *et al.* (1995)
tipA	Novel 83-kDa protein	Sorting of prestalk cells	Tight aggregate, no elongation	Stege *et al.* (1997)
spnA (*spalten*)	Novel phosphatase	Cell-type differentiation	Mound stage	Aubry and Firtel (1998)
nosA	Putative E4	Ubiquitination	Tight aggregate	Pukatzki *et al.* (1998)
talB	Talin-like protein	Tip formation	Tight aggregate	Tsujioka *et al.* (1999)

The table lists the majority of mutants that are blocked at this period, but the list is increasing rapidly. The number of known genes is less than that expected from complementation analysis of mutant collections by parasexual genetic methods.

(or slightly beyond), it is clear that we have recovered only a minority of the genes that act during this period.

The events that occur as the cells convert from a loose aggregate to a tight aggregate include the elevation of intercellular cAMP, induction of G box-binding factor, activation of the earliest cell type-specific markers, and the establishment of basic adhesion systems. During this time some of the constituents used during aggregation, notably the cAMP receptor and the secreted cyclic nucleotide phosphodiesterase, are transcribed from new promoters which do not adapt to conditions in which cAMP is permanently high. The initial differentiation of cells within the mound is influenced by a cell-autonomous bias related to conditions during the time of starvation. The later pattern of the aggregate results from sorting of the highly motile cells. There may be a contribution to the ultimate pattern from a positional information mechanism deployed among the prestalk cell subtypes. The sorting of cells is driven by signals from a new morphological structure – the tip. The movements of cells in the close confines of the mound are extensive and follow various paths. Waves of cAMP signaling sweep around the mound and elicit movement responses from the cells. The molecules involved in cell sorting include cAMP, but it is possible to create a situation in which cAMP synthesis is reduced and the cells still sort to the apical tip. This result implies sorting to low levels of cAMP or to other, as yet undefined, molecules. The adhesion systems employed by the cells are activated at this time, but there is much left to learn about this subject.

10

The Behavior of Cells in the Slug

10.1   The tight aggregate elongates under control of the tip

The tight aggregate is at first a hemisphere, but soon elaborates a tip that becomes populated with prestalk cells. A large body of evidence indicates that prestalk cells move apically in the tight aggregate to establish the pattern, sorting out from the prespore cells as they go. The first sign of a tip is a signaling center in a mound, in which cells move radially around a center that does not yet form a morphologically distinct structure (Rietdorf *et al.*, 1996; Siegert and Weijer, 1995; Sucgang *et al.*, 1997). Often there will be more than one such center, but eventually, one predominates or, in large mounds, two tips form and the cell mass is subdivided. There is lingering confusion about what constitutes the tip because people use the word differently. In the text that follows, the tip refers to the most anterior part of the prestalk zone, which is usually suspended above the substratum. The word *tip* is also used to denote a signaling entity, and the morphological tip certainly fulfills such a role. We do not know whether the tip contains a specific group of cells, smaller than the physical structure, from which organizing signals emanate. The cells in the front of the slug are in constant motion, and it is hard to imagine how a few key cells at the tip could remain as a group. Signaling seems to be a property of the region, which controls a number of important responses of the newly organized tissue. Laser technology has been used to destroy the tip, but these techniques have not recently been used (Klaus and George, 1974).

Under the influence of the tip the mound elongates, as more prestalk cells enter the apical region, and forms a standing structure called the finger or first finger. Normally such a large change in form requires a change in cell shape and a rigid extracellular matrix on which cells can gain a purchase, but little is known about the mechanical forces that participate in the elongation of the mound. During the migration of prestalk cells into the tip, certain prestalk cells remain behind and are found at the base of the aggregate. Called pstB cells by Williams, because they express *ecmB,* these cells will eventually form part of the basal disc (Williams *et al.*, 1989). Other prestalk-like cells linger among the prespore cells, to form a population of anterior-like cells (Sternfeld and David, 1981, 1982). These cells will move later, during culmination, to form the cups that support the spore mass and part of the basal disc. By the time the first finger forms, the basic pattern, with most prestalk cells in the front and pre-spore cells in the rear, has been established by sorting of the differentiated prestalk cells, which are more motile than prespore cells.

The cells in the elongated mound have now transformed themselves into a cooperating organism and are now faced with an important decision. The upright finger-like structure can culminate and form a fruiting body on the spot, or it can fall over, become horizontal, and migrate. The conditions that control the choice are the environment, as well as the direction of the light that shines on the finger (Bonner *et al.*, 1950; Newell *et al.*, 1969; Poff and Hader, 1984). Well-buffered substrates lead to fruiting on the spot, probably by neutralizing weak bases such as ammonia. Overhead light causes the first finger to skip the migratory slug stage. In the event that culmination proceeds in place, the next structure formed is called a Mexican hat. The movement of cells in the Mexican hat gives rise to the fruiting body and will be described in Chapter 11. The other alternative – a migrating structure called the slug or pseudoplasmo-dium – is induced by metabolite accumulation and light that does not come from overhead, but comes from the side. The mechanical details involved in falling over and adhering to the substratum remain unknown. The slug stage is often thought of as a period of suspended animation, when there are conveniently demarcated prestalk and prespore zones. John Bonner has called it a motile embryo. It is a convenient source of prestalk and prespore cells for a variety of experiments, but to think of it as static is an oversimplification.

10.2   Tips inhibit the formation of other tips

The tip, as Raper (1940b) showed, controls and entrains the movement of cells, as is shown in Fig. 2.5. Removing a tip blocks all developmental progress until a new tip forms, which happens spontaneously after a delay of about 30 minutes (Farnsworth, 1973). If tips can form spontaneously, why are there not many of them? The answer is that there is a gradient of inhibition that emanates from each tip and suppresses the formation of new tips, assuring a single axis. The inhibitory effects were elegantly shown by Durston and others in a series of tip transplantation experiments (Durston, 1976; Farnsworth, 1973). Tips transplanted along the length of a normal slug entrain cells of the host poorly at the front and better toward the rear. A decapitated *Dictyostelium* slug will accept a new tip transplanted into its anterior and

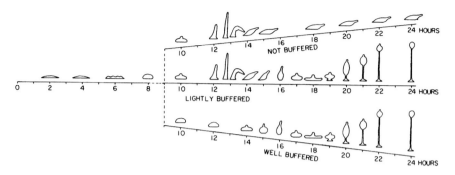

Figure 10.1 Controlling the different fates of aggregates. When aggregated amoebae were shifted to unbuffered, to lightly buffered, or to well-buffered environments, their responses varied (Newell *et al.*, 1969). The absence of buffer maintained the slug, modestly buffered environments led to transient slug formation, and well-buffered substrata caused the slug stage to be skipped altogether. (Reprinted with permission of the American Society for Microbiology.)

will not then spontaneously form its own tip. If a segment from the rear of the donor slug is used, there is no inhibition of the donor's propensity to form a spontaneous tip. Thus the idea arose that there is a gradient of inhibition. The regulation of signaling centers within the *Dictyostelium* slug is analogous to the regulation of hypostome formation in hydra (MacWilliams, 1991).

Rubin and Robertson (1975) performed heterochronous tip transplantation experiments, to determine whether tips from one stage would, when transplanted to an earlier or later stage, cause development to occur according to the history of the donor or of the recipient. The developmental stage of the recipient tissue mass is always the important one. If a slug tip is transplanted into a culminating fruiting body, it pulls in cells, but these form a fruiting body and not a slug. If a tip from a fruiting body is transplanted into a tipped aggregate, two indistinguishable slugs result. Late tips can also organize fields of cells that are ready to aggregate, themselves becoming the center of aggregation. These experiments show that the tip acts as an organizer from the late aggregate stage until culmination and that the nature of the signal does not change. For a review of the extensive literature that was available in 1984, see MacWilliams and David (1984).

If we ask more specific questions about the nature of the tip inhibitor, we find ourselves in a quandary. Is there a gradient of some unidentified molecule or does inhibition of tip formation arise from the oscillatory signaling properties of the very anterior of the prestalk zone? Some clues arise from inhibition experiments. Caffeine, known to be a relatively specific inhibitor of adenylyl cyclase in *D. discoideum* (Brenner and Thoms, 1984), allows new tips to form along the spine of the slug, so ultimately fruiting bodies spring from the slug along its length, as shown in Fig. 10.2 (MacWilliams, 1991; Siegert and Weijer, 1989). The tip dominance properties of the slug are also disrupted when slugs are placed on cAMP. New organizing centers appear along the length of the

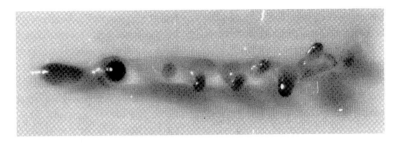

Figure 10.2 Destruction of tip inhibition by caffeine. A caffeine-treated slug develops supernumerary tips (MacWilliams, 1991). The tips are stained with neutral red, which accumulates in the autophagic vacuoles of prestalk cells. Caffeine inhibits adenylyl cyclase activation. The results suggest that polarity along the slug is controlled by a cAMP-based mechanism. (Reprinted from MacWilliams (1991), with permission from Academic Press Ltd, London, UK.)

slug and fruiting bodies arise along its length (Nestle and Sussman, 1972; Wang and Schaap, 1985). Adenosine, produced by hydrolysis of cAMP and the action of a $5'$ nucleotidase known to exist in the prestalk region (Armant and Rutherford, 1982; Armant et al., 1980), causes bigger slugs to form. This may result from tip suppression so that mounds that are producing tips do not subdivide (Schaap and Wang, 1986). Adenosine inhibits cAMP relay (Newell and Ross, 1982b; Theibert and Devreotes, 1984). Thus, an oscillatory signal beginning in the front and creating waves of cell movement as it travels, may also be responsible for tip inhibition, perhaps through secondary production of adenosine. A model that postulates direct control of organizing center formation by relayed signals requires that the period of oscillations in the anterior be more frequent than in the posterior. Otherwise, the tip cells would be entrained by the posterior cells.

A difficulty with models of tip inhibition based on cAMP or its metabolites occurs when mutants that lack a major adenylate cyclase (ACA), and hence most extracellular cAMP, manage to make slugs that have tips and migrate (Wang and Kuspa, 1997). Whether the total tissue controlled by a single tip (and hence the degree of tip dominance), varies in these mutants, has not been examined. There is only one transformant that has managed development without cAMP, because of overexpression of PKA, but its existence creates the possibility that tip inhibition, among other developmentally important events, is mediated by something in addition to relayed cAMP signals. It is also possible that low levels of cAMP, below the levels described by Wang and Kuspa, continue to act. A new adenylyl cyclase activity has been described which may provide this function (Kim et al., 1998). The matter remains to be settled.

Before leaving the subject of tip control of polarity, we should mention the elaborate tips produced by *Polysphondylium pallidum*, which forms a more complex pattern than does *D. discoideum*. *P. pallidum* produces a fruiting body with whorls of branches spaced evenly along the primary stalk. At the

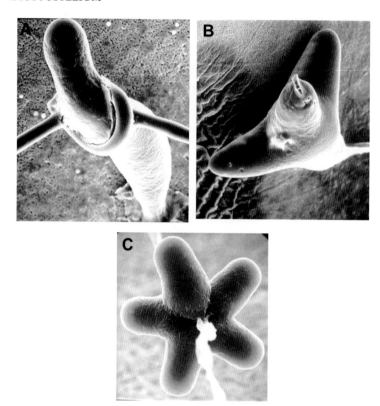

Figure 10.3 Tip inhibition in *P. pallidum*. During fruiting body formation, *P. pallidum* extends on a stalk with a primary tip and releases small masses that give rise to secondary whorls at regular intervals (see Fig. 1.4). Additional primary tips are suppressed. If the primary tip is removed by ligation (panel A), new primary tips form below the site of ligation (panel B) within 2 hours (Byrne *et al.*, 1982; Cox *et al.*, 1988; Gregg *et al.*, 1996). A developing whorl of a normal *P. pallidum* fruiting body is shown in panel C. The stalk runs through the middle. (Courtesy of E. Cox, Princeton University.)

end of each branch is a ball of spores. The distance between successive whorls along the stalk is extraordinarily regular. The equal spacing of the whorls suggests that the ascending tip produces an inhibitor, such that new whorls are inhibited until the primary cell mass has extended relatively far away (see Chapter 1). Cutting off the primary tip permits whorl formation at the base of the developing sorogen – something that does not happen if the initial extending tip is left intact (Byrne *et al.*, 1982; Gregg *et al.*, 1996; McNally and Cox, 1989). Each whorl produces five stalks that extend off the main stalk, and these are also very regularly spaced. They form from equally spaced tips, each of which must also produce a gradient of inhibition (Bonner and Cox, 1995; Cox, 1993; Gregg *et al.*, 1996). Whether the *Polysphondylium* tip inhibition mechanism is the same as that of *Dictyostelium* is not known. The effects of removing the *P. pallidum* tip are shown in Fig. 10.3. The results are best explained in terms of an inhibitory gradient emanating from the tip in the primary sorogen. Tip inhibition varies in other species and has a profound influence on the final morphology. *D. caveatum*, where tip inhibition is low, forms a small aggregate that has many tips, and these eventually give rise to a bushy fruiting structure with many stalks.

10.3 Slugs move toward light and heat

Slugs of any size have a variety of behaviors and are astonishingly sensitive to heat and to light (Bonner *et al.*, 1950; Bonner, 1998). The sensory capacity that a slug has to distinguish overhead from slanting light may have evolved as a signal that the slugs have reached the surface of the soil, where dispersal is more likely (Newell *et al.*, 1969). The movement of slugs in light can be easily observed by applying a line of washed amoebae on agar, usually made dark with a little charcoal. If the Petri dish is wrapped with foil and a small pinhole is made on an edge, the cells see low angle light as soon as they aggregate. Once aggregated, the cells begin a period of directed migration, which may go on for days and cover a distance of 10–20 cm. The migration of slugs toward light is density sensitive – at high densities the slugs tend to disperse. This has led to the idea of a secreted factor, called slug turning factor, that causes slugs to turn away from each other, even when confronted with unidirectional light, as shown in Fig. 10.4 (Fisher *et al.*, 1981; Fisher, 1997).

The action spectrum of the light recognized by the slugs is bimodal, with peaks at 425 and 550 nm (Francis, 1964; Poff and Häder, 1984). Attempts have been made to isolate the photoreceptor, but have not been successful (Poff and Butler, 1974; Poff *et al.*, 1974; Staples and Gregg, 1967). The migration of slugs results in a reasonably direct path, although deviations vary among strains (Haser and Häder, 1992). Mutants that do not perform phototaxis have been isolated, but as yet no genes have been characterized (Darcy *et al.*, 1993, 1994; Fisher, 1997; Wilczynska and Fisher, 1994). This would require REMI or another form of insertional mutagenesis, but a screen of thousands of colonies for migration defects is not a small undertaking.

The slugs are very sensitive to shallow temperature gradients and migrate toward warmth, as long as it is in the range at which amoebae normally grow (Bonner *et al.*, 1950; Poff and Skokut, 1977). The tracks of migrating slugs are shown in Fig. 10.5. When the heat source exceeds this range, the slugs exhibit negative thermotaxis. The mechanism of positive thermotaxis is so sensitive that it can detect a difference of $0.009°C/cm$ (Fig. 10.5). The mechanism of heat detection has elements in common with phototaxis because some mutants that are defective in phototaxis are also defective in thermotaxis (Fisher, 1997). Slugs exhibit negative phototaxis to ultraviolet light and to noxious compounds such as ammonia. It is not known how slugs turn, except that the turning is mediated by the front of the slug.

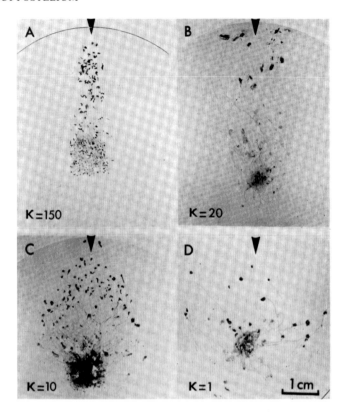

Figure 10.4 Slugs migrate toward lateral light. Overhead light causes fruiting body formation without migration. The panels show the tracks of slugs migrating toward light that comes from the direction of the arrow. K values indicate the precision of migration. Panel A shows accurate migration of a wild-type strain, while panel B shows the less accurate migration of wild-type cells placed at higher density. Panels C and D show the behavior of mutants that are defective in phototaxis (Fisher *et al.*, 1981). (Reproduced with permission from Cell Press.)

Non-phototactic mutant slugs will turn toward light if a wild-type anterior is grafted onto them (Fisher *et al.*, 1984). Light beams directed at the slug only have an effect if they are directed at the front of the slug (Häder and Burkart, 1983; Poff and Loomis, 1973). Solitary amoebae are phototactic, but we do not know if this occurs by the same mechanism that is used by cells in the slug (Schlenkrich *et al.*, 1995). Recently it has been shown that light affects cAMP signaling (Miura and Siegert, 2000).

In their migrations slugs are very robust and are capable of crossing rough terrain (Kessin *et al.*, 1996). As they move, the pseudoplasmodia use reserves, both for energy and for synthesis of the slime sheath that we discussed in Chapter 9. Some cells are left in the slime sheath that trails behind the moving slugs. As a result, the slugs become smaller and the fruiting bodies

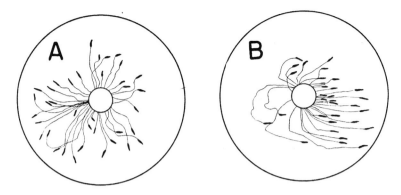

Figure 10.5 Slugs migrate up shallow heat gradients: Amoebae were deposited in the ring of the Petri dish and allowed to form slugs in the dark. The plate labeled A had no heat gradient across it. The plate labeled B was placed in a heat gradient of 0.009°C/cm. The drawings show the tracks left by migrating slugs (Bonner *et al.*, 1950). (Reprinted with permission of Wiley-Liss, a division of John Wiley & Sons, Inc.)

that eventually result are also smaller. The proportions of the cell types are maintained, which raises problems that we will discuss below.

10.4   The unanticipated complexity of cells in the slug

Early workers thought that there were only three cell types in the slug – rear-guard cells that would form part of the attachment to the substratum during culmination, prespore cells, and prestalk cells. However, subtypes of prestalk cells were then discovered, and when prestalk-specific markers became available other distinct prestalk populations were found. The first of these previously unsuspected populations were called anterior-like cells (ALCs) by Sternfeld and David, and are prestalk cells that do not reside in the anterior compartment, but rather remain in the rear of the slug, in the prespore zone (Devine and Loomis, 1985; Sternfeld and David, 1982). There are many of these cells, perhaps equal in number to the cells in the prestalk region of the slug. Like all prestalk cells, the ALCs have acidic autophagic vacuoles which accumulate and protonate the dye neutral red, and this is how they were detected. There is now a suggestion of heterogeneity within the ALC population, based on the observation that some cells express EcmA, some express EcmB, some express both, and not all of the latter stain with neutral red (Gaskel *et al.*, 1992; Williams, 1997). The various new classes of ALC may represent variation in expression of individual cells and may not have different fates. ALCs are evenly distributed along the anterior to posterior axis in the rear of the slug, but they are concentrated on the ventral side of the slugs near the substrate (Sternfeld and David, 1982).

How do the ALCs behave during normal morphogenesis? As the slug transforms into a fruiting body, the ALCs move and eventually form a sheet of amoeboid cells on the top and the bottom of the spore mass. These regions are called the upper and lower cups and cradle the spore mass. We do not know why it is that a portion of the prestalk cell population fails to sort to the tip of an aggregate and remains in the main body of the organism when the tipped aggregate transforms into a slug.

In the soil, slugs may often be severed or partially eaten by microarthropods, many of which are about as big as a slug. Being able to regulate and to carry on with the severed remainder is highly adaptive. If the slug is severed, there is another behavior of ALCs. When the front (the prestalk zone) is removed, the ALCs move toward the front of the prespore zone and re-establish a prestalk zone. When culmination occurs and a stalk is made, the cells that converted from anterior-like to anterior cells form a stalk and not the upper and lower cups to which they were originally fated. The fate of ALCs is different if they have had a chance to be in the anterior, and therefore the signals that they receive there must be different from the signals that cells receive when they are ALCs in the rear. What if the slug is severed again? Is there a reserve of ALCs or can spores converting to form these cells, now perform the reconstitution of a front again? This problem has recently been re-investigated by Ratner (personal communication), who purified prespore cells by flow cytometry and then asked whether the pure population could form a prestalk zone. The pure population of prespore cells converts to prestalk cells and, eventually, stalk. This result is consistent with a trans-differentiation of prespore into prestalk cells, as are the results of Abe *et al.* (1994).

10.4.1 The prestalk region in the slug can be subdivided

A more detailed analysis of the complexity of prestalk cells resulted from the isolation of reliable prestalk-specific genes. These genes were described in Chapter 8 when we were concerned with the establishment of the prestalk–prespore pattern. Neutral red, first used by Bonner, stains all cells of the prestalk class and does not reveal any heterogeneity (Bonner, 1952b; Bonner *et al.*, 1955). In the search for cell type-specific markers several genes have been recovered (see Table 9.1). These include *cprB,* which codes for the cysteine proteinase CP2 (Datta *et al.*, 1986; Gomer *et al.*, 1986). The more extensively used markers are *ecmA* and *ecmB*, which are induced by DIF, and are never expressed in a prespore precursor. The product of *ecmA* is an extracellular matrix protein called EcmA (ST310) and *ecmB* codes for a related protein called EcmB (ST430; see Table 9.1). Both products are part of the sheath material and later of the outer covering of the stalk (Ceccarelli *et al.*, 1991; Gaskell *et al.*, 1992). These proteins are also enriched in the trails of sheath left behind migrating slugs, but are not present in prespore cells or between cells

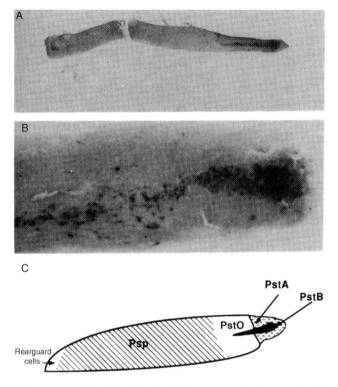

Figure 10.6 The zones of the slug. Panel A shows a whole slug marked to show the pstB cells. Panel B shows a magnified pstB region, and panel C shows a diagram of all of the zones of the slug. Rearguard cells populate the rear of the slug and express *ecmB*. Within the zone of cells fated to be spores, there are a large number of ALCs that maintain prestalk cell characteristics. In the prestalk region there are three zones – a pstO zone that expresses *ecmA* at a low level, a pstA zone (now called AO) that expresses *ecmA* strongly, and a pstB zone (now called pstAB) that forms a cone at the top of the cells and expresses both genes. The pstA and then the pstO cells migrate through the cone during culmination, transforming into stalk cells that express *ecmB* (Jermyn *et al.*, 1989). Not shown are the ALCs that populate the prespore region of the slug. (Reprinted with permission of *Nature*, slightly modified for clarity.)

(McRobbie *et al.*, 1988). Disruption of *ecmA* alters the morphology of slugs (Morrison *et al.*, 1994).

Analysis of where these genes are expressed, first by antibody studies and then by expression of β-galactosidase, lead to the discovery of sub-types of prestalk cells and redefined the anatomy of the slug (Early *et al.*, 1993; Gomer *et al.*, 1986; Jermyn *et al.*, 1989; McRobbie *et al.*, 1988). The promoter of *ecmA* is expressed in the anterior of the prestalk cells in a region called the pstA zone. Behind this is a region that is still in the prestalk zone, but expresses *ecmA* at a low level and is called the pstO region. Gomer calls this the null-cell region (Gomer and Ammann, 1996). There is much more expression of *ecmA* in the very front, especially soon after the slug forms, as shown in Fig. 10.6. The nomenclature of the zones has evolved since the original discovery to reflect the expression patterns of the prestalk genes. Thus pstB cells were originally thought not to express *ecmA*, but now it is known that they do. To reflect this fact they are now called pstAB cells.

Other genes are expressed exclusively in the prestalk region. The late promoter of the secreted phosphodiesterase is expressed in the pstA zone (Hall *et al.*, 1993), as is the cAMP receptor cAR2 (Saxe *et al.*, 1996; Yu and Saxe, 1996). A putative homeobox gene (*Wariai*) is also expressed in the pstA zone. The proportion of cells within the prestalk zone is changed by its mutation (Han and Firtel, 1998). In *Wariai* mutants there is an expanded ecmO region, although the gene is maximally expressed in pstA cells and in ALCs. *Wariai* has sequence homology to a homeobox region, but there is no evidence that the homeobox domain of *Wariai* binds to a specific DNA sequence. *Wariai* is a strictly prestalk-specific sequence and, like other prestalk-specific sequences, it depends on prior expression of GBF. Other mutants with altered patterns of prestalk and prespore cells have been isolated, but with the exception of *Wariai*, not in a manner that allows recovery of the affected genes (Blaschke *et al.*, 1986).

Another gene expressed in the prestalk zone is *rasD*, and overexpression of an activated form results in multitipped structures (Esch *et al.*, 1992; Reymond *et al.*, 1984, 1986, 1989). The expression of *cprB* and *rasD* differs from that of *ecmA* in that it is initially expressed in prespore and prestalk cells and then, as the slug forms, it becomes enriched in prestalk cells. Another indication of independent regulation is that *ecmA* is induced by DIF, whereas *rasD* is not (Jermyn and Williams, 1995). There is now evidence that overexpression of *rasD* has a role in cell fate decisions because in the multitipped mutants that overexpress the activated form of RasD, prestalk-specific genes are induced while prespore-specific genes are repressed (Louis *et al.*, 1997). The presence of RasD as an important participant in prestalk and prespore decisions means that the other proteins, such as MAP kinases and the ligands and receptors that transduce the signals associated with ras pathways, probably also play a role. Several MAP kinase activities have been detected, as described in Chapter 8.

The gradation of expression of EcmA-β-gal in pstA and pstO cells is consistent with a morphogen gradient acting from the extreme tip. MacWilliams has asked whether cells removed from the pstO region and transplanted into the pstA region will move back to the pstO zone. They do, within an hour. When cells from the extreme anterior zone, presumably expressing *ecmA* and *ecmB*, are put in the pstO zone, they rapidly move forward. Put back into their zone of origin, cells do not move (Buhl and MacWilliams, 1991). These results can be explained by a gradient of attractant to which the various types of prestalk cells respond differently (Early *et al.*, 1995).

The complexity of the prestalk population, beginning with ALCs and expanding into pstA, pstB, and pstAB cells, has increased. The number of cell type-specific genes expressed in the pstA zone has also increased and may continue to do so. The definition of these zones results from our improved ability to express region-specific marker genes rather than to perform all experiments with vital dyes like neutral red.

10.4.2 Cellular traffic and cell-type conversion during slug migration

Cells are gradually lost from the rear of migrating slugs as they move. These cells are pstAB cells that form in a cone of the migrating slug and gradually move rearward. The ratio of prespore cells to prestalk cells is maintained by converting a few of the prespore cells to ALCs. During normal slug migration, there is movement of some ALCs forward into the prestalk zone where they differentiate into pstO and then into pstA cells. Those cells that are lost to the ALC population are replaced by prespore cells that convert to ALCs (Abe *et al.*, 1994; Harwood *et al.*, 1991; Kakutani and Takeuchi, 1986). Thus, during long periods of migration, cells that had expressed prespore-specific genes convert to express an ALC-specific promoter and then gradually move into the prestalk zone. This conversion is necessary to maintain the ratio of prestalk and prespore cells (Fig. 10.7). Prespore cells labeled with β-galactosidase convert into ALCs and move into the prestalk zones as slugs move (Detterbeck *et al.*, 1994; Williams 1995). Mutants accelerate this conversion process of conversion to ALCs (Bichler and Weijer, 1994). The fate of the converting cells can be followed because of the stability of the β-galactosidase or other markers first expressed in the prespore cells. The slug is a dynamic structure, and its zones are not impervious to movement of cells, especially if the slug migrates for some time. This trans-boundary migration indicates that there is no physical barrier and stands in contrast to the compartment boundaries in *Drosophila*, which are not crossed. The movement of cells in a gradual manner during slug migration mimics what happens on a greater scale during culmination, as we will discuss in Chapter 11 (Williams, 1997).

10.4.3 How do prespore segments restore their severed prestalk tips?

The isolated prespore region restores its anterior prestalk-like region over several hours, but this restoration is not entirely due to redifferentiation of prespore cells. The ALCs sort to the front to re-establish a prestalk zone and a normally proportioned slug (Devine and Loomis, 1985; Sternfeld and David, 1982). How they do this is a mystery because if the tip is removed, there is no source of signals to direct them to the front. This observation also begs the question of whether prespore cells can convert to prestalk cells. Older evidence indicates that this is the case. Sakai and Takeuchi studied prespore vesicle removal using specific antibodies and showed that conversion occurs (Sakai and Takeuchi, 1971; Takeuchi and Sakai, 1971). As mentioned before, this

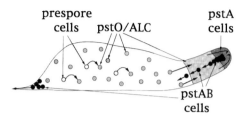

Figure 10.7 The flow of cells and their trans-differentiation in the moving slug. The pstA cells in the anterior express *ecmA* at a high level (Early *et al.*, 1993). At a low frequency, pstA cells prematurely express *ecmB* , and embark on the stalk pathway, becoming pstAB cells. These fall back through the slug and are extruded at the rear, becoming deposited in the slime trail. In other species, such as *D. mucoroides*, these cells adhere and form a stalk primordium, which looks like a rod that extends the length of the slug. The depleted pstA population is replenished by pstO cells, which engage additional transcriptional sites in their *ecmA* promoters. The decline in pstO cells is made up by conversion of anterior-like cells (here labeled pstO/ALC), and this population in turn is sustained by the conversion of prespore cells, which form the ultimate reservoir (Abe *et al.*, 1994). (Reproduced with permission of Cell Press. The figure has been relabeled for clarity.)

problem has recently been re-examined by Ratner and colleagues who showed that highly purified populations of prespore cells re-differentiate into stalk cells.

10.5   How do the cells in the slug move?

Because of advances in optics and the software to interpret cell movement, it has become clear that cells move in organized ways in the mound and in the slug. In the slug, Weijer, Siegert and their colleagues have shown that the cells in the front move in a scroll wave, twisting around the area that is roughly defined as the pstAB cone (Abe *et al.*, 1994; Siegert and Weijer, 1992, 1993). This circumferential movement continues in the area that would be defined as the pstO region. Once behind that area, in the prespore zone, the cells move straight forward, in columns, with periodic increases and decreases of speed, as if they are still responding to cAMP signals emanating from the front, as shown in Fig. 10.8 (Siegert and Weijer, 1989). Such scroll wave behavior of cells also exists in *D. mucoroides* where the stalk is formed continuously and not at the end of migration, as is the case in *D. discoideum* (Dormann *et al.*, 1997). The scroll wave noted in the front of *D. discoideum* extends through the *D. mucoroides* slug, probably due to rotational signaling around the core of pstAB cells.

Are the scroll waves necessary for correct slug formation and movement? The waves of cell movement described by Weijer and Siegert are probably due to cAMP relay, so that a system used during aggregation is also used during later development. The prestalk region contains secreted cAMP phosphodiesterase, that is the result of a specific promoter that is expressed only in this

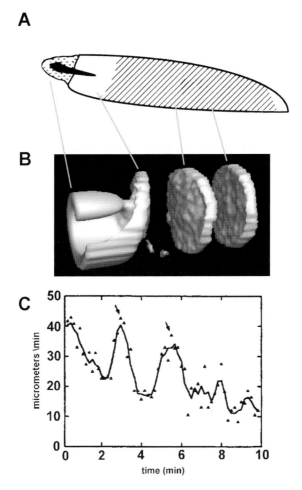

Figure 10.8 Cells in the prestalk and the prespore regions of the slug are in constant movement. Panel A shows the microanatomy of a slug (see Fig. 10.6), with the prestalk region on the left. PstAB cells are shown in black. Panel B shows that the cells in the anterior of the slug move in a spiral, or scroll wave, around the pstAB cell core. Cells in the posterior and hence prespore region, move directly forward (Dormann *et al.*, 1998; Siegert and Weijer, 1992). In *D. mucoroides*, the pstAB population extends to the rear and the spiral motion occurs throughout the slug (Dormann *et al.*, 1997). Panel C shows that a cell within the prespore zone of a slug undergoes periodic increases and decreases of velocity. A fluorescent cell among many unlabeled cells was followed with confocal microscopy. When the cell is moving fast, it is elongated, when it is moving slowly, it extends lateral pseudopods, and has a less polarized shape. Arrows show two periods when the cell is moving rapidly (Siegert and Weijer, 1991, 1993). Prestalk cells also move in a periodic fashion (Abe *et al.*, 1994). (Panel A is reproduced courtesy of *Nature*; panels B and C courtesy of C. J. Weijer, University of Dundee.)

region (Faure *et al.*, 1990; Hall *et al.*, 1993). If the gene that codes for a specific inhibitor of the phosphodiesterase is expressed in the prestalk region, aberrant slugs form that cannot migrate and fail to culminate. Expressing the inhibitor everywhere in the slug blocks slug formation entirely (Wu *et al.*, 1995a). These results are consistent with the idea that a cAMP signal is modified by the phosphodiesterase in the prestalk region, perhaps altering the sensitivity of the prestalk cells. Interfering with the signaling system may also, through adaptation, cause an inactivation of PKA.

A finding that is relevant to the requirement for signal relay in slug movement has been reported by Bonner, who invented a way to make two-dimensional slugs (Bonner, 1998). If cells are removed from slugs with a micropipette and inserted under mineral oil, the cells crawl out of the pipette and form an elongated and flat group of cells. Astonishingly, the mass of a few hundred cells is only one cell layer thick, but it moves. Those cells that are in the front move in chaotic patterns, as if they are trying to imitate the scroll waves in a three-dimensional slug. The cells in the rear move forward, but do not move in a periodic fashion. The fact that the whole group of a few hundred cells moves like a slug, argues against a signal relay system that is essential for movement, but is consistent with a cAMP relay system that makes movement more efficient or rapid. All cells in the tip were in motion, making the idea that one or a few cells that act as a master regulator less likely.

The question of how cells gain traction when they move, presents a mystery. The internal mobilization of the cytoskeleton has been examined, but what concerns us here is the substratum on which the cells in the slug move toward light or heat. The slugs move on an immobile matrix of sheath, which is thickest between the substratum and the most ventral cells, yet when the cells move forward in their periodic way, cells at all depths in the slug move together. If the cells were crawling over each other, each using the cell below to gain a foothold, there would be no coordinated movement toward the front. Rather, cells at the ventral surface would move first. This does not occur and therefore there may be secretion of extracellular matrix from the tip between the cells. The adherence of cells in the slug to an extracellular matrix is loose, because they can be easily dissociated. The secretion of matrix does not depend on cellulose, because a cellulose synthase mutant makes no cellulose but does make migrating slugs, although it is not yet established that these migrate at the normal speed (R. L. Blanton, personal communication). Siegert and colleagues have reviewed cell movement during aggregation and mound formation, and cell movement in the slug (Siegert and Weijer, 1997).

10.6   Gene regulation within the prestalk and prespore zones

What promoter sequences and signal transduction pathways lead to expression of prestalk genes like *ecmA* and *ecmB*, or prespore genes like *cotB*? Is there any predictable sequence of a prestalk or prespore gene that would allow a new gene to be assigned with high probability to either class? We have already seen that known post-aggregation genes potentially bind GBF, but this is a prerequisite for prestalk as well as prespore genes. Without GBF, cells do not pass the loose aggregate stage. Binding of GBF does not distinguish prestalk gene and prespore gene expression in any simple way. What then does? One possibility suggested by Gollop and Kimmel is that it is the combination of two or more GBF-binding sites that determines cell-type expression. In the cAR3 promoter, 350 base pairs is adequate for normal expression, but only one of the three potential GBF-binding regions is necessary for prespore-specific expression. By contrast, two such elements are required for prestalk cell expression, and one of these is in a transcribed region at +50 after the start of transcription (Gollop and Kimmel, 1997). The number of GBF-binding regions may contribute to specificity.

10.6.1 *The* ecmA *and* ecmB *promoters are highly regulated*

Analysis of the *ecmA* and *ecmB* promoters has been extensive and revealing. The fundamental difference between these two prestalk promoters is that *ecmA* is expressed throughout the prestalk region, although more at the anterior, while the *ecmB* promoter has the much more restricted expression pattern shown in Fig. 10.6. How can we account for these differences in expression? The laboratory of Jeffrey Williams has systematically examined the promoters of *ecmA* and *ecmB*. This analysis has led to a theory that explains the absence of EcmB in pstA and pstO cells – until the moment during culmination, when the pstA and pstO cells pass into the developing stalk tube on their way to becoming stalk. The next chapter examines culmination, but for the moment it is enough to realize that when the slug culminates, movement and conversion of prestalk cells begins. The movement is first upward and then down through the now vertically oriented stalk tube primordium formed by the pstAB cells (see Plates 4 and 5 and Chapter 11 for greater detail). As the pstA cells and then the pstO cells pass through an open annulus, they express *ecmB*, vacuolate, and become stalk cells. The continuous pile-up of vacuolizing prestalk cells inside a tube of extracellular matrix material forms a stalk and lifts the presumptive spore mass from the substratum.

 A fundamental regulatory problem of the several prestalk genes is to explain how *ecmB* can be kept off in pstA and pstO cells in the slug, until they are called upon to form stalk. There are presumably many genes like *ecmB* that have to be repressed. Dissection of the *ecmB* promoter revealed both activating and inhibitory elements (Ceccarelli *et al.*, 1991; Harwood *et al.*, 1993; Kawata *et al.*, 1996). Two inhibitory elements are close to the cap site, where transcription begins; the activating domains are more distal. Removal of the two redundant inhibitory elements in the *ecmB* promoter, which have the sequence TTGA in an inverted repeat, allows expression of a reporter gene in the pstA zone. Moving the inhibitory sequences into the *ecmA* promoter restricts its zone of expression to the pstAB region. There is good evidence that this proximal sequence acts in concert with a repressor protein to prevent expression of the *ecmB* gene in the pstA zone.

10.6.2   Dd-STATa binds to the ecmB promoter inhibitory sequences

What is the nature of the repressor that binds the *ecmB* promoter and prevents its expression in pstA cells? A protein that binds to the TTGA sequence within the *ecmB* promoter has been purified, sequenced, and identified as a transcriptional activator of the STAT family (Signal Transducers and Activators of Transcription) (Kawata *et al.*, 1997). These important proteins were first described as intermediates in interferon induction, and are well known in higher organisms. In mammalian cells, cytokines interact with transmembrane receptors. Cytokines lack intrinsic tyrosine kinase activity, but are associated with the Janus family of tyrosine kinases. The binding of ligand causes dimerization of the receptors and activates the Janus kinases, which phosphorylate the receptors on tyrosine residues. The STAT proteins bind the phosphorylated receptors through their SH2 domains and are also phosphorylated by the Janus kinase. STATs phosphorylated on tyrosine dimerize, and when this happens, they migrate rapidly to the nucleus where they act as transcriptional activators and bind the sequence TTGA (Darnell, 1997).

It came as a surprise that a protein purified from *Dictyostelium* that binds the TTGA sequence was homologous to the mammalian STATs. No such proteins had been discovered in lower eukaryotes, and the STATs do not exist in budding yeast. The *Dictyostelium* STATs are made throughout growth and development and localize to the nucleus during aggregation (Araki *et al.*, 1998). Following aggregation, as the slug forms, STATa becomes localized to pstA cells (Araki *et al.*, 1998). Dd-STATa seems to function in the manner of mammalian STATs, except that we do not know the equivalents of the Janus tyrosine kinases or the cytokine receptors. *Dictyostelium* has several tyrosine kinases, but these are not known to be the enzymes that phosphorylate STATa (Adler *et al.*, 1996; Tan and Spudich, 1990). The inducer of nuclear localization is the ever-present cAR1. Cells stimulated with cAMP induce the nuclear migration of Dd-STATa within 10 seconds. Cells that lack cAR1 do not localize STATa to the nucleus in response to exogenous cAMP. The nuclear localization of Dd-STATa occurs predominantly in the prestalk zone.

The idea that the repressor sequences in the proximal part of the *ecmB* promoter bind Dd-STATa, and that this accounts for the restricted expression of *ecmB*, leads to a hard prediction. If this model is correct, then deletion of the Dd-STATa gene should result in a wider expression of *ecmB* within the pstA zone. This is apparently the case. The *Dd-STATa*-null cells have a defective culmination, indicating that this repressor works widely to restrain culmination and not simply in the regulation of *ecmB*. Why Dd-STATa should be made at all times and localize to nuclei at the time of aggregation only to assume a more restricted deployment in pstA cells, is not known. Deletion of *dd-STATa* does

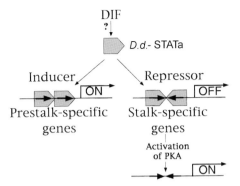

Figure 10.9 Dd-STATa and the repression of *ecmB*. Dd-STATa binds TTGA sequences in promoters that are specifically transcribed in the prestalk pathway. TTGA sequences can be arranged head-to-tail or head-to-head. When Dd-STATa binds to a head-to-tail dimer, it activates, as in the case of *ecmA*. When it binds head-to-head sequences it represses transcription, as in the case of *ecmB* during slug migration. Within the part of the *ecmB* promoter that directs expression in the stalk tube, there is an activator region of about 130 nucleotides. When the repressor elements (head-to-head TTGA sequences) are removed from the *ecmB* promoter sequence, the modified sequence promotes general prestalk expression, not confined to pstAB cells (Kawata *et al.*, 1996). (Reprinted with permission from the *EMBO Journal*, Oxford University Press, and slightly modified for clarity.)

not block development at an early time, so there is no critical early role of this protein. Two other STAT genes have been isolated, but *dd-STATb* is expressed in prespore cells and has no known phenotype when deleted and *Dd-STATc* is not induced by cAMP or DIF to localize in nuclei (J. Williams, personal communication).

What relieves the inhibition during culmination when the pstA cells start to express *ecmB*? Relief from inhibition depends on protein kinase A. Mutants that make no active PKA do not induce *ecmB*, presumably because inhibition is not relieved (Zhukovskaya *et al.*, 1996). During culmination, a rise in intracellular cAMP is crucial for prestalk cell differentiation, leading to the model shown in Fig. 10.9 (Harwood *et al.*, 1992b; Inouye and Gross, 1993; Kubohara *et al.*, 1993; Maeda, 1988). PKA is not the only element that is involved in *ecmB* repression. GSK-3, the *Dictyostelium* homologue of glycogen synthase kinase-3, is also an important regulator. Mutants lacking GSK-3 overproduce *ecmB* and pstB cells (Harwood *et al.*, 1995).

10.6.3 Regulation of ecmA

Transcription of *ecmA* is induced by cAMP and DIF, while *ecmB* transcription is induced by DIF and repressed by extracellular cAMP (Berks and Kay, 1990). These results were obtained in dispersed cell culture and are difficult to extrapolate to the intact slug and culminant (see Williams (1997) for a discussion of this problem). If the *in vitro* results apply in the intact slug, the pstAB region should be an area of low cAMP concentration. There is a secreted PDE dedicated to the pstA zone, but not specifically to the pstAB cells in the cone at the front of the slug (Hall *et al.*, 1993). The other major regulator, which may act indirectly, is ammonia – a product of amino acid catabolism by the cells. As long as ammonia levels are high, slugs will not culminate and form a fruiting body; rather, they will continue to migrate as slugs (Schindler and Sussman, 1977). Presumably the repression mechanism for *ecmB* that we have just described depends on the balance of ammonia and cAMP.

The *ecmA* promoter has a modular construction and has been implicated in the expression in pstO cells. A distal sequence of 132 bp about 1 kb upstream of the cap site directs transcription in the pstO zone and not any other region (Early *et al.*, 1993). This region, fused to a basal promoter and β-galactosidase, has been useful in marking the pstO zone of slugs and in tracing the cells to which it gives rise (Early *et al.*, 1993). This fragment of the promoter also drives reporter-gene expression in the ALCs and was used to derive the results shown in Fig. 10.7. It is an excellent marker for the pstO region and for ALCs.

The distal sequence in the *ecmA* promoter binds Dd-STATa and acts as an inducer. The gene coding for EcmA was originally defined because it responded to DIF induction. We know that STATa binds the TTGA sequence, as shown in Fig. 10.9, but its nuclear localization does not seem to be responsive to DIF. There may be other signals that are required for this, and therefore the terminus of the DIF pathway remains incompletely known.

10.6.4 The prespore gene promoters

A family of similar proteins, including SP96, SP70, and SP60, is transcribed from coordinately regulated genes (see Table 12.1 for references). These proteins are secreted from the developing spore later in the developmental cycle, but the proteins are made at about the time the cells aggregate, and accumulate in vesicles until the spore coat is formed (Devine *et al.*, 1983). Like other prespore genes, *cotC* (SP60), *cotB* (SP70), and *cotA* (SP96) are inducible in suspension culture by high levels of exogenous cAMP. Thus, they have both structural and regulatory similarities. If regulatory elements of prespore promoters can be recognized, this family of genes is the place to do it.

The *cotC* promoter has been examined by the laboratory of Richard Firtel (Haberstroh and Firtel, 1990; Haberstroh *et al.*, 1991; Powell-Coffman *et al.*, 1994). Analysis of the *cotC* promoter revealed three CA-rich sequences which stand out from the general background of A and T sequences (Haberstroh and Firtel, 1990). When these sequences were deleted along with flanking nucleotides, the expression and the localization of the transcribed product (in this case β-galactosidase) was reduced and expressed in a gradient that left the strongest staining in the front of the prespore zone (Haberstroh and Firtel, 1990). This gradient-like expression was not observed with equivalent mutations in the promoter of *cotB* (Fosnaugh and Loomis, 1993). Although the CACA repeats observed in the coordinately regulated prespore promoters are not found in prestalk promoters, they do not bind a novel prespore-specific transcription factor. These sequences bind the previously characterized G box-binding factor, which does not confer specificity and is equally active with other G-rich sequences in prestalk promoters (Powell-Coffman *et al.*, 1994).

GBF appears to recognize a variety of G-rich sequences (Hjorth *et al.*, 1990). GBF is necessary for the transition from loose aggregate to tight aggregate and for the transcription of the prespore-specific genes, but does not by itself confer cell-type specificity. What does? By deletion analysis, Powell-Coffman has found a relatively proximal element that is AT-rich and 37-bp long, located between nucleotides −318 to 355. Point mutations within this AT-rich segment prevent expression of *cotC* (Powell-Coffman *et al.*, 1994). Attaching this sequence to a minimal promoter caused prespore-specific expression in the presence of the CA-rich elements. This AT-rich region is homologous to a region from the prespore-specific promoter of *pspB* (Powell-Coffman and Firtel, 1994). So far, no transcription factor that binds the AT-rich sequence has been purified, nor has a gene coding for one been cloned. An attempt at comparing all elements of known prespore promoters has been made, but as in the more specific cases discussed above, conclusions are limited (Miller *et al.*, 1996). In the end, we cannot look at a promoter sequence and determine whether it would be expressed in a prespore or prestalk cell. The *cis*-acting regions are too small and degenerate. *Dictyostelium* is not alone in its use of AT-rich promoters; other organisms, notably the pathogens *Plasmodium falciparum* and *Entamoeba histolytica*, present similar problems.

10.7 Slugger mutants maintain their repression of culmination

There is a class of mutants in which the repressors of culmination are not eliminated, but are constitutive, causing slugs to wander like Odysseus and not to form a fruiting body. Such mutants, called *sluggers*, have been known for many years (Gee *et al.*, 1994; Newell and Ross, 1982a; Sussman *et al.*, 1978). Conditions of high ammonia favor slug migration and after an extensive parasexual genetic analysis Newell and Ross (1982a) found that slugger mutations fall into several complementation groups. Members of some complementation groups are excessively sensitive to ammonia. It is possible to remove ammonia enzymatically (Schindler and Sussman, 1977), and under these conditions slugger mutants – which fall in three complementation groups – form fruiting bodies. Ammonia has effects at several stages in the life cycle but its role in inhibiting culmination, whether in the wild-type or in the hypersensitive mutants, depends on its properties as a weak base. Octylamine, for example, is even more efficient at preventing culmination than ammonia (Gee *et al.*, 1994). (For a discussion of how ammonia may accomplish this by raising the pH of acidic cellular compartments, see Davies *et al.* (1993, 1996).) Many of the slugger mutants isolated to date have been the result of chemical mutagenesis, and there is as yet no way to recover their genes.

The methods elucidated for isolating slugger mutants – essentially a visual screen of a mutagenized population – can be applied after REMI mutagenesis, and several genes whose mutated alleles lead to an extended slug-like state have been recovered. *CudA* (culmination defective) has no known homologies except to a sequence in *Entamoeba*. CudA is a nucleoprotein with unusual regulation – it is expressed in prespore cells and in pstA cells, but not in pstO cells. Cells that lack CudA are strong sluggers – they will migrate for long periods when slugs composed of wild-type cells would culminate. If the CudA protein is expressed in the pstA cells, this is enough to cause the slugs to culminate correctly.

A mutation in a gene called *dhkC* eliminates the sensitivity of slugs to ammonia and is thought to be part of the mechanism by which ammonia is sensed (Singleton *et al.*, 1998). Null mutations of *dhkC* culminate in concentrations of ammonia that would cause the wild-type to keep migrating. DhkC is a sensor histidine kinase, which is part of a two-component signaling system. Truncated forms of DhkC cause the slugs to migrate until they run out of reserves. DhkC effects depend on the response regulator domain of a phosphotransfer protein called RdeA and on a phosphodiesterase called RegA (Chang *et al.*, 1998; Singleton *et al.*, 1998; Thomason *et al.*, 1998). Two-component systems were described in Section 8.5.6, and play a significant role in determining how *Dictyostelium* amoebae sense their environment during growth and development (Loomis *et al.*, 1998).

A gene called *chrA* has been mentioned in Section 3.7 because of its ability to form spores rather than stalk in chimeras. When incubated without a wild-type partner, *chrA* mutants form a long slug that will not proceed to culmination and makes an enlarged prespore region, such that the prespore and prestalk cell ratios are distorted. These slugs have very small prestalk zones – 5% of the total or less (H. L. Ennis, unpublished results). ChrA is an F-box protein, which in other organisms cause the breakdown of specific target proteins by bringing them into proximity with ubiquitination enzymes. In the absence of the F-box protein, the target accumulates. The timely removal of critical proteins is essential for development, the cell cycle, and many other events. Thus the identification of the target proteins by genetic or biochemical methods may help our understanding of the transition from slug to culminant.

10.8 The stability of the differentiated slug cells

Cells in the slug have not forgotten how to aggregate. Simple experiments can sometimes tell us something about the properties of the regulatory systems that control development. We can ask whether cells, having passed a developmental stage, can still perform the events of the prior stage. Cells in the slug stage can be disrupted into single cells by gentle pipetting, and the separated cells, when placed on agar or a filter paper, rapidly re-aggregate by the same apparent chemotactic mechanisms that they used before (Newell *et al.*, 1972; Takeuchi and Sakai, 1971). The cells stop the enzyme synthesis that was occurring when they were disaggregated, and do not resume the process until they are back in an aggregate. This recapitulation of the developmental program can accomplish in a few hours what previously took 16–18 hours. In later development, cAR1 and other elements necessary for aggregation are not repressed; rather, the system adapts so that when cells are disaggregated, cAMP levels – normally high in the aggregate and the later stages – decline by dilution and the always present chemotactic system (based on cAR1) is immediately reactivated. There is direct evidence that cAMP levels in the prespore zone maintain spore cell differentiation. If cAMP phosphodiesterase bound to beads is introduced into this region, the cells surrounding the beads de-differentiate (Wang *et al.*, 1998b).

While the cells remember their developmental functions if they are dissociated and starved, they will not remember if they are fed. This loss of information is called erasure. Dissociated cells retain their developmental features, including rapid re-aggregation, only for about 80 minutes in the presence of food (Finney *et al.*, 1979; Kraft *et al.*, 1989; Soll and Waddell, 1975; Waddell and Soll, 1977). After 80 minutes, cells resuspended in dextrose buffer or nutrient medium rapidly lose the enzyme and other molecules that they need for development, and start to re-synthesize the proteins that they need to grow and divide (Finney *et al.*, 1987). The loss of development-specific proteins can be prevented by extracellular cAMP. PKA deactivation is probably the key to loss of development-specific synthesis, but there may also be an activation of specific proteolysis. The *Dictyostelium* developmental program must constantly assess nutritional status – cells can go back to feeding until late stages, indicating that there are no absolute commitment features to the program. The slug, covered by the sheath, may be prevented from feeding on bacteria, and it is often the case that slugs can be found wandering out of the colonies in which they form on to bacterial lawns without de-differentiation. The physical barrier of the sheath may provide a sort of commitment that is overcome by dissociation and feeding.

In conclusion, the slug has many unusual properties: it migrates; it is sensory; and it regulates. It also controls the number of tips and its axis. A slug regulates the movement of cells and retains a cadre of cells to form a new tip in case it is severed. Within it, cells differentiate, move, and sometimes transdifferentiate, revealing the astonishing complexity of cell types that make up this organism. The slug employs regulatory molecules thought previously to exist only in the world of animals. Thus, by maintaining a relatively stable arrangement of prespore and prestalk cells for days, it helps us to examine a number of critical problems.

11

Culmination

11.1 Deciding when migration has gone on long enough

One of the decisions that a slug must make is when to form a fruiting body. It is possible to trick slugs into migrating toward light until they expire, having used all of their reserves. The slug offers the organism escape from noxious environments, dispersal, and perhaps protection from nematodes, but ultimately the major protection is the creation of a resistant spore, placed so that it can be dispersed. This transition is accomplished late in development by a series of elaborate cell maneuvers. The slug begins with a set of partially differentiated prespore cells, all contiguous in the rear of the structure, and at the end has put these cells, fully encapsulated, into a loosely held sphere at the top of a stalk.

Ammonia, as we learned in Chapter 10, causes the slugs to refrain from culmination. They migrate away from it, so it may have a negative chemotactic effect. Sussman, White, and Schindler determined that 10^8 cells contain about 5 mg of protein, and that during the course of development about 2 mg of this protein is degraded, eventually releasing a substantial amount of ammonia (Schindler and Sussman, 1977; White and Sussman, 1961). Similar observations were made by Gregg et al. (1954). Wilson and Rutherford (1978) measured the amounts of ammonia in tissue slices and found that it accumulated at the end of development. Ammonia accumulates as cells develop and, if not allowed to evaporate, has a profound effect, causing aggregation to be delayed and slugs to migrate (Schindler and Sussman, 1977). Schindler and Sussman also showed that if slugs were maintained in oblique light and then treated in such a way that ammonia was removed from the environment by incorporation into α-ketoglutarate by the action of glutamate dehydrogenase, they went on to culmination almost immediately. The removal of ammonia overcame the effects of oblique light.

How ammonia can exert an effect is not known, but exposure to ammonium carbonate inhibits cAMP production and release (Schindler and Sussman, 1979). Gross has proposed that neutralization of acidic compartments by ammonia leads to calcium release, and that this calcium has a controlling role in late development (Cubitt *et al.*, 1995; Davies *et al.*, 1993; Jaffe, 1997; Pinter and Gross, 1995). The ammonia sensitivity of certain slugger mutants, discussed in Chapter 10, also suggests a role for ammonia sensing in the initiation of culmination (Sussman *et al.*, 1978; Zinda and Singleton, 1998).

One of the critical events in the transition to culmination is the release from inhibition of genes that were repressed in the pstA and pstO compartments. As we know, a transcription factor, Dd-STATa, plays a major role in maintaining the inhibition of *ecmB* in all cells, except the pstAB cells. Dd-STATa may also act as a general inhibitor of the expression of genes that are not needed before culmination in prestalk cells. One of the major questions that arises about the transition from slug to culminant is whether Dd-STATa has a more general role than inhibiting *ecmB*. We are limited by the fact that *ecmB* is the only gene known to be expressed in the stalk primordium. If Dd-STATa is the general repressor of culmination, it should control the expression, during the slug period, of other genes that are expressed at the same core site as *ecmB*, but not as yet expressed in the pstA zone. Before we venture into the localized expression of *ecmB* and the induction of spore maturation, it is important to deal with earlier events, and with anatomy.

11.2   Early steps in culmination

Culmination was a word coined by Bonner (1944) to denote the conversion from migrating slug to an upright spore-containing fruiting body. The first indication that a slug is about to stop migrating and undergo fruiting body formation is a cessation of movement in the front, but not in the rear (Fig. 11.1). The effect of this, which is obvious in time-lapse microscopy, is to shorten the slug as the rear catches up with the front. The tip of the slug then begins to point upward until the combination of these two motions has, over the course of 30 minutes, converted the axis of the organism from horizontal to vertical (Rand and Sussman, 1983; Raper and Fennell, 1952). Over the next 30 minutes, the aggregate compacts to form a structure called the Mexican hat in which the prespore cells occupy the rim. In the center, a cellulose primordium of the stalk tube forms (Fig. 11.2). The formation of a stalk tube initially of a diameter that is appropriate for the cell mass was first observed by Raper and Fennell (1952). At the beginning, the stalk tube forms in the upper part of the aggregate and then reaches the base. How the initial stalk tube forms and extends to the substrate at the beginning of culmination is not known, but probably involves a pathfinder effect by pre-existing pstAB cells in the front of the slug. Once the stalk tube has made contact with the

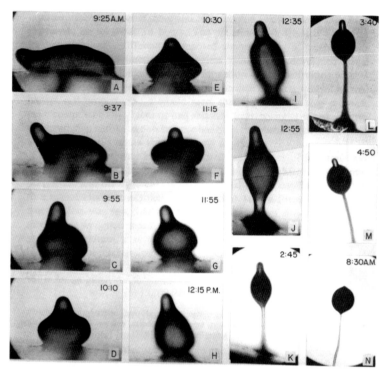

Figure 11.1 The stages of culmination. Note the shortening of the slug in panels A and B as migration ends and culmination begins. In panels C and D, the culminant assumes a vertical orientation. Panels E and F show the period at which stalk formation begins, and in G and H the basal disc has begun to form. Panels I, J and K show the beginning of the ascent of the sorus, while L and M show the stages before spore maturation. The final sorus is shown in panel N. This classical study by Raper and Fennell (1952) indicates the times between stages. The sorus in panel N had been mature for some time before it was photographed. (Reprinted with permission of Princeton University Press.)

basal disc, all growth and elongation of the fruiting structure occurs at the apex.

We know more about the sub-populations within the slug and the early culminant because we can mark them with the promoters and promoter fragments fused to β-galactosidase or other genes, as described in Chapter 10. Microscopy and image analysis have also increased our understanding of what happens as the slug converts to a fruiting structure. Several studies have shown that there is a group of cells that expresses *ecmB* in the slug and is located next to the substratum and just behind the prestalk zone (Dormann et al., 1996; Jermyn et al., 1996). When the slug begins to culminate it stops moving and this group of cells moves forward into the prestalk zone, forming the rearguard cells, as shown in Fig. 11.3. When the rear of the slug moves up, the rearguard cells and the ventral pstB cells are brought into contact with each

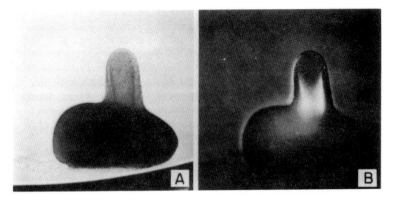

Figure 11.2 The origin of the stalk tube. An intact tipped aggregate stained with iodide reveals the stalk tube. The structure on the left was photographed in normal light, while the structure on the right was photographed in polarized light. Cells will migrate up the outer side and then into it and convert into stalk. During this period they add cellulose and other material to the tube (Raper and Fennell, 1952). (Reprinted with permission of Princeton University Press.)

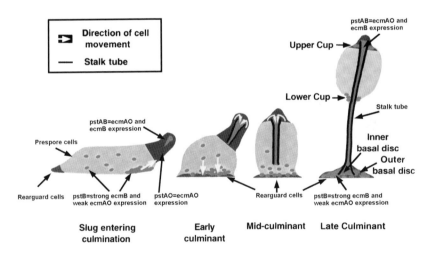

Figure 11.3 The movement of cell types within the culminating slug. The *ecmA* and *ecmB* expressing cells are marked. As the slug shifts into culmination, the *ecmB*-expressing cells are massed just behind the prestalk zone. When the slug ceases movement, these cells move up into the back half of the prestalk zone – within the pstO region but closely attached to the substratum, perhaps providing an anchor. The posterior cells move over them (the rear of the slug keeps moving) and together these two cell types begin formation of the basal disc. They are joined by ALCs that migrate down and form the outer basal disc. Meanwhile, at the apex, pstA and then pstO cells migrate into the stalk tube and create stalk. This drawing does not show the pstA and pstO compartments separately. A lower cup forms of ALCs from an *ecmB* expressing population. There is also an apical disc population at the very top that expresses *ecmA* (Jermyn *et al.*, 1996; Williams, 1997). (See Plates 5 and 6.) (Reproduced with permission of the Company of Biologists Ltd.)

other and begin to form the basal disc. Later, a part of the population of ALCs migrates downward to add to the top of the basal disc (Sternfeld and David, 1982; Sternfeld, 1992). When the movements of the cells in the early culminant were studied, it became apparent that in addition to the radial movement of cells in the prestalk zone, the cells that join to form the basal disc are also highly motile and form an independent signaling center. Thus, the basal disc forms as an early event in culmination, and the cells within it have independent signaling properties (Jermyn *et al.*, 1996). Some of these cells are separated from the basal disc and form a cup below the spore mass (Sternfeld, 1998).

The basal disc population can regulate, as shown most clearly in a mutant affected in glycogen synthase kinase-3 (GSK-3). Although this enzyme plays an important role in the biochemistry of glycogen metabolism, it is also a crucial element in signal transduction, mediating the effects of the *wnt* pathway in higher organisms (Shimizu *et al.*, 1997). Mutants of *gskA* have an increased basal disc. Harwood and colleagues have shown that the increased basal disc cells come at the expense of prespore cells, which convert from one lineage to another (Harwood *et al.*, 1995; Insall, 1995). This cell-type specification event begins much earlier, in the mound, when the initial differentiations are taking place. The fruiting bodies of *gskA* mutants have very few spores. GSK-3 is regulated by extracellular cAMP through the cAR3 receptor (Ginsburg and Kimmel, 1997; Plyte *et al.*, 1999). Other regulatory elements that activate GSK-3 include the ZAK1 kinase (A. Kimmel, personal communication). GSK-3, in turn, regulates the nuclear localization of STATa (A. J. Harwood, personal communication).

In other species, the basal disc can be much more elaborate than in *D. discoideum* (Raper and Fennell, 1967). It serves to grip the various substrates on which fruiting bodies form. Watching fruiting bodies form on autoclaved soil rather than on agar is an instructive experience. The basal disc is a critical structure because without it the fruiting body does not stand up, the spores are not projected into crevices in the soil, and dispersal – the presumed adaptive value of the stalk in this and many other organisms – is lost.

11.3 Movements at the Mexican hat stage

The compacted slug forms the Mexican hat. This is a transient structure, because very rapidly, the whole community of cells begins to gain in height and to transform in shape. This occurs by addition of stalk tube at the apex and the local formation of stalk. As culmination proceeds, the prestalk cells march into the stalk tube – first the pstA and then the pstO cells. As the cells move into the stalk tube, they synthesize and secrete EcmB, and in this sense become like the cells that formed the initial pstAB zone. How the prestalk cells make a 180 degree turn and head downward into the stalk tube is not known. We do not know, for example, whether there is some chemotactic attraction, or if the turn is based on substrate adhesion of the developing stalk cells to the stalk tube.

Once prestalk cells enter into the tube, they vacuolate, expand, deposit cellulose, and die as stalk cells. The prespore cells, contained in their sheath material and supported by a cup of anterior-like cells, are gradually lifted to the top (Chen *et al.*, 1998; Sternfeld, 1998).

11.4   The anatomic details of a culminant

Before returning to the events that control culmination, we should define the parts of the fruiting body in more detail. The stalk is a composite structure, communally formed, and always consisting of a stalk tube surrounding stalk cells. The stalk is not simply a series of cellulose-covered cells piled one on the other – it is always surrounded by an external tube of cellulose and glycoprotein. The stalk tube is critical for the formation of the stalk, because prestalk cells migrate into it. The source of the stalk tube primordium in the aggregate, as well as its extension from the apex, are not well understood (Fig. 11.2).

In the stalk shown in Fig. 11.4, the vacuolated stalk cells pass through the spore mass. At the top, there are more specializations. Below the tip, there is a narrowing called Grimson's constriction, above which there is a cone-like apex that flares out and includes an area of forming stalk tube. Prestalk cells migrate into this region. The cells within the cone are elongated and are oriented perpendicularly (Fig. 11.5).

Outside of the stalk tube, there is also heterogeneity among cell types. In Fig. 11.4, the prespore cells are stained darkly and are bounded by a lower cup of amoeboid cells. These form from a population of ALCs in the slug posterior that move upward during culmination. They may be a part of a population that also forms the top of the basal disc (Jermyn *et al.*, 1996; Williams, 1997). The upper cup is formed by another population. Before all of the prestalk O cells have passed into the stalk tube, about half of them induce *ecmB* (Bichler and Weijer, 1994; Jermyn and Williams, 1991). This pattern of expression also characterizes the cells in the stalk tube. Upper cup cells utilize certain elements of the *ecmB* promoter differently than the cells that form the stalk and the thought is that their gene regulation properties are distinct (Ceccarelli *et al.*, 1991).

An even more apical structure in culminants is called the apical disc. This has been observed with vital dyes, but is only apparent in culminants of slugs that have migrated. Cells in the apical disc express the prestalk cell markers *ecmA* and *ecmB* They also express a gene called *staB*, the function of which is unknown (Robinson and Williams, 1997). The thought that these cells have a special regulatory role, perhaps as a chemotactic beacon, has never been confirmed. Careful analysis with fine laser beams, in which these cells are destroyed, could answer these questions (Klaus and George, 1974).

Figure 11.5 shows the prestalk zone. Surprisingly, in the electron micrograph it is not possible to see exactly where the cells enter the stalk tube, nor where stalk tube is being synthesized. The pstA cells that have not yet been

Figure 11.4 An electron micrograph showing the anatomy of a mass of cells during mid to late culmination. The vacuolated stalk cells are surrounded by a stalk tube, with which their cellulose coats merge. The stalk tube is laid down at the apex and pstA and then pstO cells migrate into it. At its base, the stalk tube inserts into the basal disc, which is a composite structure. Its outer layer is formed by *ecmB*-expressing cells derived from the ALC population. These form the dark, outer layer on the basal disc. Other ALCs form a cup of amoeboid cells below the spore mass. (This electron micrograph was provided by M. Grimson, Texas Tech University.)

converted to stalk cells are in circular motion, as the experiments of Weijer, Siegert and their colleagues have shown (Dormann *et al.*, 1997; Zimmermann and Siegert, 1998).

Within the prestalk region, an even higher magnification analysis has shown that there are cell junctional specializations that were not expected in so primitive an organism. These are seen in Fig. 11.7, and resemble the junctional complexes of epithelial cells. The dark plaques at the plasma membranes of apposing cells are associated with actin microfilaments, similar to the adherens junctions of multicellular organisms.

One of the components of junctions in higher organisms is β-catenin, and this protein (called aardvark) has also been discovered in *Dictyostelium*. Deletion of the aardvark gene causes a loss of the culmination junctional complexes and the production of supernumerary stalks (A. Harwood, personal

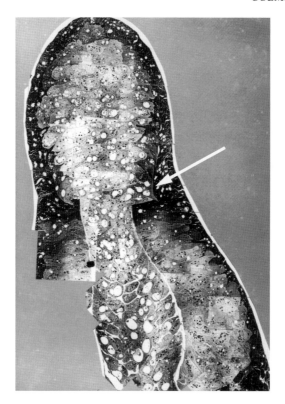

Figure 11.5 An expanded view of the upper part of the forming stalk. Note the constriction of Grimson and the stalk tube extending above it (arrow). Above this, stalk tube is being synthesized. Cells in the center of the tube are horizontally oriented and will soon convert to stalk. The stalk tube itself must be extended upward, but the mechanism by which this is accomplished is not well understood. (Courtesy of M. Grimson, Texas Tech University.)

communication). The intriguing point is that β-catenin in higher organisms also serves an important regulatory function by controlling the activity of GSK-3 (Harwood *et al.*, 1995; Plyte *et al.*, 1999). *Dictyostelium* employs this kinase for regulatory decisions, and so it will be of interest to learn whether an interaction similar to the one between β-catenin and GSK-3 also exists in *Dictyostelium*.

The entire slug and culminant is surrounded by an electron-dense cell type (Fuchs *et al.*, 1993) (Fig. 11.6). This layer is detected best with rapid freezing and other techniques, and seems to be composed of cells that lie under the extracellular matrix. Like epithelial cell layers in other organisms, these cells have a basal–apical polarity and are closely apposed. The fusion of the outer membrane leaflets of the putative epithelial cells has been observed, but no other junctional specializations were found in the peripheral layer of the slug.

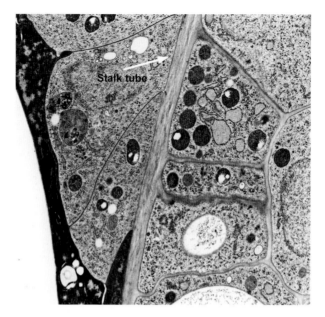

Figure 11.6 A high-resolution view of the stalk tube and apposed stalk cells. The inner layer of the stalk tube is continuous with the stalk cell wall. Cells to the left of the stalk tube are outside it and those to the right have already migrated in. (Courtesy of M. Grimson and R. L. Blanton, Texas Tech University.)

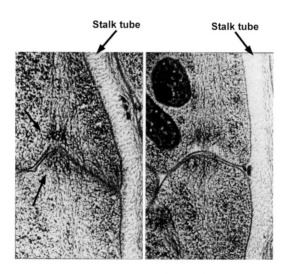

Figure 11.7 Junctional complexes in a culminant. Cells on the outside of the stalk tube appear to be joined by junctional complexes composed of actin filaments (dark arrows). Two examples are shown. (This micrograph was prepared by M. Grimson, Texas Tech University.)

11.5 Cellulose synthesis and the formation of stalk

The major component of the stalk cell wall and the stalk tube is cellulose. Glucose from UDPG is polymerized into glucan chains, presumably on the cytoplasmic side of the plasma membrane, and is spooled out through openings in the membrane (Blanton, 1997; Grimson *et al.*, 1996). The glucan chains then crystallize to form fibrils on the outside surface of the plasma membrane. While the primary structure of cellulose, 1,4-β-linked glucose, is constant, cellulose varies in its microfibril dimensions and orientation, its crystalline form, and its degree of polymerization, and can provide a number of structural roles. The type of cellulose made depends on the site and on the time of development. Some fibrils are thicker than others, some are ordered, while in other cases the fibrils form a mesh. It is surprising that the synthesis of cellulose, which is one of the most common polymers in the biosphere, has not been more thoroughly understood. Until recently, no cellulose synthases had been recovered from higher organisms. One of the reasons for this difficulty is that synthesis and extrusion from *Dictyostelium* or plant cells requires a multicomponent array of intramembranous proteins which does not survive membrane solubilization, making purification of the components difficult (Blanton, 1997; Grimson *et al.*, 1996). The absence of solubilization hindered the purification of the *Dictyostelium* synthase. In plants, the synthase cannot be assayed at all (Haigler and Blanton, 1996).

The trans-membrane specializations from which cellulose is transported to the exterior are known as terminal complexes (TCs). How they are arranged determines the type of cellulose synthesized. The stalk tube mode of cellulose synthesis involves prestalk cells that move up on the outside of the tube and produce longitudinal orientation of microfibrils of a certain size. In this case, all of the terminal complexes are arranged in a row, as shown in Fig. 11.8. After the cells have entered the tube, the synthesis of cellulose is quite different. The cells no longer move and there is a random orientation of cellulose microfibrils, while the terminal complexes are arranged in clusters. It is possible that cellulose synthesis, in the case of the prestalk cells on the outside of the tube, helps move the cells upward. For a full discussion of the role of cellulose synthesis in culmination and a comparison with plant cell wall synthesis, see Blanton (1997) and Grimson *et al.* (1996).

In crude extracts, cellulose synthase can be assayed and the enzyme activity starts at around the tight aggregate stage, when cellulose is needed for the sheath. Cellulose synthesis, when studied in dispersed cells, requires the stalk cell-inducing molecule DIF, because a mutant that is incapable of making DIF makes no cellulose until DIF is added (Blanton, 1993, 1997). Although the attempt to purify the cellulose synthase component of the terminal complexes failed, good fortune intervened in that a REMI mutant was recovered in which the disrupted gene had sequence homology to a known cellulose synthase gene from *Acetobacter xylinum* (Saxena *et al.*, 1994).

Disruption of the cellulose synthase gene caused a distinct phenotype – the structural integrity of culmination was eliminated (Blanton *et al.*, 2000). The tight aggregate formed – and so did the slug – and these seem relatively normal

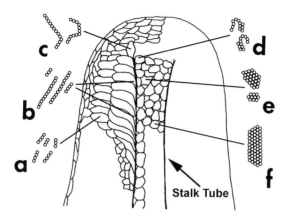

Figure 11.8 The spatial arrangement of terminal complexes determines the physical properties of the cellulose that is produced. The types of terminal complexes and their positions in the culminating fruiting body are indicated by a–f. Prestalk cells that do not abut the stalk tube have rows of 3–4 intramembranous particles (IMPs) in their terminal complexes (a). Those cells that border the stalk tube have longer rows of IMPs and contribute cellulose to the outside of the tube (b). Once they have made the turn into the stalk tube, the prestalk cells become immobile stalk cells and the IMPs collapse into clusters, leading to a different randomly arranged form of cellulose (c, d). Larger clusters can form within the stalk tube (e, f). (Grimson *et al.*, 1996.) (Courtesy of M. Grimson and R. L. Blanton, Texas Tech University. Reprinted with permission of the Company of Biologists Ltd.)

without any cellulose, indicating that cellulose is not essential until culmination. At the time of culmination, the effort to raise the spores is a failure and the structure falls down, only to try again with an equal lack of success.

11.6   Death comes to the stalk cells

When the prestalk cells head into the stalk tube, a de-repression takes place. The gene coding for EcmB is induced and the nature of the cellulose synthesized by the terminal complexes changes so that it now encases the cell and makes a cell wall (Grimson *et al.*, 1996). The cells pile up on top of one another within the stalk tube and the structure is extended. These cells now die. Although their death is programmed in the sense that they expire as part of a developmental program, there is no evidence that they die by the well-defined criteria of apoptosis. Apoptotic death includes a blebbing of the dying cell, the activation of caspases, and a particular kind of nuclear DNA fragmentation. Cell death in *Dictyostelium* is not blocked by caspase inhibitors, nor is it accompanied by DNA fragmentation (Cornillon *et al.*, 1994; Olie *et al.*, 1998). We do not know whether the cells die through a specific signal in common with apoptotic cells, or from the exertion and strangulation of cellulose synthesis and vacuolization. One approach to this question, now that a cellulose synthase mutant is available, will be to ask whether the cells that make no cellulose die at the same rate as do normal cells. Genetic approaches are also available if we can select cells that move into the stalk but do not die.

11.7   Participation of the prespore cells in culmination

As an increasing number of prestalk cells move into the tube, the mass of prespore cells lifts from the substratum. The mass is wrapped in the slug sheath, and gradually pulled upward. It is unclear what provides the motive force to lift the mass of prespore cells. Although not tested, it is possible that the individual cells retain motility and are in part responsible for late movements. One clue is that prespore cells, during their ascent of the stalk, move in a spiral pattern – a behavior that cannot be passive. To examine the question, Rex Chisholm and his colleagues disengaged the myosin II-based motility system of the cells by disrupting the myosin regulatory light chain gene. Such cells are capable of some motion during chemotaxis, but this is severely restricted because myosin II is not engaged (Chen *et al.*, 1998). Prespore cells that lack myosin II function do not successfully mount the stalk in chimeras with wild-type cells. The ability to culminate is restored by transforming the mutant cells with the RLC sequence expressed from a prespore-specific promoter, so that it is expressed during culmination. The prespore cells, which do not encapsulate until reaching the top of the stalk, need to move to get to the top of the stalk. These results suggest that they are not simply hauled up. The movement and integration of prestalk cells into the forming stalk also require myosin II function (Chen *et al.*, 1998).

11.8   The final step in spore formation is regulated exocytosis

There are major transitions at culmination. The first of these occurs with relief of the repression caused by ammonia or oblique light when the slug converts to a fruiting body, but the last – the encapsulation of the spores and the death of the stalk cells – are the final acts of development. Final they may be, but they are certainly highly coordinated. It is about them that recent genetic experiments have been most revealing. Before asking how the process of encapsulation is regulated, we should follow morphological developments a little further in the spore lineage.

As the prespore cells are elevated to the top of the stalk, a wave of encapsulation sweeps downward through the 80,000 cells of the spore mass and a three-layered spore coat is created around each cell (Richardson *et al.*, 1994). The final wave of encapsulation begins at about the stage of the culminant shown in Fig. 11.4. It is accompanied by the transcription of a spore-specific gene called *spiA* (Richardson *et al.*, 1994). Spores that lack Dd31, the product of *spiA*, are killed by hydration (Richardson and Loomis, 1992). The nature of the inductive signal that passes down the spore mass, potentially from the prestalk cells in the tip, is not known but several candidates are being evaluated. Soaking the developing fruiting bodies in high concentrations of 8-Br-cAMP causes encapsulation and transcription of *spiA* through the whole spore mass, eliminating the gradient effect (Richardson *et al.*, 1994).

The prespore cells have long been preparing the components of the spore walls and carrying it with them in prespore vesicles (PSVs) since the tight aggregate stage. At the time of encapsulation, the vesicles fuse with the plasma membrane by regulated exocytosis. This releases a number of spore coat proteins, some of which have been described in previous chapters and will be revisited in Chapter 12 on the mature spore and its germination. It is expected that cell fusion events are mediated by small G proteins, and one of these has recently been detected in purified PSVs (Srinivasan *et al.*, 1999, 2000). A number of small GTPase proteins have been described (Buczynski *et al.*, 1997; Bush *et al.*, 1993, 1996), but other than to say that the fusion process is PKA-dependent, the regulatory events of exocytosis are not well known.

The origin of the PSVs is probably the Golgi apparatus, although this idea has not been pursued since PSVs were described (Hohl and Hamamoto, 1969; Maeda and Takeuchi, 1969). The PSV contains a unit membrane and a small electron-lucent gap and then, on the inside an electron-dense layer that will form the initial spore wall. The material inside the membrane is polysaccharide in nature. Several antibodies that recognize the polysaccharides of *Dictyostelium* glycoproteins stain the PSVs (Takeuchi, 1963). A galactose-rich polysaccharide called galuran is also present in the PSVs (Zhang *et al.*, 1998). The presence of spore coat proteins and galuran raises the question of synthesis – how for example, do the nine major spore coat proteins get into the vesicles? Do the enzymes involved in polysaccharide synthesis, also work within the PSVs? UDPG-polysaccharide transferase is one of several enzymes that mediate polysaccharide synthesis, and was the first enzyme used to show specific gene induction during development (Sussman and Lovgren, 1965; Sussman and Osborn, 1964). At late stages of culmination this enzyme is released into the extracellular space, as expected if the enzyme functions within a PSV that fuses with the cell membrane.

During the course of development the number of PSVs increases and they fill up, although if the cells are caused to trans-differentiate, the vesicles disappear. How they are degraded is not known, but it probably involves a form of autophagy, which is more thoroughly studied in budding yeast (Jentsch and Ulrich, 1998). At the end of culmination, the PSVs accumulate near the periphery of the cells, which have begun to elongate. At the appropriate time, the numerous PSVs coordinately fuse with the plasma membrane and the electron-dense layer now becomes an outer layer. Gradually, the whole plasma membrane is replaced with membrane from the PSVs, and as a consequence the cells shrink. A large amount of electron-dense material is left between the nearly mature spore cells. The series of events is shown in Fig. 11.9.

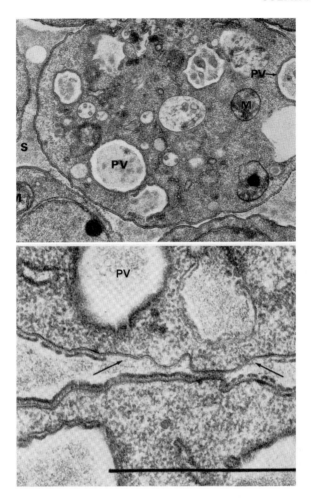

Figure 11.9 The PSVs are formed before encapsulation. Synthesis of extracellular matrix proteins of the spore coat begins in the tight aggregate stage. These proteins reside in the vesicles until, at the moment the spores have nearly reached the apex of the stalk, they are induced to undergo regulated exocytosis. PV, prespore vesicles; S, slime matrix; M, mitochondria; N, nuclei. The bar length is 1 µm. The lower panel shows various stages of PSV vesicle fusion with the plasma membrane. There are areas of plasma membrane (arrows) that have not yet been replaced by the double layered PSV membrane (Hohl and Hamamoto, 1969). (Reprinted with permission of Academic Press, Inc.)

11.9 The coordination of spore and stalk formation

The fundamental coordination problem of culmination is to make sure that the prespore cells, loaded with PSVs, do not try to make a spore coat before they are well up the stalk. What signals pass to the prespore cells at the critical moment as they move up the stalk, and what is the origin of the signal? How is it passed from the apex of the spore mass to the base of the spore mass?

Much of what we know about the final steps of spore and stalk formation has been learned from culture experiments *in vitro* in which prespore and prestalk cells have been dispersed (Kay, 1987). Both cell types can be made to differentiate in culture at low densities such that there is no communication between the dispersed cells and all differentiation is dependent on factors added by the experimenter. Under dispersed conditions, both prestalk and prespore cell differentiation in culture requires a period of induction by cAMP. This part of the treatment regime mimics what the cells would ordinarily see during aggregation. As we saw in Section 9.2, a role of cAMP at this time is to induce the transcription factor GBF, which is essential for the induction of a number of cell type-specific genes as the cells pass from loose aggregate to tight aggregate. Stalk cells can be induced by further addition of DIF.

Normally, the final encapsulation of the spores, including the exocytosis of prespore vesicles, does not occur in dispersed cells in culture. In cells dispersed in culture, prespore-specific genes can be induced by activation of PKA. One way to do this is to introduce many copies of a gene coding for the catalytic subunit, under the control of its own promoter (Anjard *et al.*, 1998). Other methods create the same effect: deletion of the phosphotransfer protein RdeA (Chang *et al.*, 1998), elimination of the intracellular phosphodiesterase RegA, or removal of the regulatory subunit activate the kinase. If a cell permeant analogue of cAMP, 8-Br-cAMP, is added to the culture at high concentrations, the cells encapsulate. The encapsulation of such cells can be increased by several factors, which are discussed below. These genetic and pharmacological results led to the idea that a sudden rise in cAMP during culmination activates PKA, and this leads to encapsulation (Kay, 1989; Maeda, 1988, 1992). A constitutively active PKA allows them to cross several normal regulatory barriers in development. Experiments in which PKA is inhibited block all of the events of development – from aggregation to encapsulation.

One could get the idea that PKA regulates everything – from the growth to development transition (Chapter 7), the synthesis of a cAMP pulse during aggregation (Chapter 8), the loss of inhibition by Dd-STATa during early culmination (Chapter 10), and now encapsulation. This is to an extent true, and represents the re-utilization of signal circuitry during development. Yet, there must be a second lock on all of these events, or else no coordinated morphogenesis would be possible – cells would simply form spores in a disorganized way, rushing unrestrained through all of development, like an orchestra with no conductor. Between developmental transitions, PKA must be regulated or there must be other controls that gate the effects of PKA. This role could fall to the sensor histidine kinases. We have discussed the importance of these kinases at several earlier times during the development of *Dictyostelium*. Table 11.1 re-emphasizes the point that these molecules play a major role and that there remain several expressed sequence tags (ESTs), with unknown roles, which probably represent histidine kinase genes.

One view of how histidine kinases control transitions in development is shown in Fig. 11.10. In this view, the cAMP receptors (cAR1–4) function through the G proteins and other components discussed in Chapter 8 to acti-

Table 11.1 *Histidine kinases of* Dictyostelium

Gene name	Phenotype of null mutants	Proposed function	Reference
dokA	Osmo-sensitive, late defect in development, see Chapter 5	Osmosensor	Schuster *et al.* (1996)
dhkA	Long stalk, few spores, cell autonomous	Culmination	Wang *et al.* (1996)
dkhB	Normal development, premature spore germination; see Chapter 12	Receptor for spore germination inhibitor?	Zinda and Singleton (1998)
dhkC	Rapid aggregation, normal fruiting, insensitive to ammonia	Ammonia receptor?	Singleton *et al.* (1998)
dhkD	Unknown	Unknown	C. Singleton (personal communication)
6 EST candidates	Unknown	Unknown	http://www.csm.biol. tsukuba.ac.jp

Source: These data were compiled by P. Thomason, D. Traynor and R. Kay, MRC, Cambridge, to whom I am grateful (Thomason *et al.*, 1999).

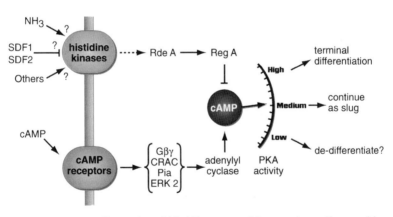

Figure 11.10 The controlling roles of histidine sensor kinases. According to this formulation, two systems compete for the soul of PKA. Active RegA is a phosphodiesterase controlled by a response regulator (RR) domain that is responsive to a histidine kinase through a phosphotransfer protein called RdeA. Removal of RegA, RdeA, or one of the histidine kinases all cause rapid development by removing a control on cAMP levels (Thomason *et al.*, 1999). For a discussion of the proteins involved in the transduction of the cAMP signal, see Chapter 8. Another source of cAMP may be the newly discovered, late-acting adenylate cyclase ACB (Kim *et al.*, 1998). (Courtesy of R. Kay. Medical Research Council, Cambridge, UK. Reproduced from *Trends in Genetics*, with permission from Elsevier Science.)

vate adenylyl cyclase, which in turn drives up cAMP activity and activates PKA. PKA can have several levels of activation and does not run free (Thomason *et al.*, 1998, 1999). The phosphodiesterase RegA inhibits activation of PKA by degrading cAMP. RdeA is a phosphotransfer protein that is controlled by a histidine kinase – DhkA – during culmination. The histidine kinases are constitutively active, and their phosphotransfer activity is reduced by binding any of the several ligands. Unphosphorylated RegA is inactive, allowing cAMP levels to rise and then encapsulation takes place (Thomason *et al.*, 1999).

11.10 Small peptides and the timing of encapsulation

Given that an elevation of intracellular cAMP causes spores to encapsulate, we are still left with the problem of how this induction takes place just at the right moment. What signals stimulate the activity of PKA, and where do they come from? To study this problem, the *in vitro* system has been invaluable. Any compound that causes encapsulation of appropriately prepared prespore cells in culture is a candidate for the inducer. These studies are based on a cell assay system in which a transformant overexpresses the catalytic subunit of PKA and encapsulates in dispersed culture. The efficiency of encapsulation increases at high cell density. At low density, the cells remain as prespores and do not have the ellipsoid shape or refractile nature of true spores. This behavior of sporogenous mutants – the dependence on cell density – suggested to Anjard and coworkers that encapsulation was driven by an extracellular factor that activates the PKA pathway (Anjard *et al.*, 1997; Thomason *et al.*, 1999). Such a factor was found and purified many-fold and is now known to be a heat-stable peptide that can be phosphorylated by PKA. Its molecular mass is about 1,100 Da, but its sequence is not known. The effect of this peptide, which is now called SDF-1 (Spore Differentiation Factor-1), is to cause encapsulation of the prespore cells, which is to say, that exocytosis of PSVs is stimulated. In the case of SDF-1, this takes about 75 minutes and requires protein synthesis (Anjard *et al.*, 1997). SDF-1 works at sub-nanomolar concentrations – but no receptor is yet known. The test cells in these experiments are always sporogenous mutant cells, which by virtue of their overexpression of PKA, are capable of sporulating in culture.

A second spore-differentiation factor has been isolated. The sporogenous *RegA*-null cells produce a factor that is secreted into the medium, presumably as a result of their advanced state of development. This factor, SDF-2, is produced by prestalk cells and is also a small peptide (1,300 Da), but it differs from SDF-1 in that it is not phosphorylated (Anjard *et al.*, 1998). It causes the PKA overexpressing test cells to encapsulate in seconds, not 75 minutes, and its effects do not require protein synthesis. SDF-2 has been partially purified, but its sequence is not yet known.

11.11   Genetic experiments suggest the source of the inducer of spore encapsulation

One of the mutants isolated by REMI affected a gene called *tagB* (tight aggregate), the stage at which these mutants were blocked. The gene has an interesting structure. At the amino terminus, it codes for sequences that resemble a protease and at the carboxyl terminus there is an entity that looks like a transporter of the ABC class. Two additional *tag* genes, *tagC* and *tagD*, were found to be located in a cluster with *tagB* (see Fig. 4.1). The current model is that the products of *tagB*, *C* and *D* form heterodimers and that loss of any one of them causes a block at the tight aggregate stage (Shaulsky *et al.*, 1995).

The biology of the *tagB*, *C*, *D* complex has two interesting aspects. The first is the possibility that the genes are prestalk-specific but do not require DIF for their induction. In fact, they are not stimulated by DIF. The effect of this result on ideas about prestalk cell specification has been discussed in Section 9.6. The second aspect of these prestalk-specific genes is that they may also be involved in generating a signal that causes prespore cells to encapsulate. Cells with defective *tagB* or *tagC* make no early prestalk markers, but in chimeras with wild-type cells, they make spores. This failure of *tagB* mutant cells to make spores when developing without a wild-type partner derives from their inability to make stalk cells (Shaulsky *et al.*, 1995). Signals from stalk cells induce the last stage of spore maturation. In the true culminant, this induction probably accounts for the fact that spores only encapsulate when they reach the top of the stalk.

The first indication of a pathway in which Tag proteins participated came from genetic suppressor analysis. *Tag* mutants do not make spores, but a second round of REMI mutagenesis, followed by a screen for sporulating colonies, led to suppressors of *tagB*. These double mutants made structures that were more advanced than the tight aggregate *tagB* mutants – and they made viable spores. Some of these suppressor mutants were affected in the phosphodiesterase gene *regA* (Shaulsky *et al.*, 1996). *RegA* was also recovered in other selections (Thomason *et al.*, 1998). Eliminating RegA has the same result as eliminating the PKA regulatory subunit – cells develop rapidly and without a need for the normal architecture of culmination. This is presumably because intracellular cAMP levels are driven up due to the absence of PDE. The fact that RegA contains an RR domain implies the involvement of a histidine kinase that phosphorylates it after sensing a change in conditions at the cell surface. The several histidine kinase receptors, like the four cAMP receptors, are expressed at different stages of development (Table 11.1).

A mutant allele of *dhkA* was recovered from a REMI mutant in which prespore cells fail to form spore cells, even when mixed with wild-type cells. DhkA is a membrane-spanning protein with an extracellular loop and an intracellular domain that is similar to other histidine kinases that work in two-component systems (Thomason *et al.*, 1999; Wang *et al.*, 1996). The null mutant makes fruiting structures with long stalks and very few spores. The block to sporulation in *dhkA*-null cells can be suppressed by mutations in *pkaA*, *regA*, or *rdeA*.

Kay and his colleagues suggest, with direct biochemical evidence, that the role of phosphorylation of the RR region in RegA is to stimulate PDE activity (Thomason *et al.*, 1998). The RR region of RegA has two sites for phosphorylation – one is stimulatory and one is inhibitory. During aggregation an aspartate residue is phosphorylated by transfer from RdeA, which is a phosphotransfer protein (Chang *et al.*, 1998). This increases the activity of RegA and removes cAMP, reducing the activity of PKA catalytic subunit. The activity of RegA is inhibited by the MAP kinase Erk2, as described in Section 8.5.5. Without Erk2 to control it, RegA destroys all of the cAMP bound for export or for activation of PKA (A. Kuspa, personal communication). The site of phosphorylation by Erk2 has been determined and altering it prevents Erk2 from controlling RegA. Current thinking is that DhkA, which accumulates after aggregation, may activate Erk2 and then cause the phosphodiesterase to be turned off, driving cAMP levels up. This would activate PKA and drive late events in culmination. These results and speculations are summarized in Fig. 11.11.

Another element that may be crucial to elevating intracellular cAMP levels during culmination has recently been discovered. A gene coding for a putative adenylyl cyclase has been cloned and called *acrA* (W. F. Loomis, personal communication). The null mutant has long thin stalks that resemble those of DhkA mutants. Although *acrA* codes for a large membrane-bound protein (243 kDa), nothing is yet known about its regulation by ligands or signal transduction pathways. *AcrA* almost certainly codes for ACB, the protein described by Kim *et al.* (Kim *et al.*, 1998; W. F. Loomis, personal communication).

11.12 Ligands for two-component sensors

The two-component histidine kinase systems now include DhkA, DhkB, DhkC, and DokA, but a number of other possible genes of this class exist (Thomason *et al.*, 1999). These elements are thought to span the membrane and function as receptors for unknown ligands. Can a relation be made between the membrane-spanning histidine kinases and the factors SDF-1 or SDF-2, described above? SDF-1 is not a ligand for DhkA because cells which are mutant for *dhkA*, still encapsulate when exposed to SDF-1. Thus, the hope that SDF-1 was the ligand that causes the final events of sporulation by binding DhkA was not realized. However, SDF-2, which has a very rapid effect on encapsulation, requires the wild-type *dhkA* allele. It does not cause encapsulation of test cells carrying a mutant *dhkA* allele. SDF-2 has an autocatalytic capacity – addition to cells in culture leads to a priming effect so that large amounts are secreted (Anjard *et al.*, 1998). This signaling system, which contains untested assumptions, is shown in Fig. 11.11.

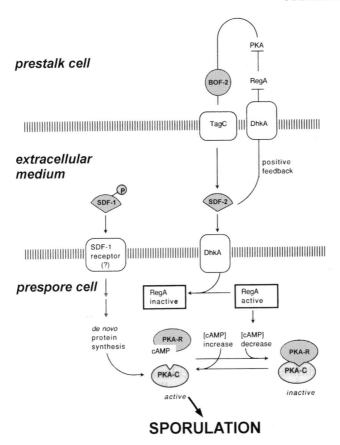

Figure 11.11 A model to explain the encapsulation of prespore cells at the top of the culminant. Certain predictions of the model have support. Encapsulation, the final action in the formation of a spore, is caused by SDF-1 and SDF-2 through different pathways. DhkA is needed for encapsulation induced by SDF-2. As yet, there are no structures for SDF-1 or SDF-2. Spatial localization is inferred. There is evidence for a pathway in which SDF-2 binds to DhkA on prestalk cells and stimulates its own release by an autocrine mechanism. Histidine kinases other than DhkA may play roles (Anjard *et al.*, 1998). (Courtesy of William F. Loomis, University of California at San Diego. Reprinted with permission of Academic Press, Inc.)

The difficulty with results that depend on dispersed cell experiments, is that encapsulation normally happens in the three-dimensional architecture of the culminant that was described at the beginning of this chapter. How can we decide whether any of these models is real? One answer may come from immunocytochemistry, as soon as SDF-1 and SDF-2 are purified and characterized. SDF-1 should be found in prestalk cells. SDF-2 should spread over the nearly mature spore mass in a wave, perhaps mimicking the wave of encapsulation noticed by Richardson *et al.* (1994). These molecules are presumably part of larger precursors, and these too may be detected by immunocytochemical methods. Production of appropriate antisera awaits complete purification and structural analysis of SDF-1 and SDF-2. Receptor binding and other biochemical studies are essential to test the model.

11.13 The targets of PKA

The models in Figs 11.10 and 11.11 predict that all pathways converge on the catalytic subunit of PKA, and this is consistent with what is known from PKA overexpression experiments and from experiments that employ 8-Br-cAMP. However, PKA has to phosphorylate target proteins that are essential for exocytosis and other aspects of the encapsulation of the spores. There are several known very late-acting genes in spore development. One target, SpiA, is a spore coat protein (Richardson and Loomis, 1992). A second target is a potential transcription factor called DdSRF, which resembles the serum-response factors of mammalian cells. It is expressed late in spore formation and may regulate late events. Cells that lack DdSRF make misshapen and unstable spores. *SpiA* is not expressed when DdSRF is absent (Escalante and Sastre, 1998). We can predict that membrane-fusion proteins will be among the targets of PKA. At least for SDF-2, whose action occurs over seconds, all targets must be present, and no *de novo* synthesis is required for encapsulation. There may be further events needed for spore maturation, such as those mediated by DdSRF – but at least the initial production of spore coats does not require it.

11.14 Culmination-defective mutants

There are several mutants which affect the culmination process, but cannot yet be associated with one of the pathways discussed above. One of the most interesting of these mutants is *stalky*, which makes stalk cells at the expense of spore cells (Chang *et al.*, 1996). As a consequence, the stalks are extremely long. During development, *stalky* mutants aggregate and form slugs in the normal manner, but when culmination occurs only stalk cells are formed and these may form within the stalk tube or outside of it. Like the *gskA* mutants, the *stalky* mutants represent a trans-differentiation. This is most evident when the β-galactosidase gene is fused to a prespore-specific promoter. This gene is expressed normally in the prespore cells of the *stalky* slug, but as the cells convert to prestalk and stalk cells during culmination, the β-galacto-sidase is retained in the newly converted prestalk and stalk cells. The critical point is that the developing cells are not committed and can convert from one cell type to another, as we saw previously with the transplantation experiments of Raper and with *in vitro* experiments performed by Sakai (1973). The *stkA* product is apparently essential for the final steps in spore formation and without it, there is a default program – to make stalk. When precisely during the culmination process the cells do this, and what becomes of the PSVs, is not known (Sakai and Takeuchi, 1971).

Stalky mutants were first described by Morrissey and made up only one complementation group, defining the *stkA* allele (Morrissey *et al.*, 1981; Morrissey and Loomis, 1981). The advent of REMI mutagenesis methods made it possible to isolate new *stkA* mutants and then to determine the nature of the affected gene. This was done by Chang, Newell and Gross (1996), who determined that *stkA* might code for a GATA type transcription factor. Sequence analysis showed two potential zinc-finger domains that had homology with GATA transcription factors. Several other facts are consistent with StkA being a transcription factor. First, it is nuclear. Second, the inability of the cells to complete spore differentiation without it is not overcome by over-expressing PKA, which means that it is downstream of PKA in the spore encapsulation program. PKA may activate StkA – there is a PKA phosphorylation site in one of the putative zinc fingers.

Elimination of StkA may relieve a repression of prestalk and stalk cell differentiation. *EcmB*, which we know is repressed under most conditions, is prematurely expressed in *stalky* mutants. The *ecmB* promoter contains several potential GATA sites. *Stalky* is developmentally regulated, and is transcribed after aggregation. The very late spore-specific gene *spiA* is not transcribed without StkA. This may be because the trans-differentiation of the prespore cells takes place before *spiA* is expressed. There is no direct evidence that StkA binds DNA or interacts directly with promoters. The events of culmination seem, in *stalky* mutants, to involve not only a trans-differentiation to prestalk cells, but also a continued migration into the stalk tube, until a tall stalk is produced.

The morphology of culmination is complex, requiring a re-alignment of the cells in the slug, synthesis of an internal stalk tube, movement of cells into the tube, and at the same time an extension of the tube. We do not know how the prespore cells move up the elongating stalk or how the cells that make the stalk are pulled into it. Nonetheless, some of the morphological requirements are becoming clear and we are assembling a number of mutants that affect the process. These mutants will allow us to test what are now speculative models to explain the elaboration of morphological structures and the encapsulation of the spore cells. The events of culmination also force us to confront the problem of trans-differentiation – the conversion of prespore cells to basal disc cells in the case of *gskA* mutants and the conversion of spore cells to stalk cells in the case of *stalky*. These are problems that are critical to *Dictyostelium* and to developmental biology in general, and they have yet to be answered.

12

Formation and Germination of Spores

The fusion of the prespore vesicles (PSVs) with the plasma membrane creates an immature spore. A number of further events must occur before the cell can be assured of protection from desiccation, osmotic shock, or the digestive tracts of soil creatures. The spore proteins, galuran, and cellulose must be organized and cross-linked in such a way that they protect the delicate amoebae within. The formation of the coat, about which we know quite a lot, still presents a number of problems (Lydan and Cotter, 1994; West and Erdos, 1990; West et al., 1996). We know that a number of the prespore proteins and the polysaccharide galuran are synthesized and modified in the early and intermediate stages of the secretory pathway, and then stored in a regulated secretory compartment, the PSV (Srinivasan et al., 2000). We do not know how the proteins of the spore coat interact with cellulose. There are alternative views regarding the extent of pre-assembly of proteins into specific complexes in the PSV. There is no information on how the spore coat is anchored to the plasma membrane. We are beginning to understand the mechanism that the spores use to maintain their dormancy, but new components in these pathways remain to be discovered. We do not know a great deal about how the spores detect the appearance of nutrients and launch the developmental program that leads to germination and the re-establishment of amoeboid life.

12.1   The spore coat has a complex architecture

The complete coat is 210 nm thick and forms a permeability barrier to molecules that are the size of proteins, but it does not constrain the passage of water or other small molecules (Cotter, 1981; Fosnaugh et al., 1994; West and Erdos,

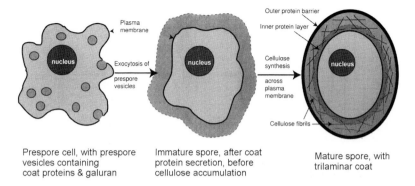

Figure 12.1 The formation of the spore coat. Prespore vesicles fuse with the plasma membrane by regulated exocytosis in response to a signal that issues from prestalk cells during culmination. The fusion depends on PKA and can be induced prematurely with 8-Br-cAMP. Galuran is a repeating polymer of galactose and *N*-acetylgalactosamine. Two protein layers form, separated by a layer of cellulose. Much of the structural organization occurs after exocytosis. (Courtesy of C. M. West, University of Florida.)

1990). In this it differs from the cell walls of plants, which are permeable to proteins. The spore coat is arranged in the layers shown in Figs 12.1 and 12.2. The coat has an outer protein layer, which provides the barrier to proteins and perhaps toxins. A middle layer consists of cellulose, and is in fact two layers – one with fibers arranged to run around the spore, and one in which the fibers run randomly (Hemmes *et al.*, 1972). An inner protein layer surrounds the plasma membrane (West *et al.*, 1996). The polysaccharide galuran, present at less than 10% of the level of cellulose, is concentrated near the plasma membrane.

Coat formation is initiated by secretion of specific proteins or protein complexes into an initially soluble pool – spore coat proteins made by one cell may find their way into the coats of another cell (West and Erdos, 1992). Disulfide bonds stabilize the position of the proteins in the coat by cross-linking them. Cellulose is extruded from the cells and also binds specifically to certain of the proteins to create a highly organized structure. The marriage of cellulose and coat glycoproteins occurs after the fusion of the PSVs with the plasma membrane. Proteins that fail to incorporate into the coat become part of the fluid interspore matrix outside of the boundary of the coat.

12.2 The proteins of the spore coat

When purified spore coats are eluted with denaturing and reducing agents, a limited number of major proteins is recovered (Devine *et al.*, 1983; Orlowski and Loomis, 1979). By two-dimensional gel analysis, there are nine major proteins and a number of minor ones (West *et al.*, 1996). Each spore coat

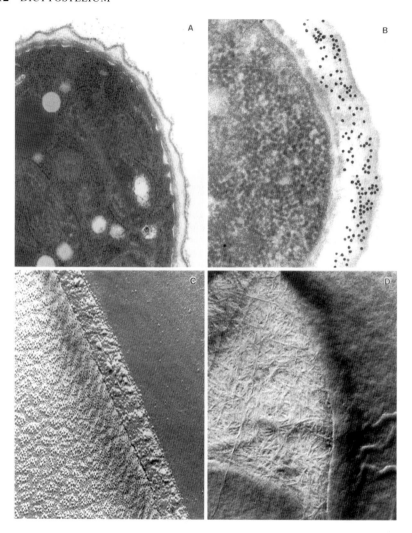

Figure 12.2 Electron microscopy of the spore wall. Panel A shows a conventional osmium-stained thin section counter-stained with uranyl acetate and lead citrate. Note the triple layer of the coat, which has inner and outer electron-dense layers and an electron-lucent middle layer. Panel B shows that the electron-lucent middle layer contains cellulose, which in this case is marked with a cellulase/gold complex. Panel C shows a spore that has been freeze-fractured, deep etched, and shadowed. The fracture plane split the leaflets of the plasma membrane and then traveled perpendicularly across the thickness of the coat. Note the intramembranous particles in the membrane and the fibrils in the coat. Panel D shows an isolated fragment of the inner middle layer of the coat. The fragment was facing outward and shows randomly oriented cellulose fibrils. See also Erdos and West (1989) and Hemmes *et al.* (1972). (Courtesy of G. W. Erdos and C. M. West, University of Florida.)

Table 12.1 *The spore coat proteins, their mutations, and their localization (see West et al. (1996) for a list of additional minor proteins)*

Protein (M_r)	Gene	Localization	References
SP96	*cotA*	Outer layer	Fosnaugh and Loomis (1989a); Tasaka *et al.* (1990)
SP87	*pspD*	Outer layer	West *et al.* (1996); Yoder *et al.* (1994)
SP85 (PsB)	*pspB*	Cellulose layer, inner layer	Power-Coffman and Firtel (1994); West *et al.* (1996)
SP75	DP87/*cotD*	Outer layer	Nakao *et al.* (1994)
SP70	*cotB*	Outer layer	Fosnaugh and Loomis (1989b)
SP65		ND	West *et al.* (1996)
SP60	*cotC*	Outer layer	Fosnaugh and Loomis (1989b)
SP52	?	ND	West *et al.* (1996)
SP35	*psvA*	ND	Hildebrandt *et al.* (1991); West *et al.* (1996)

ND, not determined.

contains about 5 million protein molecules that were stored in prespore vesicles during the last half of development (West *et al.*, 1996). The synthesis of the spore coat proteins, and the synthesis of cellulose and galuran, is a considerable commitment of resources.

A number of the genes that code for spore coat proteins have been cloned and these are all prespore-specific – they are not transcribed in stalk cells. One of these prespore vesicle proteins, SP35, has a curious mode of regulation in that it is controlled by antisense transcripts (Hildebrandt and Nellen, 1992). The coat protein genes are coordinately regulated, but no sequence within the promoters has been dissected enough to determine the elements which give rise to coordinate regulation, as discussed in Chapter 10. A list of spore coat proteins is provided in Table 12.1. All of the proteins contain mucin repeats and other sequence motifs rich in cysteine, and stem from duplication of an ancestral gene; most of the homologues remain closely linked on chromosome 2 (West *et al.*, 1996).

All of the spore coat proteins have leader sequences, which is consistent with the fact that they are secreted. SP96 and SP75 are phosphorylated and all, except possibly SP60, are either N- or O-glycosylated, with the latter predominating (West *et al.*, 1996). An analysis of the sequences suggests that the proteins are folded into a series of small cysteine-rich domains similar to vertebrate extracellular matrix proteins such as fibronectin. These domains may mediate specific interactions with other coat molecules (West *et al.*, 1996). The mucin repeats, if they prove to be glycosylated and phospho-glycosylated as early evidence suggests, would make the outer surface of the spores hydro-

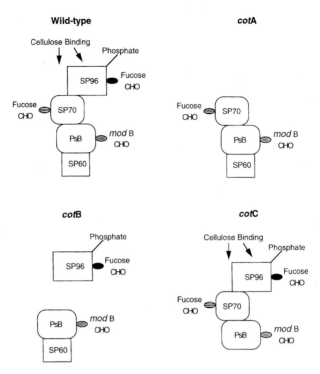

Figure 12.3 The prespore protein complex psB is thought to assemble in PSVs. The wild-type complex is shown in the upper left. SP96 is phosphorylated and is thought to bind cellulose. In several mutants of spore coat proteins, components of the complex are removed as shown (McGuire and Alexander, 1996). (Reprinted with permission of the *Journal of Biological Chemistry*; © The American Society for Biochemistry and Molecular Biology.)

philic and negatively charged. Immunocytochemistry shows one protein, SP85, to be concentrated in the inner and middle region, where it may bind to cellulose (Erdos and West, 1989; Zhang *et al.*, 1999). Other studies suggest that SP96 is a cellulose-binding protein (McGuire and Alexander, 1996).

Assembly of the spore coat with galuran occurs as proteins and the galuran intermix in the secretory pathway. Electron microscopy shows a concentration of material apposed to the inner surface of the PSV membrane, suggesting that segments of the coat may be preformed and later patched together after exocytosis. However, further study showed that the organization of this material did not match that of the spore coat and was probably a preparation artifact. Using immunoprecipitation methods, Loomis and colleagues (Devine *et al.*, 1983) detected a large multiprotein complex which is found in the PSVs, and Alexander and colleagues have suggested that this complex is secreted as a unit (McGuire and Alexander, 1996; Srinivasan *et al.*, 1999; Watson *et al.*, 1993, 1994). This proposed complex, shown in Fig. 12.3, contains proteins identified as spore coat proteins by other means. The components of the complex are held together by covalent and non-covalent bonds. Work with mutants of some of the genes listed in Table 12.1 gave rise to the model shown in Fig. 12.3. Alexander's group has proposed that the complex binds cellulose through its SP96 component (McGuire and Alexander, 1996).

The spore coat proteins affect permeability. In mutants that have deletions in one or more spore coat genes, or are defective in O-glycosylation, permeability of the spores to protein probes is increased (Fosnaugh et al., 1994; West and Erdos, 1990). These findings led to the idea that the outer protein layers of the spore coat block the entry of large molecules from the environment.

Final assembly follows secretion of the proteins and the galuran to the cell surface. Initially, there is an exchange of spore coat proteins or protein complexes, such that proteins made and ejected by one cell find their way into neighboring spore coats (West and Erdos, 1992). The characteristic tri-laminar organization of the coat is not seen until the cellulose appears. The effect of cellulose may be mediated via the pre-coat complex described above, or by individual proteins, which have been shown to bind cellulose after isolation from the coat (McGuire and Alexander, 1996; Zhang et al., 1998). The attachment of proteins to the spore coat is stabilized by disulfide bonds, and the proteins can only released by denaturing and reducing agents. There is some disagreement regarding the significance of the protein complex observed *in vitro* prior to secretion. West suggests that associations may occur after cell lysis, and observes that several proteins of the complex appear to localize to different layers of the coat (West et al., 1996).

12.3   The synthesis of cellulose in spores

In contrast to galuran and the glycoproteins of the coat, cellulose is extruded by the forming spores after the fusion of prespore vesicles with the plasma membrane. This presents a problem, as the plasma membrane is gradually replaced with the membrane of the prespore vesicles. The machinery for synthesis of cellulose – the terminal complex discussed in Section 11.5 – would have to be present for the cellulose of the maturing spore to be made. It may be that these protein complexes are present in remaining plasma membrane and that they diffuse laterally into the newly inserted membrane.

Little is known about the binding of cellulose to the proteins of the matrix *in vivo*. The coat proteins cannot be released from cellulose except by reducing agents and denaturation. The discovery of a cellulose synthase mutant (Section 11.5) and the resulting reagents, should tell us much more about the synthesis of cellulose at the time of final spore coat maturation (R. L. Blanton, personal communication). We should also be able to recover the components of the terminal complexes and learn where they are in the re-engineered membrane of the spore cell.

12.4 The formation of the spore entails changes in the cytoplasm

During the formation of spores the shape of the mitochondria change, but there is no obvious change in the nucleus – it does not become heterochromatic or lose its nucleolus. Polysomes break down to single ribosomes. The cytoplasm is densely packed because the everted prespore vesicles reduce the total volume. There is a major change in shape to form oval spores.

The spore, having made the coating that will protect it against osmotic swelling and predation by animals, would still be a vulnerable structure if it had not stored energy reserves. The cells produce high levels of trehalose toward the end of development, when glycogen decreases (Rutherford and Jefferson, 1976). The amount of trehalose can be as much as 5% of the spore mass, and it is likely that this trehalose is used during germination (Ceccarini and Filosa, 1965). The trehalase gene has not been cloned and therefore no knockouts are available to test this idea. Trehalase has been purified, and its role may be understood once the gene is cloned (Killick, 1983; Temesvari and Cotter, 1997). Another potential energy source is polyphosphates. These are long polymers of high-energy phosphates and are synthesized in large amounts during the later stages of *Dictyostelium* development (Al-Rayess *et al.*, 1979; Gezelius, 1974; M. Sims and E. R. Katz, personal communication). Glycogen is made during *Dictyostelium*'s development, but its monomers seem to be dedicated to the formation of the cellulose of the stalk and spore (Favis *et al.*, 1998; Rutherford *et al.*, 1997).

12.5 Sorocarps contain inhibitors of germination

The imprisoned amoeba is not dormant in the sense of total inactivity. Life depends on whether it senses food, and whether it can determine that it has been dispersed from the fruiting body in which it formed. To understand what the spores confront, we have to ask about conditions in the sorocarp. Spores are extremely sensitive to osmolarity. High concentrations of osmolytes in the sorus keep them as spores and prevent germination (Fig. 12.4). The osmolytes include the soluble contents of the prespore vesicles and a relatively high concentration of ammonium phosphate, which can reach 200–300 mM in dry sori (D. Cotter, personal communication). The concentration of ammonium phosphate that blocks germination is about 70 mM. The sorus is essentially a sphere, and its volume depends critically on the humidity – which in turn determines the concentration of osmolytes. The liquid portion of the sorus is roughly one-half to one-third of its volume. Osmotic control of germination of spores and seeds is a general phenomenon.

In addition to the effects of osmolarity on the inhibition of spore germination, there is a specific inhibitor, called discadenine, which resembles plant cytokinins. The structure of discadenine is shown in Fig. 12.5. Discadenine synthesis rises dramatically during culmination, and the modified purine can

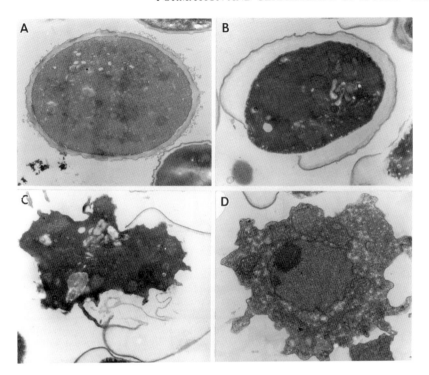

Figure 12.4 The structure of the mature spore and its germination. Panel A shows a dormant spore, with densely packed ribosomes. Panel B shows a swollen spore, 90 minutes after activation. The spore coat has begun to rupture. Panel C shows an amoeba emerging, 2.5 hours after activation. After 3 hours the mature amoeba has escaped from the spore coat. (Ennis *et al.*, 1988). (Courtesy of Herbert L. Ennis, Columbia University. Reprinted with permission of Wiley-Liss, Inc., a subsidiary of John Wiley & Sons, Inc.)

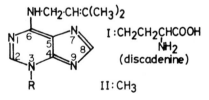

Figure 12.5 The structure of discadenine. Discadenine and the enzymes responsible for its synthesis are highly regulated, appearing only as spores form in the sorus (Abe *et al.*, 1976).

be isolated from sorocarps (Ihara *et al.*, 1980). Its synthesis begins with $5'$-AMP and requires two enzymes, which are also developmentally regulated (Ihara *et al.*, 1980, 1986). Discadenine acts at low concentration (about 50–100 nM) to inhibit the germination of spores from some species, but not from others (Abe *et al.*, 1981).

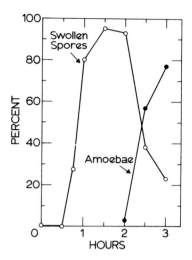

Figure 12.6 Amoebae emerge with high synchrony. In this experiment, spores were activated with DMSO for 30 minutes in phosphate buffer. Activation replaces the need for nutrients. Emergence is preceded by swelling. Protein synthesis begins a rapid increase after 1 hour as polysomes appear (Giri and Ennis, 1977; Yagura and Iwabuchi, 1976). Cycloheximide blocks emergence (Cotter *et al.*, 1969; Cotter and Raper, 1970). (Courtesy of H. L. Ennis, Columbia University. Reprinted with permission of Academic Press, Inc.)

Spores germinate if they are removed from the inhibition of high osmolarity and discadenine, as would happen if the sorus were dispersed into the environment. The spores require the amino acid part of their food supply and a short heat shock to germinate. Certain chemicals such as dimethyl sulfoxide (DMSO) also induce germination, and it is likely that these disparate treatments serve to inactivate an inhibitor of germination. When induced with heat-shock or DMSO, emergence can be remarkably synchronous, as shown in Fig. 12.6. The heat-activated spores have a lag phase, during which they can reverse the process, under certain conditions, and return to dormancy. After about an hour, they start to swell and then emerge from the spore coat, which is opened by the action of cellulases.

12.6 Amoebae emerge with caution

What mediates the response of the cells to high osmolarity and to discadenine? Spores that lack the unusual adenylyl cyclase ACG (adenylyl cyclase of germination) germinate under high osmolar conditions when the parental strain does not. Unlike other adenylyl cyclases (but rather like some guanylyl cyclases), this protein spans the membrane only once (Pitt *et al.*, 1992). The gene coding for ACG is expressed only in spores. Disruption of *acgA* by homologous recombination leads to spores which have lost the ability to respond to high osmolarity, and germinate when they should not. Otherwise, the amoebae that lack ACG have normal development (van Es *et al.*, 1996). Mutant *acgA* cells do not require heat activation to stimulate germination, and when germinating they cannot be forced to return to dormancy by high osmolarity. How the osmotic conditions stimulate the activation of this adenylyl cyclase is not clear, nor is it clear why there should be a need for a membrane-spanning protein to detect hyperosmolar conditions, unless the extracellular domain functions as a receptor. When expressed at other times of development, ACG also responds to hypertonic conditions. Thus, activation of an adenylyl cyclase in the spore is thought to activate PKA, which prevents germination. Mutants that overexpress the catalytic subunit of PKA germinate very poorly, which supports the idea that PKA activity controls dormancy. Spores with defective *acgA* alleles do not germinate in the sorus, where they are still subject to inhibition by discadenine (van Es *et al.*, 1996).

The cells contain at least one other mechanism to prevent premature germination. The two-component histidine kinase gene, *dhkB*, was isolated in a PCR based search for two-component signaling systems (Zinda and Singleton, 1998). We have established that *Dictyostelium* has a number of these two-component systems and that they may represent a major signal transduction mechanism by which the amoebae receive and process information about their environment. A gene called *dhkB* codes for what appears to be a membrane-spanning protein that is not developmentally regulated. DhkB has all of the motifs expected of a sensor histidine kinase. Mutants with defective *dhkB* genes develop well, but once the spores have formed in the sorocarp, they germinate spontaneously because they are impervious to the signals generated through high osmolarity and discadenine. The amoebae in the sorus dehydrate because of the high osmolarity, and then lyse. This pathway, too, seems to converge on PKA because ectopic activation of PKA protects the cells from germination and death. Cells that lack DhkB have less cAMP than normal spores. The combination of active DhkB and ACG seem to keep the spores in dormancy. DhkB can maintain an active PKA in the absence of ACG, but ACG alone is not sufficient to maintain dormancy. It would be reasonable to think that DhkB is the receptor for discadenine, although there is no direct evidence for this idea. The report by Zinda and Singleton (1998) contains a thorough discussion of two-component systems in fungi and in plants. Readers should also consult the review by W. F. Loomis on two-component systems and the organisms that employ them (Loomis *et al.*, 1998).

12.7 The program of spore germination

Once spores have been dispersed and removed from the effects of high osmolarity and of discadenine, they begin a program of germination. Transmission electron micrographs were shown in Fig. 12.4; a scanning electron micrograph is shown in Fig. 12.7 (George *et al.*, 1972). The spores respond to germination according to their age; spores that are just a few days old will not germinate if

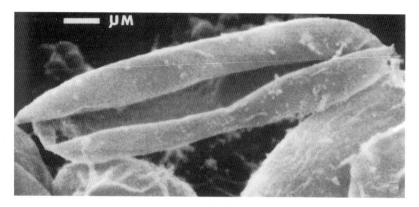

Figure 12.7 A view of germination through the scanning electron microscope. The spores swell after being induced by heating or by feeding. Amoebae emerge along a longitudinal fissure (George *et al.*, 1972). (Reprinted with permission of the American Society for Microbiology.)

dispersed into buffer without nutrients. Those that have been in sorocarps for several weeks undergo a process called auto-activation. Such old spores germinate if put into buffer; they have lost their requirement for nutrients when dispersed, but do not germinate as long as they are in a sorocarp. An auto-activator of germination has been postulated, but not purified (Dahlberg and Cotter, 1977; Lydan and Cotter, 1995).

When germination is induced by DMSO, the first event that is detected is a swelling, as shown in Fig. 12.7. The swelling is reversible and does not require protein synthesis. A challenge to swollen spores with 250 mM sucrose convinces them to return to their dormant state (van Es *et al.*, 1996). Once past the swollen stage, a new program of transcription is engaged and protein synthesis is required (Dowbenko and Ennis, 1980; Giri and Ennis, 1977, 1978). A number of new proteins appear and disappear (Dowbenko and Ennis, 1980). Several families of transcripts have been studied, some of which remain unidentified (Kelly *et al.*, 1985). Many proteins resume synthesis to assure the ability of the emergent amoeba to survive, but several genes are induced that are specifically required at the time of germination.

A number of germination-specific cDNAs have been isolated and sequenced (Giorda and Ennis, 1987; Giorda *et al.*, 1989; Kelly *et al.*, 1983). One group of germination-specific proteins is defined by the presence of a common tetrapeptide repeat of threonine–glutamic acid–threonine–proline. Two members of the family, *celA* and *celB*, are expressed solely during spore germination, the level of the respective mRNAs being low or absent in dormant spores, rising during germination to a peak at about 2 hours, and then rapidly declining as amoebae are released from spores (Giorda *et al.*, 1990). The mRNAs disappear with a half-life of about 42 minutes, as shown in Fig. 12.8. The transcripts are not found in growing cells or during multicellular development. The rapidity with which these transcripts accumulate and then disappear during germination

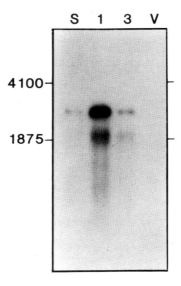

Figure 12.8 The cellulase genes are exquisitely regulated as part of the germination program. This Northern blot shows that the mRNAs of *celA* (upper band), and the related *celB*, appear after 1 hour of induction with DMSO and then disappear. Lanes: S = spores, 1 and 3 hours past germination; V = vegetative amoebae. (Reprinted with permission of Wiley-Liss, Inc., a subsidiary of John Wiley & Sons, Inc.)

implies that the respective gene products may be important for exit of spores from the spore coat. No gene knockouts are available – since there are several cellulases, the elimination of one gene might not produce a distinct phenotype.

CelA has homology with a number of cellulases in the databases, and the purified protein has cellulase activity. The cellulases produced by these genes are secreted and are presumed to digest the spore wall, thus allowing the amoebae to emerge (Blume and Ennis, 1991). Deletion analysis shows that CelA is composed of three functional domains: the amino terminus is the catalytic domain; the carboxy terminus is the cellulose-binding domain; and the region between these two is a tetrapeptide repeat, which is O-glycosylated. CelB has a similar structure, although its catalytic activity has not yet been determined (Ramalingam and Ennis, 1997).

The transcription of *celA* is controlled by AT-rich 5′ sequences (Ramalingam *et al.*, 1995). Analysis of reporter gene expression driven by *celA* promoter deletions indicates that *celA* gene expression is controlled by two distinct *cis*-acting elements located between −664/−620 and −619/−584 upstream of the translation initiation AUG codon, as shown in Fig. 12.9. The −664/−620 sequence contains a canonical Myb-response element (TAACTG), which is essential for the function of the element. The −619/−584 sequence is exclusively AT, and is sufficient by itself for *celA* gene expression. It does not contain any known transcription factor-binding site. The two *cis*-elements can function independently to activate *celA* transcription, but at

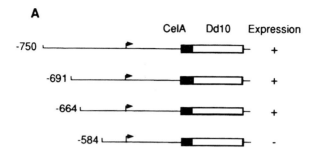

-664 AATTTTCAAATTTTTTTTTTTTA<u>TAACTG</u>ATTTCTTTTTTTTTT

ATTTTAATTTTTTTTTTAATTTTTTTTTTTATTTAAAAAAT **-584**

Figure 12.9 The functional regions of the promoter of cellulase. The promoter of *celA* has transcription factor binding regions at −664 to −620, which controls germination-specific transcription. This defined sequence binds a transcription factor of the Myb class. A different Myb is active during aggregation (Otsuka and van Haastert, 1998). The arrows mark the start of transcription. Dd10 is a reporter sequence. See text for references. (Reprinted with permission of Oxford University Press.)

least one of them is required. In addition, the *celA cis*-elements confer spore germination-specific expression on a heterologous promoter, confirming the physiological relevance of these upstream activating sequences. Factors present in germinating spores form three complexes with the −664/−620 *cis*-element, and the ability of these factors to bind this sequence is developmentally regulated. Deletions or mutations of the Myb response element alter factor binding to the *celA cis*-element *in vitro* and reduce *celA* gene expression, reinforcing the idea that these factors regulate the expression of *celA in vivo*.

Spore germination is as closely controlled as the rest of development. It would be counter-productive, after all, to go through the elaborate steps of development and then have the spores die on the vine. We know very little about spore germination, although we are beginning to understand that spores continuously sample their environment through two-component histidine kinases and cAMP signaling systems. They are not so inactive as to be totally passive. The strategy that they employ to exit their cellulose prison is highly regulated, constituting a developmental program of its own. The precise induction of genes within this program – notably two coordinately regulated cellulases – is impressive in its onset and equally decisive in its shutdown as the liberated amoebae enter vegetative growth. Part of this program must be a reactivation of vegetative genes, about which we know almost nothing. The programs of germination probably began long before fruiting bodies had evolved, when amoebae were confronted with the problem of emerging from microcysts and macrocysts.

13

Resources

13.1 Books

A number of books have been produced previously on *Dictyostelium* and its relatives. Some of these remain important sources of information on techniques and earlier work, and should be available in university libraries.

John Bonner has written a classic volume on *Dictyostelium* (1967), and this is an important source for early experiments and details that might get overlooked.

Kenneth Raper has also written well on *Dictyostelium*, and his 1984 book contains much material that is not covered elsewhere.

L. S. Olive has written a book on the mycetozoans, for those interested in the relatives of *Dictyostelium*, or at least in species with a similar life style.

William Loomis has been a longtime contributor and has written and edited several books that are listed below.

As a technical help, a *Methods in Cell Biology* book, edited by J. Spudich has much value. This provides recipes for media and detailed methods for growing and developing *Dictyostelium*.

An excellent book by D. Wessels and D. R. Soll describes microscopy and imaging methods for the study cell movement. Several chapters concern *Dictyostelium*.

The most recent book dedicated to *Dictyostelium* is the result of a meeting held in Sendai in Japan, and is an excellent source of recent reviews. It can be ordered through the Kyoto *Dictyostelium* Group Website, which is listed below. A relatively unknown monograph by Hagiwara deserves more attention.

- Bonner, J. T. (1967). *The Cellular slime Molds*, 2nd edition. Princeton University Press, Princeton, New Jersey.
- Goldbeter, A. (1996). *Biochemical oscillations and cellular rhythms. The molecular bases of periodic and chaotic behavior*. Cambridge University Press.
- Hagiwara, H. (1989). *The Taxonomic Study of Japanese Dictyostelid Slime Molds*. National Science Museum, Tokyo, Japan.
- Loomis, W. F. (1975). *Dictyostelium discoideum: A Developmental System*. Academic Press, New York.
- Loomis, W. F. (ed.) (1982). *The Development of Dictyostelium discoideum*. Academic Press, New York.
- Maeda, Y., Inouye, K., and Takeuchi, I. (eds.) (1997). *Dictyostelium – a Model System for Cell and Developmental Biology*. Universal Academy Press, Inc., Tokyo.
- Maeda, M. and Maeda, Y. (1978). *Biology of Slime Molds* (in Japanese). University of Tokyo Press, Tokyo.
- Olive, L. S. (1975). *The Mycetozoans*. Academic Press, New York.
- Raper, K. B. (1984). *The Dictyostelids*. Princeton University Press, Princeton, N.J.
- Segel, L. A. (1984). *Modeling dynamic phenomena in molecular and cellular biology*. Cambridge, U.K.: Cambridge University Press.
- Spudich, J. A. (ed.) (1987). *Dictyostelium discoideum: Molecular Approaches to Cell Biology*. Methods in Cell Biology, vol. 28. Academic Press, Orlando, Florida.

John Tyler Bonner has written a number of other books, most of which stretch beyond *Dictyostelium*, but usually contain something of his favorite organism. John's books are always worth reading:

Bonner, J. T. (1965). *Size and cycle. An essay on the structure of biology*. Princeton, NJ: Princeton University Press.

Bonner, J. T. (1980). *The evolution of culture in animals*. Princeton, NJ: Princeton University Press.

Bonner, J. T. (1988). *The evolution of complexity by means of natural selection*. Princeton, NJ: Princeton University Press.

Bonner, J. T. (1993). *Life cycles: reflections of an evolutionary biologist*. Princeton, NJ: Princeton University Press.

Bonner, J. T. (1996). *Sixty years of biology. Essays on evolution and development*. Princeton, NJ: Princeton University Press.

13.2 Articles for the non-scientist

Several articles have been published that are useful to the non-scientist or the young scientist.

Bonner, J. T. (1959). Differentiation in social amoebae. *Sci. Am.* **201** (December), 152–162.

An article by the author appeared several years ago in *American Scientist*. It has the virtue of excellent color photographs:

Kessin, R. H., and van Lookeren Campagne, M. M. (1992). The development of a social amoeba. *Am. Scientist* **80**, 556–565.

13.3   Films and videos

A number of striking films have been made over the years by Gerisch, Bonner and others. Two films are commercially available:

1. Higuchi, G., Yamada, T., and Ishii, S. (1993). *Behavior and Differentiation of Cellular Slime Molds.* Cine Document, Tokyo.
2. Higuchi, G., Yamada, T., and Ishii, S. (1982). *The Life Cycle of Cellular Slime Molds.* Cine Document, Tokyo.

The address is: 1-12-12 Kitami, Setagaya-ku, Tokyo 157 Japan. Phone: + 81-3-3416-7255 and Fax: + 81-3-3417-2365. Consult also the Website of the Kyoto *Dictyostelium* laboratory: http://cosmos.bot.kyoto-u.ac.jp/index.html.

A film called *A Dozen Eggs: Time Lapse Microscopy of Normal Development* includes an excellent view of *Dictyostelium*. It is edited by Rachel Fink for The Society of Developmental Biology, and is available from Sinauer Associates.

13.4   The Franke bibliographic database

A searchable database of every *Dictyostelium* reference is maintained by Jakob Franke of Columbia University as a service to the community. It can be reached through the Website maintained by Rex Chisholm listed below. This is an invaluable resource. It can be used with great creativity and is an enormous time saver.

13.5   Websites

There are a number of extraordinarily useful Websites. All of these are linked to the site maintained by Rex Chisholm of Northwestern University: http://dicty.cmb.nwu.edu/dicty/dicty.html. This site provides news items, meeting announcements, links to genomic DNA and cDNA databases, recent abstracts, the cellular slime mold newsletter, and the Franke database of *Dictyostelium* literature. The site also provides a list of laboratories on the web, information on techniques and more.

Dictyostelium is listed on an NIH Website for model organisms: http://www.nih.gov/science/models/. This provides access to a number of specific websites, including: http://dicty.cmb.nwu.edu/dicty/pnd. This site has a number of excellent images and videos of *Dictyostelium* cell biology and development.

DictyDB, the AceDB database for *Dictyostelium* was developed and is maintained by Doug Smith and Bill Loomis of The University of California at San Diego, as part of the *Dictyostelium* developmental gene program. It provides chromosome maps, a list of genes disrupted by REMI, movies, and links to the Tsukuba cDNA and other sites:
http://glamdring.ucsd.edu/others/dsmith/dictydb.html.

Genome projects can be reached through DictyDB at:
http://genome.cornell.edu/cgi-bin/WebAce/webace?db = dictydb and
http://dicty.cmb.nwu.edu/dicty/genomeseq.htm.

The Japanese cDNA sequencing project provides a superb resource for searching and requesting developmental and vegetative cDNA sequences at:
http://www.csm.biol.tsukuba.ac.jp/cDNAproject.html.

A number of personal Websites have valuable information. Most of these can be accessed from the Chisholm site listed above.

References

Abe, H., Uchiyama, M., Tanaka, Y., and Saito, H. (1976). Structure of discadenine, a spore germination inhibitor from the cellular slime mold *Dictyostelium discoideum*. *Tetrahedron Lett.* **42**, 3807–3810.

Abe, H., Hashimoto, K., and Uchiyama, M. (1981). Discadenine distribution in cellular slime molds and its inhibitory activity on spore germination. *Agr. Biol. Chem.* **45**, 1295–1296.

Abe, K., and Yanagisawa, K. (1983). A new class of rapid developing mutants in *Dictyostelium discoideum*: implications for cyclic AMP metabolism and cell differentiation. *Dev. Biol.* **95**, 200–210.

Abe, T., Early, A., Siegert, F., Weijer, C., and Williams, J. (1994). Patterns of cell movement within the *Dictyostelium* slug revealed by cell type-specific, surface labeling of living cells. *Cell* **77**, 687–699.

Abu-Elneel, K., Karchi, M., and Ravid, S. (1996). *Dictyostelium* myosin II is regulated during chemotaxis by a novel protein kinase C. *J. Biol. Chem.* **271**, 977–984.

Adachi, H., Hasebe, T., Yoshinaga, K., Ohta, T., and Sutoh, K. (1994). Isolation of *Dictyostelium discoideum* cytokinesis mutants by restriction enzyme-mediated integration of the blasticidin S resistance marker. *Biochem. Biophys. Res. Commun.* **205**, 1808–1814.

Adachi, H., Takahashi, Y., Hasebe, T., Shirouzu, M., Yokoyama, S., and Sutoh, K. (1997). *Dictyostelium* IQGAP-related protein specifically involved in the completion of cytokinesis. *J. Cell Biol.* **137**, 891–898.

Adessi, C., Chapel, A., Vincon, M., Rabilloud, T., Klein, G., Satre, M., and Garin, J. (1995). Identification of major proteins associated with *Dictyostelium discoideum* endocytic vesicles. *J. Cell Sci.* **108**, 3331–3337.

Adler, K., Gerisch, G., von Hugo, U., Lupas, A., and Schweiger, A. (1996). Classification of tyrosine kinases from *Dictyostelium discoideum* with two distinct, complete or incomplete catalytic domains. *FEBS Lett.* **395**, 286–292.

Aeckerle, S., and Malchow, D. (1989). Calcium regulates cAMP-induced potassium ion efflux in *Dictyostelium discoideum. Biochim. Biophys. Acta* **1012**, 196–200.

Aeckerle, S., Wurster, B., and Malchow, D. (1985). Oscillations and cyclic AMP induced changes of the K$^+$ concentration in *Dictyostelium discoideum. EMBO J.* **4**, 39–43.

Aerts, R.J., de Wit, R.J.W., and van Lookeren Campagne, M.M. (1987). Cyclic AMP induces a transient alkanization in *Dictyostelium. FEBS Lett.* **220**, 366–370.

Aiba, K., Fang, H., Yamaguchi, N., Tanaka, Y., and Urushihara, H. (1997). Isoforms of gp138, a cell-fusion related protein in *Dictyostelium discoideum. J. Biochem.* **121**, 238–243.

Aizawa, H., Sutoh, K., Tsubuki, S., Kawashima, S., Ishii, A., and Yahara, I. (1995). Identification, characterization, and intracellular distribution of cofilin in *Dictyostelium discoideum. J. Biol. Chem.* **270**, 10923–10932.

Aizawa, H., Fukui, Y., and Yahara, I. (1997). Live dynamics of *Dictyostelium* cofilin suggests a role in remodeling actin latticework into bundles. *J. Cell Sci.* **110**, 2333–2344.

Albe, K. R., and Wright, B. E. (1994). Carbohydrate metabolism in *Dictyostelium discoideum*. 2. Systems' analysis. *J. Theor. Biol.* **169**, 243–251.

Alcantara, F., and Bazill, G. W. (1976). Extracellular cyclic AMP-phosphodiesterase accelerates differentiation in *Dictyostelium discoideum. J. Gen. Microbiol.* **92**, 351–368.

Alcantara, F., and Monk, M. (1974). Signal propagation during aggregation in the slime mould *Dictyostelium discoideum. J. Gen. Microbiol.* **85**, 321–334.

Alexander, H., Lee, S. K., Yu, S. L., and Alexander, S. (1996). *repE* – The *Dictyostelium* homolog of the human xeroderma pigmentosum group E gene is developmentally regulated and contains a leucine zipper motif. *Nucleic Acids Res.* **24**, 2295–2301.

Alexander, S., Sydow, L. M., Wessels, D., and Soll, D. R. (1992). Discoidin proteins of *Dictyostelium* are necessary for normal cytoskeletal organization and cellular morphology during aggregation. *Differentiation* **51**, 149–161.

Al-Rayess, H., Ashworth, J. M., and Younis, M. S. (1979). The phosphate and polyphosphate content of *Dictyostelium discoideum* cells during their growth and subsequent differentiation. *J. Univ. Kuwait (Sci)* **6**, 103–108.

Amagai, A. (1992). Induction of heterothallic and homothallic zygotes in *Dictyostelium discoideum* by ethylene. *Devel. Growth Differ.* **34**, 293–300.

Amagai, A., and Maeda, Y. (1992). The ethylene action in the development of cellular slime molds – an analogy to higher plants. *Protoplasma* **167**, 159–168.

André, B., Noegel, A. A., and Schleicher, M. (1996). *Dictyostelium discoideum* contains a family of calmodulin-related EF-hand proteins that are developmentally regulated. *FEBS Lett.* **382**, 198–202.

André, E., Brink, M., Gerisch, G., Isenberg, G., Noegel, A., Schleicher, M., Segall, J. E., and Wallraff, E. (1989). A *Dictyostelium* mutant deficient in severin and F-actin fragmenting protein shows normal motility and chemotaxis. *J. Cell Biol.* **108**, 985–996.

Angata, R., Kuroe, K., Yanagisawa, K., and Tanaka, Y. (1995). Codon usage, genetic code and phylogeny of *Dictyostelium discoideum* mitochondrial DNA as deduced from a 7.3-kb region. *Curr. Genet.* **27**, 249–256.

Anjard, C., Pinaud, S., Kay, R. R., and Reymond, C. D. (1992). Overexpression of DdPK2 protein kinase causes rapid development and affects the intracellular cAMP pathway of *Dictyostelium discoideum. Development* **115**, 785–790.

Anjard, C., Etchebehere, L., Pinaud, S., Véron, M., and Reymond, C. D. (1993). An unusual catalytic subunit for the cAMP-dependent protein kinase of *Dictyostelium discoideum*. *Biochemistry* **32**, 9532–9538.

Anjard, C., van Bemmelen, M., Véron, M., and Reymond, C. D. (1997). A new spore differentiation factor (SDF) secreted by *Dictyostelium* cells is phosphorylated by the cAMP dependent protein kinase. *Differentiation* **62**, 43–49.

Anjard, C., Zeng, C., Loomis, W. F., and Nellen, W. (1998). Signal transduction pathways leading to spore differentiation in *Dictyostelium discoideum*. *Dev. Biol.* **193**, 146–155.

Arakane, T., and Maeda, Y. (1997). Relevance of histone H1 kinase activity to the G2/ M transition during the cell cycle of *Dictyostelium discoideum*. *J. Plant Res.* **110**, 81–85.

Araki, T., Nakao, H., Takeuchi, I., and Maeda, Y. (1994). Cell-cycle-dependent sorting in the development of *Dictyostelium* cells. *Dev. Biol.* **162**, 221–228.

Araki, T., Abe, T., Williams, J. G., and Maeda, Y. (1997). Symmetry breaking in *Dictyostelium* morphogenesis: evidence that a combination of cell cycle stage and positional information dictates cell fate. *Dev. Biol.* **192**, 645–648.

Araki, T., Gamper, M., Early, A., Fukuzawa, M., Abe, T., Kawata, T., Kim, E., Firtel, R. A., and Williams, J. G. (1998). Developmentally and spatially regulated activation of a *Dictyostelium* STAT protein by a serpentine receptor. *EMBO J.* **17**, 4018–4028.

Ardlie, K. G. (1998). Putting the brake on drive: meiotic drive of t haplotypes in natural populations of mice. *Trends Genet.* **14**, 189–193.

Armant, D. R., and Rutherford, C. L. (1982). Properties of a 5′-AMP specific nucleotidase which accumulates in one cell type during development of *Dictyostelium discoideum*. *Arch. Biochem. Biophys.* **216**, 485–494.

Armant, D. R., Stetler, D. A., and Rutherford, C. L. (1980). Cell surface localization of 5′-AMP nucleotidase in prestalk cells of *Dictyostelium discoideum*. *J. Cell Sci.* **45**, 119–129.

Armstrong, D. P. (1984). Why don't cellular slime molds cheat. *J. Theor. Biol.* **109**, 271–283.

Arndt, A. (1937). Rhizopodenstudien III. Untersuchungen über *Dictyostelium mucoroides* Brefeld. *W.R. Arch. Entw. Mech. Org.* **136**, 681–744.

Ashworth, J. M., and Quance, J. (1972). Enzyme synthesis in myxamoeba of the cellular slime mould *Dictyostelium discoideum* during growth in axenic culture. *Biochem. J.* **126**, 601–608.

Aubry, L., and Firtel, R. A. (1998). Spalten, a protein containing Gα-protein-like and PP2C domains, is essential for cell-type differentiation in *Dictyostelium*. *Genes Dev.* **12**, 1525–1538.

Aubry, L., Klein, G., and Satre, M. (1997). Cytoskeletal dependence and modulation of endocytosis in *Dictyostelium discoideum* amoebae. In *Dictyostelium – A model system for cell and developmental biology*. (Y. Maeda, K. Inouye, and I. Takeuchi, Eds.), pp. 65–74. Universal Academy Press, Tokyo, Japan.

Bacon, R. A., Cohen, C. J., Lewin, D. A., and Mellman, I. (1994). *Dictyostelium discoideum* mutants with temperature-sensitive defects in endocytosis. *J. Cell Biol.* **127**, 387–399.

Baldauf, S. L., and Doolittle, W. F. (1997). Origin and evolution of the slime molds (Mycetozoa). *Proc. Natl. Acad. Sci. USA* **94**, 12007–12012.

Barkley, D. S. (1969). Adenosine-3′,5′-phosphate: Identification as acrasine in a species of cellular slime mold. *Science* **165**, 1133–1134.

Barondes, S. H., Springer, W. R., and Cooper, D. N. (1982). Cell adhesion. In *The development of Dictyostelium discoideum*. (W. F. Loomis, Ed.), pp. 195–231. Academic Press, New York.

Barondes, S. H., Cooper, D. N. W., and Springer, W. R. (1987). Discoidins I and II: endogenous lectins involved in cell-substratum adhesion and spore coat formation. *Methods Cell Biol.* **28**, 387–409.

Barra, J., Barrand, P., Blondelet, M. H., and Brachet, P. (1980). *pdsA*, a gene involved in the production of active phosphodiesterase during starvation of *Dictyostelium discoideum* amoebae. *Mol. Gen. Genet.* **177**, 607–613.

Barth, A., Müller-Taubenberger, A., Taranto, P., and Gerisch, G. (1994). Replacement of the phospholipid-anchor in the contact site A glycoprotein of *D. discoideum* by a transmembrane region does not impede cell adhesion but reduces residence time on the cell surface. *J. Cell Biol.* **124**, 205–215.

Bauer, C. B., Kuhlman, P. A., Bagshaw, C. R., and Rayment, I. (1997). X-ray crystal structure and solution fluorescence characterization of Mg $2'(3')$-O-(N-methylanthraniloyl) nucleotides bound to the *Dictyostelium discoideum* myosin motor domain. *J. Mol. Biol.* **274**, 394–407.

Bear, J. E., Rawls, J. F., and Saxe III, C. L. (1998). SCAR, a WASP-related protein, isolated as a suppressor of receptor defects in late *Dictyostelium* development. *J. Cell Biol.* **142**, 1325–1335.

Berks, M., and Kay, R. R. (1990). Combinatorial control of cell differentiation by cAMP and DIF-1 during development of *Dictyostelium discoideum*. *Development* **110**, 977–984.

Beug, H., Gerisch, G., Kempff, S., Riedel, V., and Cremer, G. (1970). Specific inhibition of cell contact formation in *Dictyostelium* by univalent antibodies. *Exp. Cell Res.* **63**, 147–158.

Beug, H., Katz, F. E., Stein, A., and Gerisch, G. (1973). Quantitation of membrane sites in aggregating *Dictyostelium* cells by use of tritiated univalent antibody. *Proc. Natl. Acad. Sci. USA* **70**, 3150–3154.

Bichler, G., and Weijer, C. J. (1994). A *Dictyostelium* anterior-like cell mutant reveals sequential steps in the prespore prestalk differentiation pathway. *Development* **120**, 2857–2868.

Blaauw, M., Linskens, M. H. K., and van Haastert, P. J. M. (2000). Efficient control of gene expresson by a tetracycline-dependent transactivator in single *Dictyostelium discoideum* cells. *Gene*, **252**, 71–82.

Blanton, R. L. (1990). Phylum Acrasea. In *Jones & Bartlett Series in Life Sciences: Handbook of Protoctista: The structure, cultivation, habitats and life histories of the eukaryotic microorganisms and their descendents exclusive of animals, plants and fungi: a guide to algae, ciliates, foraminifera, sporozoa, slime molds, and the other protoctista*. (L. Margulis *et al.*, Eds.), pp. 75–87. Jones and Bartlett, Boston.

Blanton, R. L. (1993). Prestalk cells in monolayer cultures exhibit two distinct modes of cellulose synthesis during stalk cell differentiation in *Dictyostelium*. *Development* **119**, 703–710.

Blanton, R. L. (1997). Cellulose biogenesis in *Dictyostelium discoideum*. In *Dictyostelium – A model system for cell and developmental biology*. (Y. Maeda, K. Inouye, and I. Takeuchi, Eds.), pp. 379–391. Universal Academy Press, Tokyo, Japan.

Blanton, R. L., Fuller, D., Iranfar, N., Grimson, M. J., and Loomis, W. F. (2000). The cellulose synthase gene of *Dictyostelium*. *Proc. Natl. Acad. Sci. USA* **97**, 2395–2396.

Blaschke, A., Weijer, C., and MacWilliams, H. (1986). *Dictyostelium discoideum*: Cell-type proportioning, cell-differentiation preference, cell fate, and the behavior of anterior-like cells in Hs1/Hs2 and G+/G− mixtures. *Differentiation* **32**, 1–9.

Blumberg, D. D., and Lodish, H. F. (1981). Changes in the complexity of nuclear RNA during development of *Dictyostelium discoideum*. *Dev. Biol.* **81**, 74–80.

Blume, J. E., and Ennis, H. L. (1991). A *Dictyostelium discoideum* cellulase is a member of a spore germination-specific gene family. *J. Biol. Chem.* **266**, 15432–15437.

Blusch, J. H., and Nellen, W. (1994). Folate responsiveness during growth and development of *Dictyostelium* – separate but related pathways control chemotaxis and gene regulation. *Mol. Microbiol.* **11**, 331–335.

Böhme, R., Bumann, J., Aeckerle, S., and Malchow, D. (1987). A high affinity plasma membrane Ca2+-ATPase in *Dictyostelium discoideum*: its relation to cAMP-induced Ca2+ fluxes. *Biochim. Biophys. Acta* **904**, 125–130.

Bonner, J. T. (1944). A descriptive study of the development of the slime mold *Dictyostelium discoideum*. *Am. J. Bot.* **31**, 175–182.

Bonner, J. T. (1947). Evidence for the formation of aggregates by chemotaxis in the development of the slime mold *Dictyostelium discoideum*. *J. Exp. Zool.* **106**, 1–26.

Bonner, J. T. (1952a). *Morphogenesis. An essay on development.* Princeton University Press, Princeton, NJ.

Bonner, J. T. (1952b). The pattern of differentiation in amoeboid slime molds. *Am. Naturalist* **86**, 79–89.

Bonner, J. T. (1960). Development in the cellular slime molds: The role of cell division, cell size, and cell number. In *Developing cell systems and their control.* (D. Rudnick, Ed.), pp. 3–20. Ronald Press, New York.

Bonner, J. T. (1967). *The cellular slime molds. Second edition.* Princeton University Press, Princeton, NJ.

Bonner, J. T. (1982). Evolutionary strategies and developmental constraints in the cellular slime molds. *Am. Naturalist* **119**, 530–552.

Bonner, J. T. (1992). The fate of a cell is a function of its position and vice-versa. *J. Biosci.* **17**, 95–114.

Bonner, J. T. (1998). A way of following individual cells in the migrating slugs of *Dictyostelium discoideum*. *Proc. Natl. Acad. Sci. USA* **95**, 9355–9359.

Bonner, J. T., and Adams, M. S. (1958). Cell mixtures of different species and strains of cellular slime molds. *J. Embryol. Exp. Morphol.* **6**, 346–356.

Bonner, J. T., and Cox, E. C. (1995). Pattern formation in Dictyostelids. *Semin. Dev. Biol.* **6**, 359–368.

Bonner, J. T., and Dodd, M. R. (1962). Evidence for gas-induced orientation in the cellular slime molds. *Dev. Biol.* **5**, 344–361.

Bonner, J. T., and Frascella, E. B. (1952). Mitotic activity in relation to differentiation in the slime mold *Dictyostelium discoideum*. *J. Exp. Zool.* **121**, 561–571.

Bonner, J. T., Clarke Jr, W. W., Neely Jr, C. H. L., and Slifkin, M. K. (1950). The orientation to light and the extremely sensitive orientation to temperature gradients in the slime mold *Dictyostelium discoideum*. *J. Cell. Compar. Physiol.* **36**, 149–158.

Bonner, J. T., Chiquoine, A. D., and Kolderie, M. Q. (1955). A histochemical study of differentiation in the cellular slime molds. *J. Exp. Zool.* **130**, 133–158.

Bonner, J. T., Barkley, D. S., Hall, E. M., Konijn, T. M., Mason, J. W., O'Keefe III, G., and Wolfe, P. B. (1969). Acrasin, acrasinase and the sensitivity to acrasin in *Dictyostelium discoideum*. *Dev. Biol.* **20**, 72–87.

Boy-Marcotte, E., Vilaine, F., Camonis, J., and Jacquet, M. (1984). A DNA sequence from *Dictyostelium discoideum* complements *ura3* and *ura5* mutations of *Saccharomyces cerevisiae*. *Mol. Gen. Genet.* **193**, 406–413.

Bozzaro, S., and Gerisch, G. (1978). Contact sites in aggregating cells of *Polysphondylium pallidum*. *J. Mol. Biol.* **120**, 265–279.

Bozzaro, S., and Ponte, E. (1995). Cell adhesion in the life cycle of *Dictyostelium*. *Experientia* **51**, 1175–1188.

Bozzone, D. M. (1983). A comparison of the macrocyst and fruiting body developmental pathways in *Dictyostelium discoideum*. *J. Gen. Microbiol.* **129**, 1371–1380.

Bozzone, D. M., and Bonner, J. T. (1982). Macrocyst formation in *Dictyostelium discoideum*: mating or selfing. *J. Exp. Zool.* **220**, 391–394.

Bracco, E., Peracino, B., Noegel, A. A., and Bozzaro, S. (1997). Cloning and transcriptional regulation of the gene encoding the vacuolar/H+ ATPase B subunit of *Dictyostelium discoideum*. *FEBS Lett.* **419**, 37–40.

Brandon, M. A., and Podgorski, G. J. (1997). Gα3 regulates the cAMP signaling system in *Dictyostelium*. *Mol. Biol. Cell* **8**, 1677–1685.

Brandon, M. A., Voglmaier, S., and Siddiqi, A. A. (1997). Molecular characterization of a *Dictyostelium* G-protein α-subunit required for development. *Gene* **200**, 99–105.

Brar, S. K., and Siu, C. H. (1993). Characterization of the cell adhesion molecule gp24 in *Dictyostelium discoideum* – mediation of cell-cell adhesion via a Ca2+-dependent mechanism. *J. Biol. Chem.* **268**, 24902–24909.

Brazill, D. T., Gundersen, R., and Gomer, R. H. (1997). A cell-density sensing factor regulates the lifetime of a chemoattractant-induced Gα-GTP conformation. *FEBS Lett.* **404**, 100–104.

Brazill, D. T., Lindsey, D. F., Bishop, J. D., and Gomer, R. H. (1998). Cell density sensing mediated by a G protein-coupled receptor activating phospholipase C. *J. Biol. Chem.* **273**, 8161–8168.

Brefeld, O. (1869). *Dictyostelium mucoroides*. Ein neuer Organismus aus der Verwandtschaft der Myxomyceten. *Abhandlungen der Senckenbergischen Naturforschenden Gesellschaft Frankfurt* **7**, 85–107.

Brefeld, O. (1884). *Polysphondylium violaceum* und *Dictyostelium mucoroides* nebst Bemerkungen zur Systematik der Schleimpilze. *Untersuchungen aus dem Gesammtgebiet der Mykol.* **6**, 1–34.

Brenner, M. (1978). Cyclic AMP levels and turnover during development of the cellular slime mold *Dictyostelium discoideum*. *Dev. Biol.* **64**, 210–223.

Brenner, M., and Thoms, S. D. (1984). Caffeine blocks activation of cyclic AMP synthesis in *Dictyostelium discoideum*. *Dev. Biol.* **101**, 136–146.

Bretschneider, T., Vasiev, B., and Weijer, C. J. (1997). A model for cell movement during *Dictyostelium* mound formation. *J. Theor. Biol.* **189**, 41.

Brock, D. A., and Gomer, R. H. (1999). A cell-counting factor regulating structure size in *Dictyostelium*. *Genes Dev.* **13**, 1960–1969.

Brock, D. A., Buczynski, G., Spann, T. P., Wood, S. A., Cardelli, J., and Gomer, R. H. (1996). A *Dictyostelium* mutant with defective aggregate size determination. *Development* **122**, 2569–2578.

Brookman, J. J., Town, C. D., Jermyn, K. A., and Kay, R. R. (1982). Developmental regulation of stalk cell differentiation-inducing factor in *Dictyostelium discoideum*. *Dev. Biol.* **91**, 191–196.

Brown, J. M., and Firtel, R. A. (2000). Just the right size: cell counting in *Dictyostelium*. *Trends Genet.* **16**, 191–193.

Brown, J. M., Briscoe, C., and Firtel, R. A. (1997). Control of transcriptional regulation by signal transduction pathways in *Dictyostelium* during multicellular development. In *Dictyostelium – A model system for cell and developmental biology.* (Y. Maeda, K. Inouye, and I. Takeuchi, Eds.), pp. 245–265. Universal Academy Press, Tokyo, Japan.

Brown, R. J., van Beek, E., Watts, D. J., Lowik, C. W. G. M., Papapoulos, S. E. (1998). Differential effects of aminosubstituted analogs of hydroxy bisphosphonates on the growth of *Dictyostelium discoideum. J. Bone Miner. Res.* **13**, 253–258.

Buczynski, G., Bush, J., Zhang, L. Y., Rodriguez-Paris, J., and Cardelli, J. (1997). Evidence for a recycling role for Rab7 in regulating a late step in endocytosis and in retention of lysosomal enzymes in *Dictyostelium discoideum. Mol. Biol. Cell* **8**, 1343–1360.

Bühl, B., and MacWilliams, H. K. (1991). Cell sorting within the prestalk zone of *Dictyostelium discoideum. Differentiation* **46**, 147–152.

Bukenberger, M., Horn, J., Dingermann, T., Dottin, R. P., and Winckler, T. (1997). Molecular cloning of a cDNA encoding the nucleosome core histone H3 from *Dictyostelium discoideum* by genetic screening in yeast. *Biochim. Biophys. Acta* **1352**, 85–90.

Bumann, J., Malchow, D., and Wurster, B. (1986). Oscillations of Ca^{++} concentration during the cell differentiation of *Dictyostelium discoideum. Differentiation* **31**, 85–91.

Burki, E., Anjard, C., Scholder, J. C., and Reymond, C. D. (1991). Isolation of two genes encoding putative protein kinases regulated during *Dictyostelium discoideum* development. *Gene* **102**, 57–65.

Bush, J., Franek, K., Daniel, J., Spiegelman, G. B., Weeks, G., and Cardelli, J. (1993). Cloning and characterization of five novel *Dictyostelium discoideum* Rab-related genes. *Gene* **136**, 55–60.

Bush, J., Richardson, J., and Cardelli, J. (1994). Molecular cloning and characterization of the full-length cDNA encoding the developmentally regulated lysosomal enzyme beta-glucosidase in *Dictyostelium discoideum. J. Biol. Chem.* **269**, 1468–1476.

Bush, J., Temesvari, L., Rodriguez-Paris, J., Buczynski, G., and Cardelli, J. (1996). A role for a Rab4-like GTPase in endocytosis and in regulation of contractile vacuole structure and function in *Dictyostelium discoideum. Mol. Biol. Cell* **7**, 1623–1638.

Buss, L. W. (1982). Somatic cell parasitism and the evolution of somatic tissue compatibility. *Proc. Natl. Acad. Sci. USA* **79**, 5337–5341.

Byrne, G., and Cox, E. C. (1986). Spatial patterning in *Polysphondylium*: monoclonal antibodies specific for whorl prepatterns. *Dev. Biol.* **117**, 442–455.

Byrne, G. W., Trujillo, J., and Cox, E. C. (1982). Pattern formation and tip inhibition in the cellular slime mold *Polysphondylium pallidum. Differentiation* **23**, 103–108.

Callahan, S. M., Cornell, N. W., and Dunlap, P. V. (1995). Purification and properties of periplasmic $3':5'$-cyclic nucleotide phosphodiesterase. A novel zinc-containing enzyme from the marine symbiotic bacterium *Vibrio fischeri. J. Biol. Chem.* **270**, 17627–17632.

Cappello, J., Handelsman, K., Cohen, S. M., and Lodish, H. F. (1985a). Structure and regulated transcription of DIRS-1: an apparent retrotransposon of *Dictyostelium discoideum. Cold Spring Harbor Symp. Quant. Biol.* **50**, 759–767.

Cappello, J., Handelsman, K., and Lodish, H. F. (1985b). Sequence of *Dictyostelium* DIRS-1: an apparent retrotransposon with inverted terminal repeats and an internal circle junction sequence. *Cell* **43**, 105–115.

Carrin, I., Murgia, I., McLachlan, A., and Kay, R. R. (1996). A mutational analysis of *Dictyostelium discoideum* multicellular development. *Microbiology* **142**, 993–1003.

Caterina, M. J., and Devreotes, P. N. (1991). Molecular insights into eukaryotic chemotaxis. *FASEB J.* **5**, 3078–3085.

Caterina, M. J., Milne, J. L. S., and Devreotes, P. N. (1994). Mutation of the third intracellular loop of the cAMP receptor, cAR1, of *Dictyostelium* yields mutants impaired in multiple signaling pathways. *J. Biol. Chem.* **269**, 1523–1532.

Cavender, J. C. (1990). Phylum Dictyostelida. In *Jones Bartlett Series in Life Sciences: Handbook of Protoctista: The structure, cultivation, habitats and life histories of the eukaryotic microorganisms and their descendents exclusive of animals, plants and fungi: a guide to algae, ciliates, foraminifera, sporozoa, slime molds, and the other protoctista.* (L. Margulis, *et al.*, Eds.), pp. 88–101. Jones and Bartlett, Boston.

Ceccarelli, A., Mahbubani, H., and Williams, J. G. (1991). Positively and negatively acting signals regulating stalk cell and anterior-like cell differentiation in *Dictyostelium. Cell* **65**, 983–989.

Ceccarelli, A., Mahbubani, H. J., Insall, R., Schnitzler, G., Firtel, R. A., and Williams, J. G. (1992). A G-rich sequence element common to *Dictyostelium* genes which differ radically in their patterns of expression. *Dev. Biol.* **152**, 188–193.

Ceccarini, C., and Filosa, M. (1965). Carbohydrate content during development of the slime mold *Dictyostelium discoideum. J. Cell. Compar. Physiol.* **66**, 135–140.

Chadwick, C. M., Ellison, J. E., and Garrod, D. R. (1984). Dual role for *Dictyostelium* contact sites B in phagocytosis and developmental size regulation. *Nature* **307**, 646–647.

Chandrasekhar, A., Wessels, D., and Soll, D. R. (1995). A mutation that depresses cGMP phosphodiesterase activity in *Dictyostelium* affects cell motility through an altered chemotactic signal. *Dev. Biol.* **169**, 109–122.

Chang, W. T., Newell, P. C., and Gross, J. D. (1996). Identification of the cell fate gene stalky in *Dictyostelium. Cell* **87**, 471–481.

Chang, W. T., Thomason, P. A., Gross, J. D., and Newell, P. C. (1998). Evidence that the RdeA protein is a component of a multistep phosphorelay modulating rate of development in *Dictyostelium. EMBO J.* **17**, 2809–2816.

Chang, Y. Y. (1968). Cyclic 3',5'-adenosine monophosphate phosphodiesterase produced by the slime mold *Dictyostelium discoideum. Science* **161**, 57–59.

Chen, M. Y., Insall, R. H., and Devreotes, P. N. (1996). Signaling through chemoattractant receptors in *Dictyostelium. Trends Genet.* **12**, 52–57.

Chen, M. Y., Long, Y., and Devreotes, P. N. (1997). A novel cytosolic regulator, pianissimo, is required for chemoattractant receptor and G protein-mediated activation of the 12 transmembrane domain adenylyl cyclase in *Dictyostelium. Genes Dev.* **11**, 3218–3231.

Chen, P. X., Ostrow, B. D., Tafuri, S. R., and Chisholm, R. L. (1994). Targeted disruption of the *Dictyostelium* RMLC gene produces cells defective in cytokinesis and development. *J. Cell Biol.* **127**, 1933–1944.

Chen, T. L., Kowalczyk, P. A., Ho, G. Y., and Chisholm, R. L. (1995). Targeted disruption of the *Dictyostelium* myosin essential light chain gene produces cells defective in cytokinesis and morphogenesis. *J. Cell Sci.* **108**, 3207–3218.

Chen, T. L., Wolf, W. A., and Chisholm, R. L. (1998). Cell-type-specific rescue of myosin function during *Dictyostelium* development defines two distinct cell movements required for culmination. *Development* **125**, 3895–3903.

Chia, C. P. (1996). A 130-kDa plasma membrane glycoprotein involved in *Dictyostelium* phagocytosis. *Exp. Cell Res.* **227**, 182–189.

Chisholm, R. L. (1997). Cytokinesis: A regulatory role for Ras-related proteins? *Curr. Biol.* **7**, R648–R650.

Chisholm, R. L., Barklis, E., and Lodish, H. F. (1984). Mechanism of sequential induction of cell type specific mRNAs in *Dictyostelium* differentiation. *Nature* **310**, 67–69.

Clarke, M., and Gomer, R. H. (1995). PSF and CMF, autocrine factors that regulate gene expression during growth and early development of *Dictyostelium*. *Experientia* **51**, 1124–1134.

Clarke, M., and Heuser, J. (1997). Water and ion transport. In *Dictyostelium – A model system for cell and developmental biology*. (Y. Maeda, K. Inouye, and I. Takeuchi, Eds.), pp. 75–91. Universal Academy Press, Tokyo, Japan.

Clarke, M., Kayman, S. C., and Riley, K. (1987). Density-dependent induction of discoidin-I synthesis in exponentially growing cells of *Dictyostelium discoideum*. *Differentiation* **34**, 79–87.

Clarke, M., Dominguez, N., Yuen, I. S., and Gomer, R. H. (1992). Growing and starving *Dictyostelium* cells produce distinct density-sensing factors. *Dev. Biol.* **152**, 403–406.

Cockburn, A. F., Newkirk, M. J., and Firtel, R. A. (1976). Organization of the ribosomal RNA genes of *Dictyostelium discoideum*: mapping of the nontranscribed spacer regions. *Cell* **9**, 605–613.

Cockburn, A. F., Taylor, W. C., and Firtel, R. A. (1978). *Dictyostelium* rDNA consists of nonchromosomal palindromic dimers containing 5S and 36S coding regions. *Chromosoma* **70**, 19–29.

Cohen, C. J., Bacon, R., Clarke, M., Joiner, K., and Mellman, I. (1994). *Dictyostelium discoideum* mutants with conditional defects in phagocytosis. *J. Cell Biol.* **126**, 955–966.

Cohen, M. H., and Robertson, A. (1971a). Chemotaxis and the early stages of aggregation in cellular slime molds. *J. Theor. Biol.* **31**, 119–130.

Cohen, M. H., and Robertson, A. (1971b). Wave propagation in the early stages of aggregation of cellular slime molds. *J. Theor. Biol.* **31**, 101–118.

Cole, R. A., and Williams, K. L. (1994). The *Dictyostelium discoideum* mitochondrial genome: A primordial system using the universal code and encoding hydrophilic proteins atypical of metazoan mitochondrial DNA. *J. Mol. Evol.* **39**, 579–588.

Cole, R. A., Slade, M. B., and Williams, K. L. (1995). *Dictyostelium discoideum* mitochondrial DNA encodes a NADH:ubiquinone oxidoreductase subunit which is nuclear encoded in other eukaryotes. *J. Mol. Evol.* **40**, 616–621.

Condeelis, J. (1992). Are all pseudopods created equal? *Cell Motil. Cytoskel.* **22**, 1–6.

Condeelis, J. (1998). The biochemistry of animal cell crawling (Ch. 5). In *Motion analysis of living cells*. (D. R. Soll, and D. Wessels, Eds.), pp. 85–100. Wiley-Liss, New York.

Cooper, D. N. W., and Barondes, S. H. (1984). Colocalization of discoidin-binding ligands with discoidin in developing *Dictyostelium discoideum*. *Dev. Biol.* **105**, 59–70.

Cornillon, S., Foa, C., Davoust, J., Buonavista, N., Gross, J. D., and Golstein, P. (1994). Programmed cell death in *Dictyostelium*. *J. Cell Sci.* **107**, 2691–2704.

Cotter, D. A. (1981). Spore activation. In *The fungal spore: morphogenetic controls*. (G. Turian, and H. R. Hohl, Eds.), pp. 385–411. Academic Press, London.

Cotter, D. A., and Raper, K. B. (1970). Spore germination in *Dictyostelium discoideum*: Trehalase and the requirement for protein synthesis. *Dev. Biol.* **22**, 112–128.

Cotter, D. A., Miura-Santo, L. Y., and Hohl, H. R. (1969). Ultrastructural changes during germination of *Dictyostelium discoideum* spores. *J. Bacteriol.* **100**, 1020–1026.

Coukell, M. B. (1975). Parasexual genetic analysis of aggregation-deficient mutants of *Dictyostelium discoideum*. *Mol. Gen. Genet.* **142**, 119–135.

Coukell, M. B., and Cameron, A. M. (1985). Genetic locus (*stmF*) associated with cyclic GMP phosphodiesterase activity in *Dictyostelium discoideum* maps in linkage group II. *J. Bacteriol.* **162**, 427–429.

Coukell, M. B., and Cameron, A. M. (1986). Characterization of revertants of *stmF* mutants of *Dictyostelium discoideum*: evidence that *stmF* is the structural gene of the cGMP-specific phosphodiesterase. *Dev. Genet.* **6**, 163–177.

Coukell, M. B., and Walker, I. O. (1973). The basic nuclear proteins of the cellular slime mold *Dictyostelium discoideum*. *Cell Differ.* **2**, 87–95.

Coukell, M. B., Lappano, S., and Cameron, A. M. (1983). Isolation and characterization of cAMP unresponsive (frigid) aggregation deficient mutants of *Dictyostelium discoideum*. *Dev. Genet.* **3**, 283–297.

Coukell, M. B., Moniakis, J., and Cameron, A. M. (1997). The *patB* gene of *Dictyostelium discoideum* encodes a P-type H^+-ATPase isoform essential for growth and development under acidic conditions. *Microbiology* **143**, 3877–3888.

Cox, D., Condeelis, J., Wessels, D., Soll, D., Kern, H., and Knecht, D. A. (1992). Targeted disruption of the ABP-120 gene leads to cells with altered motility. *J. Cell Biol.* **116**, 943–955.

Cox, D., Wessels, D., Soll, D. R., Hartwig, J., and Condeelis, J. (1996). Re-expression of ABP-120 rescues cytoskeletal, motility, and phagocytosis defects of ABP-120⁻ *Dictyostelium* mutants. *Mol. Biol. Cell* **7**, 803–823.

Cox, E. C. (1993). Periodic spatial patterns in the cellular slime mold *Polysphondylium pallidum*. In *Oscillations and Morphogenesis*. (L. Rensing, Ed.), pp. 437–452. M. Dekker, New York.

Cox, E. C., Spiegel, F. W., Byrne, G., McNally, J. W., and Eisenbud, L. (1988). Spatial patterns in the fruiting bodies of the cellular slime mold *Polysphondylium pallidum*. *Differentiation* **38**, 73–81.

Cox, E. C., Vocke, C. D., Walter, S., Gregg, K. Y., and Bain, E. S. (1990). Electrophoretic karyotype for *Dictyostelium discoideum*. *Proc. Natl. Acad. Sci. USA* **87**, 8247–8251.

Crowley, T. E., Nellen, W., Gomer, R. H., and Firtel, R. A. (1985). Phenocopy of discoidin I-minus mutants by antisense transformation in *Dictyostelium*. *Cell* **43**, 633–641.

Cubitt, A. B., Firtel, R. A., Fischer, G., Jaffe, L. F., and Miller, A. L. (1995). Patterns of free calcium in multicellular stages of *Dictyostelium* expressing jellyfish apoaequorin. *Development* **121**, 2291–2301.

Dahlberg, K. R., and Cotter, D. A. (1979). Autoactivation of spore germination in mutant and wild type strains of *Dictyostelium discoideum*. *Microbios* **23**, 153–166.

Dallon, J. C., and Othmer, H. G. (1997). A discrete cell model with adaptive signalling for aggregation of *Dictyostelium discoideum*. *Phil. Trans. R. Soc. Lond. B* **352**, 391–417.

Dao, D. N., Kessin, R. H., and Ennis, H. L. (2000). Developmental cheating and the evolutionary biology of *Dictyostelium* and *Myxococcus*. *Microbiology*, **146**, 1505–1512.

Darcy, P. K., Wilczynska, Z., and Fisher, P. R. (1993). Phototaxis genes on linkage group-V in *Dictyostelium discoideum. FEMS Microbiol. Lett.* **111**, 123–127.

Darcy, P. K., Wilczynska, Z., and Fisher, P. R. (1994). Genetic analysis of *Dictyostelium* slug phototaxis mutants. *Genetics* **137**, 977–985.

Darnell, J. E. (1997). Phosphotyrosine signaling and the single cell:metazoan boundary. *Proc. Natl. Acad. Sci. USA* **94**, 11767–11769.

Datta, S., Gomer, R. H., and Firtel, R. A. (1986). Spatial and temporal regulation of a foreign gene by a prestalk-specific promotor in transformed *Dictyostelium discoideum. Mol. Cell. Biol.* **6**, 811–820.

Davies, L., Satre, M., Martin, J. B., and Gross, J. D. (1993). The target of ammonia action in *Dictyostelium. Cell* **75**, 321–327.

Davies, L., Farrar, N. A., Satre, M., Dottin, R. P., and Gross, J. D. (1996). Vacuolar H^+-ATPase and weak base action in *Dictyostelium. Mol. Microbiol.* **22**, 119–126.

De Chastellier, C., and Ryter, A. (1977). Changes on the cell surface and of the digestive apparatus of *Dictyostelium discoideum* during the starvation period triggering aggregation. *J. Cell Biol.* **75**, 218–236.

De Chastellier, C., and Ryter, A. (1980). Characteristic ultrastructural transformations upon starvation of *Dictyostelium discoideum* and their relations with aggregation. Study of wild type amoebae and aggregation mutant. *Biol. Cellulaire* **38**, 121–128.

De Gunzburg, J., Part, D., Guiso, N., and Véron, M. (1984). An unusual adenosine $3',5'$-phosphate dependent protein kinase from *Dictyostelium discoideum. Biochemistry* **23**, 3805–3812.

De Gunzburg, J., Franke, J., Kessin, R. H., and Véron, M. (1986). Detection and developmental regulation of the mRNA for the regulatory subunit of the cAMP-dependent protein kinase of *Dictyostelium discoideum* by cell-free translation. *EMBO J.* **5**, 363–367.

de Hostos, E. L., Bradtke, B., Lottspeich, F., Guggenheim, R., and Gerisch, G. (1991). Coronin, an actin binding protein of *Dictyostelium discoideum* localized to cell surface projections, has sequence similarities to G protein β subunits. *EMBO J.* **10**, 4097–4104.

de Hostos, E. L., Rehfuess, C., Bradtke, B., Waddell, D. R., Albrecht, R., Murphy, J., and Gerisch, G. (1993). *Dictyostelium* mutants lacking the cytoskeletal protein coronin are defective in cytokinesis and cell motility. *J. Cell Biol.* **120**, 163–173.

de Hostos, E. L., McCaffrey, G., Sucgang, R., Pierce, D. W., and Vale, R. D. (1998). A developmentally regulated kinesin-related motor protein from *Dictyostelium discoideum. Mol. Biol. Cell* **9**, 2093–2106.

De Lozanne, A., and Spudich, J. A. (1987). Disruption of the *Dictyostelium* myosin heavy chain gene by homologous recombination. *Science* **236**, 1086–1091.

de Wit, R. J. W., van Bemmelen, M. X. P., Penning, L. C., Pinas, J. E., Calandra, T. D., and Bonner, J. T. (1988). Studies of cell-surface glorin receptors, glorin degradation, and glorin-induced cellular responses during development of *Polysphondylium violaceum. Exp. Cell Res.* **179**, 332–343.

Deasey, M. C., and Olive, L. S. (1981). Role of Golgi apparatus in sorogenesis by the cellular slime mold *Fonticula alba. Science* **213**, 561–563.

Deering, R. A. (1968). *Dictyostelium discoideum*: a gamma-ray resistant organism. *Science* **162**, 1289–1290.

Deering, R. A. (1988a). DNA repair in *Dictyostelium. Dev. Genet.* **9**, 483–494.

Deering, R. A. (1988b). Use of *Dictyostelium discoideum* to study DNA repair. In *DNA Repair: A Laboratory Manual of Research Procedures.* (E. C. Friedberg, and P. C. Hanawalt, Eds.), pp. 39–76. Marcel Dekker, New York.

Deering, R. A., Smith, M. S., Thompson, B. K., and Adolf, A. C. (1970). Gamma-ray-resistant and -sensitive strains of slime mold (*Dictyostelium discoideum*). *Radiat. Res.* **43**, 711–728.

Deering, R. A., Guyer, R. B., Stevens, L., and Watsonthais, T. E. (1996). Some repair-deficient mutants of *Dictyostelium discoideum* display enhanced susceptibilities to bleomycin. *Antimicrob. Agents Chemother.* **40**, 464–467.

Deery, W. J., and Gomer, R. H. (1999). A putative receptor mediating cell-density sensing in *Dictyostelium*. *J. Biol. Chem.* **274**, 34476–34482.

Dembinsky, A., Rubin, H., and Ravid, S. (1996). Chemoattractant-mediated increases in cGMP induce changes in *Dictyostelium* myosin II heavy chain-specific protein kinase C activities. *J. Cell Biol.* **134**, 911–921.

Depraitère, C., and Darmon, M. (1978). Croissance de l'amibe sociale *Dictyostelium discoideum* sur différentes espèces bactériennes. *Ann. Microbiol. (Institute Pasteur)* **129B**, 451–461.

Detterbeck, S., Morandini, P., Wetterauer, B., Bachmair, A., Fischer, K., and MacWilliams, H. K. (1994). The 'prespore-like cells' of *Dictyostelium* have ceased to express a prespore gene: Analysis using short-lived β-galactosidases as reporters. *Development* **120**, 2847–2855.

Devine, K. M., and Loomis, W. F. (1985). Molecular characterization of anterior-like cells in *Dictyostelium discoideum*. *Dev. Biol.* **107**, 364–372.

Devine, K. M., Bergmann, J. E., and Loomis, W. F. (1983). Spore coat proteins of *Dictyostelium discoideum* are packaged in prespore vesicles. *Dev. Biol.* **99**, 437–446.

Devreotes, P. N. 1983. Cyclic nucleotides and cell-cell communication in *Dictyostelium discoideum*. *Adv. Cyclic Nucleotide Res.* **15**, 55–96.

Devreotes, P. N., and Steck, T. L. (1979). cAMP relay in *Dictyostelium discoideum*. II. Requirements for the initiation and termination of the response. *J. Cell Biol.* **80**, 300–309.

Devreotes, P. N., and Zigmond, S. H. (1988). Chemotaxis in eukaryotic cells: A focus on leucocytes and *Dictyostelium*. *Annu. Rev. Cell Biol.* **4**, 649–686.

Devreotes, P. N., Derstine, P. L., and Steck, T. L. (1979). Cyclic $3',5'$-AMP relay in *Dictyostelium discoideum*. I. A technique to monitor responses to controlled stimuli. *J. Cell Biol.* **80**, 291–299.

Dharmawardhane, S., Warren, V., Hall, A. L., and Condeelis, J. (1989). Changes in the association of actin-binding proteins with the actin cytoskeleton during chemotactic stimulation of *Dictyostelium discoideum*. *Cell Motil. Cytoskel.* **13**, 57–63.

Dharmawardhane, S., Cubitt, A. B., Clark, A. M., and Firtel, R. A. (1994). Regulatory role of the Gα1 subunit in controlling cellular morphogenesis in *Dictyostelium*. *Development* **120**, 3549–3561.

Dimond, R. L., Mayer, M., and Loomis, W. F. (1976). Characterization and developmental regulation of β-galactosidase isozymes in *Dictyostelium discoideum*. *Dev. Biol.* **52**, 74–82.

Dinauer, M. C., MacKay, S. A., and Devreotes, P. N. (1980a). Cyclic $3',5'$-AMP relay in *Dictyostelium discoideum* III. The relationship of cAMP synthesis and secretion during the cAMP signaling response. *J. Cell Biol.* **86**, 537–544.

Dinauer, M. C., Steck, T. L., and Devreotes, P. N. (1980b). Cyclic $3',5'$-AMP relay in *Dictyostelium discoideum* IV. Recovery of the signaling response after cAMP stimulation. *J. Cell Biol.* **86**, 545–553.

Dinauer, M. C., Steck, T. L., and Devreotes, P. N. (1980c). Cyclic $3',5'$-AMP relay in *Dictyostelium discoideum*. V. Adaptation of the cAMP signaling response during cAMP stimulation. *J. Cell Biol.* **86**, 554–561.

Dingermann, T., Bertling, W., Brechner, T., Nerke, K., Peffley, D. M., and Sogin, M. L. (1986). Structure of two tRNA genes from *Dictyostelium discoideum*. *Nucleic Acids Res.* **14**, 1127.

Dingermann, T., Amon, E., Williams, K. L., and Welker, D. L. (1987). Chromosomal mapping of tRNA genes from *Dictyostelium discoideum*. *Mol. Gen. Genet.* **207**, 176–187.

Dingermann, T., Reindl, N., Werner, H., Hildebrandt, M., Nellen, W., Harwood, A., Williams, J., and Nerke, K. (1989). Optimization and *in situ* detection of *Escherichia coli* β-galactosidase gene expression in *Dictyostelium discoideum*. *Gene* **85**, 353–362.

Dingermann, T., Reindl, N., Brechner, T., Werner, H., and Nerke, K. (1990). Nonsense suppression in *Dictyostelium discoideum*. *Dev. Genet.* **11**, 410–417.

Dingermann, T., Werner, H., Schutz, A., Zundorf, I., Nerke, K., Knecht, D., and Marschalek, R. (1992). Establishment of a system for conditional gene expression using an inducible tRNA suppressor gene. *Mol. Cell. Biol.* **12**, 4038–4045.

Dormann, D., Siegert, F., and Weijer, C. J. (1996). Analysis of cell movement during the culmination phase of *Dictyostelium* development. *Development* **122**, 761–769.

Dormann, D., Weijer, C., and Siegert, F. (1997). Twisted scroll waves organize *Dictyostelium mucoroides* slugs. *J. Cell Sci.* **110**, 1831–1837.

Dormann, D., Vasiev, B., and Weijer, C. J. (1998). Propagating waves control *Dictyostelium discoideum* morphogenesis. *Biophys. Chem.* **72**, 21–35.

Dowbenko, D. J., and Ennis, H. L. (1980). Regulation of protein synthesis during spore germination in *Dictyostelium discoideum*. *Proc. Natl. Acad. Sci. USA* **77**, 1791–1795.

Drayer, A. L., van der Kaay, J., Mayr, G. W., and van Haastert, P. J. M. (1994). Role of phospholipase C in *Dictyostelium* – formation of inositol 1,4,5-trisphosphate and normal development in cells lacking phospholipase C activity. *EMBO J.* **13**, 1601–1609.

Dunbar, A. J., and Wheldrake, J. F. (1994). A calcium and calmodulin-dependent protein kinase present in differentiating *Dictyostelium discoideum*. *FEMS Microbiol. Lett.* **115**, 113–118.

Durston, A. J. (1976). Tip formation is regulated by an inhibitory gradient in the *Dictyostelium discoideum* slug. *Nature* **263**, 126–129.

Dynes, J. L., Clark, A. M., Shaulsky, G., Kuspa, A., Loomis, W. F., and Firtel, R. A. (1994). LagC is required for cell-cell interactions that are essential for cell-type differentiation in *Dictyostelium*. *Genes Dev.* **8**, 948–958.

Early, A., McRobbie, S. J., Duffy, K. T., Jermyn, K. A., Tilly, R., R., Ceccarelli, A. and Williams, J. G. (1988a). Structural and functional characterization of genes encoding *Dictyostelium* prestalk and prespore cell-specific proteins. *Dev. Genet.* **9**, 383–402.

Early, A. E., Williams, J. G., Meyer, H. E., Por, S. B., Smith, E., Williams, K. L., and Gooley, A. A. (1988b). Structural characterization of *Dictyostelium discoideum* prespore-specific gene D19 and of its product, cell surface glycoprotein PsA. *Mol. Cell. Biol.* **8**, 3458–3466.

Early, A. E., Gaskell, M. J., Traynor, D., and Williams, J. G. (1993). Two distinct populations of prestalk cells within the tip of the migratory *Dictyostelium* slug with differing fates at culmination. *Development* **118**, 353–362.

Early, A., Abe, T., and Williams, J. (1995). Evidence for positional differentiation of prestalk cells and for a morphogenetic gradient in *Dictyostelium*. *Cell* **83**, 91–99.

Early, V. E., and Williams, J. G. (1988). A *Dictyostelium* prespore-specific gene is transcriptionally repressed by DIF *in vitro. Development* **103**, 519–524.

Eddy, R. J., Han, J. H., Sauterer, R. A., and Condeelis, J. S. (1996). A major agonist-regulated capping activity in *Dictyostelium* is due to the capping protein, cap32/34. *Biochim. Biophys. Acta* **1314**, 247–259.

Eddy, R. J., Han, J., and Condeelis, J. S. (1997). Capping protein terminates but does not initiate chemoattractant-induced actin assembly in *Dictyostelium. J. Cell Biol.* **139**, 1243–1253.

Egelhoff, T. T., and Spudich, J. A. (1991). Molecular genetics of cell migration – *Dictyostelium* as a model system. *Trends Genet.* **7**, 161–166.

Egelhoff, T. T., Manstein, D. J., and Spudich, J. A. (1990). Complementation of myosin null mutants in *Dictyostelium discoideum* by direct functional selection. *Dev. Biol.* **137**, 359–367.

Egelhoff, T. T., Lee, R. J., and Spudich, J. A. (1993). *Dictyostelium* myosin heavy chain phosphorylation sites regulate myosin filament assembly and localization *in vivo. Cell* **75**, 363–371.

Eichinger, L., and Schleicher, M. (1992). Characterization of actin-binding and lipid-binding domains in severin, a Ca^{2+}-dependent F-actin fragmenting protein. *Biochemistry* **31**, 4779–4787.

Eichinger, L., Bähler, M., Dietz, M., Eckerskorn, C., and Schleicher, M. (1998). Characterization and cloning of a *Dictyostelium* Ste20-like protein kinase that phosphorylates the actin-binding protein severin. *J. Biol. Chem.* **273**, 12952–12959.

Eichinger, L., Lee, S. S., and Schleicher, M. (1999). *Dictyostelium* as model system for studies of the actin cytoskeleton by molecular genetics. *Micros. Res. Tech.* **47**, 124–134.

Ekelund, F., and Ronn, R. (1994). Notes on protozoa in agricultural soil with emphasis on heterotrophic flagellates and naked amoebae and their ecology. *FEMS Microbiol. Rev.* **15**, 321–353.

Ellison, A. M., and Buss, L. W. (1983). A naturally occurring developmental synergism between the cellular slime mold, *Dictyostelium mucoroides* and the fungus *Mucor hiemalis. Am. J. Bot.* **70**, 298–302.

Ennis, H. L., and Sussman, M. (1958). The initiator cell for slime mold aggregation. *Proc. Natl. Acad. Sci. USA* **44**, 401–411.

Ennis, H. L., Giorda, R., Ohmachi, T., and Shaw, D. R. (1988). Characterization of genes that are developmentally regulated during *Dictyostelium discoideum* spore germination. *Dev. Genet.* **9**, 303–314.

Ennis, H. L., Dao, D. N., Pukatzki, S. U., and Kessin, R. H. (2000). *Dictyostelium* amoebae lacking an F-box protein form spores rather than stalk in chimeras with wild type. *Proc. Natl. Acad. Sci. USA* **97**, 3292–3297.

Erdos, G. W., and West, C. M. (1989). Formation and organization of the spore coat of *Dictyostelium discoideum. Exp. Mycol.* **13**, 169–182.

Erdos, G. W., Nickerson, A. W., and Raper, K. B. (1972). The fine structure of macrocysts in *Polysphondylium violaceum. Cytobiology* **6**, 351–366.

Escalante, R., and Sastre, L. (1998). A serum response factor homolog is required for spore differentiation in *Dictyostelium. Development* **125**, 3801–3808.

Esch, R. K., Howard, P. K., and Firtel, R. A. (1992). Regulation of the *Dictyostelium* cAMP-induced, prestalk-specific DdrasD gene – Identification of cis-acting elements. *Nucleic Acids Res.* **20**, 1325–1332.

Etchebehere, L. C., van Bemmelen, M. X. P., Anjard, C., Traincard, F., Assemat, K., Reymond, C., and Véron, M. (1997). The catalytic subunit of *Dictyostelium* cAMP-dependent protein kinase – Role of the N-terminal domain and of the C-terminal residues in catalytic activity and stability. *Eur. J. Biochem.* **248**, 820–826.

Ettensohn, C. A., and McClay, D. R. (1988). Cell lineage conversion in the sea urchin embryo. *Dev. Biol.* **125**, 396–409.

Euteneuer, U., Gräf, R., Kube-Granderath, E., and Schliwa, M. (1998). *Dictyostelium* γ-tubulin: molecular characterization and ultrastructural localization. *J. Cell Sci.* **111**, 405–412.

Faix, J., and Dittrich, W. (1996). DGAP1, a homologue of rasGTPase activating proteins that controls growth, cytokinesis, and development in *Dictyostelium discoideum*. *FEBS Lett.* **394**, 251–257.

Faix, J., Gerisch, G., and Noegel, A. A. (1990). Constitutive overexpression of the contact site-A glycoprotein enables growth-phase cells of *Dictyostelium discoideum* to aggregate. *EMBO J.* **9**, 2709–2716.

Faix, J., Gerisch, G., and Noegel, A. A. (1992). Overexpression of the csA cell adhesion molecule under its own cAMP-regulated promoter impairs morphogenesis in *Dictyostelium*. *J. Cell Sci.* **102**, 203–214.

Farnsworth, P. A. (1973). Morphogenesis in the cellular slime mould *Dictyostelium discoideum*; the formation and regulation of aggregate tips and the specification of developmental axes. *J. Embryol. Exp. Morphol.* **29**, 253–266.

Farnsworth, P. A., and Loomis, W. F. (1975). A gradient in the thickness of the surface sheath in pseudoplasmodia in *Dictyostelium discoideum*. *Dev. Biol.* **46**, 349–357.

Faure, M., Franke, J., Hall, A. L., Podgorski, G. J., and Kessin, R. H. (1990). The cyclic nucleotide phosphodiesterase gene of *Dictyostelium discoideum* contains 3 promoters specific for growth, aggregation, and late development. *Mol. Cell. Biol.* **10**, 1921–1930.

Favis, R., McCaffery, I., Ehrenkaufer, G., and Rutherford, C. L. (1998). Transcription of the *Dictyostelium* glycogen phosphorylase-2 gene is induced by three large promoter domains. *Dev. Genet.* **23**, 230–246.

Fields, S. D., Conrad, M. N., and Clarke, M. (1998). The *S. cerevisiae CLU1* and *D. discoideum cluA* genes are functional homologues that influence mitochondrial morphology and distribution. *J. Cell Sci.* **111**, 1717–1727.

Filosa, M. F. (1962). Heterocytosis in cellular slime molds. *Am. Naturalist* **XCVI (no. 887)**, 79–92.

Finney, R., Varnum, B., and Soll, D. R. (1979). "Erasure" in *Dictyostelium*: A dedifferentiation involving the programmed loss of chemotactic functions. *Dev. Biol.* **73**, 290–303.

Finney, R., Ellis, M., Langtimm, C., Rosen, E., Firtel, R., and Soll, D. R. (1987). Gene regulation during dedifferentiation in *Dictyostelium discoideum*. *Dev. Biol.* **120**, 561–576.

Firtel, R. A. (1995). Integration of signaling information in controlling cell-fate decisions in *Dictyostelium*. *Genes Dev.* **9**, 1427–1444.

Firtel, R., and Bonner, J. (1972). Characterization of the genome of the cellular slime mold *Dictyostelium discoideum*. *J. Mol. Biol.* **66**, 339–361.

Firtel, R. A., and Chung, C. Y. (2000). The molecular genetics of chemotaxis: sensing and responding to chemoattractant gradients. *BioEssays* **22**, 603–615.

Firtel, R. A., and Jacobson, A. (1977). Structural organization and transcription of the genome of *Dictyostelium discoideum*. In *International review of biochemistry. Biochemistry of cell differentiation II*. (J. Paul, Ed.), pp. 377–429. University Park Press, Baltimore.

Firtel, R. A., Silan, C., Ward, T. E., Howard, P., Metz, B. A., Nellen, W., and Jacobson, A. (1985). Extrachromosomal replication of shuttle vectors in *Dictyostelium discoideum. Mol. Cell. Biol.* **5**, 3241–3250.

Fisher, F. R., Smith, E., and Williams, K. L. (1981). An extracellular chemical signal controlling phototactic behavior by *D. discoideum* slugs. *Cell* **23**, 799–807.

Fisher, P. R. (1990). Pseudopodium activation and inhibition signals in chemotaxis by *Dictyostelium discoideum* amoebae. *Semin. Cell Biol.* **1**, 87–97.

Fisher, P. R. (1997). Genetics of phototaxis in a model eukaryote, *Dictyostelium discoideum. BioEssays* **19**, 397–407.

Fisher, P. R., Dohrmann, U., and Williams, K. L. (1984). Signal processing in *Dictyostelium discoideum* slugs. In *Modern Cell Biology.* (B. H. Satir, Ed.), pp. 197–248. A. R. Liss, New York.

Fisher, P. R., Merkl, R., and Gerisch, G. (1989). Quantitative analysis of cell motility and chemotaxis in *Dictyostelium discoideum* by using an image processing system and a novel chemotaxis chamber providing stationary chemical gradients. *J. Cell Biol.* **108**, 973–984.

Fontana, D. R. (1993). Two distinct adhesion systems are responsible for EDTA-sensitive adhesion in *Dictyostelium discoideum. Differentiation* **53**, 139–147.

Forbes, A., and Lehmann, R. (1998). Nanos and Pumilio have critical roles in the development and function of *Drosophila* germline stem cells. *Development* **125**, 679–690.

Forman, D., and Garrod, D. R. (1977). Pattern formation in *Dictyostelium discoideum.* I. Development of prespore cells and its relationship to the pattern of the fruiting body. *J. Embryol. Exp. Morphol.* **40**, 215–228.

Fosnaugh, K. L., and Loomis, W. F. (1989a). Sequence of the *Dictyostelium discoideum* spore coat gene SP96. *Nucleic Acids Res.* **17**, 9489–9490.

Fosnaugh, K. L., and Loomis, W. F. (1989b). Spore coat genes SP60 and SP70 of *Dictyostelium discoideum. Mol. Cell. Biol.* **9**, 5215–5218.

Fosnaugh, K. L., and Loomis, W. F. (1991). Coordinate regulation of the spore coat genes in *Dictyostelium discoideum. Dev. Genet.* **12**, 123–132.

Fosnaugh, K. L., and Loomis, W. F. (1993). Enhancer regions responsible for temporal and cell-type-specific expression of a spore coat gene in *Dictyostelium. Dev. Biol.* **157**, 38–48.

Fosnaugh, K., Fuller, D., and Loomis, W. F. (1994). Structural roles of the spore coat proteins in *Dictyostelium discoideum. Dev. Biol.* **166**, 823–825.

Francis, D. W. (1964). Some studies on phototaxis of *Dictyostelium. J. Cell. Compar. Physiol.* **64**, 131–138.

Francis, D. (1975). Macrocyst genetics in *Polysphondylium pallidum*, a cellular slime mould. *J. Gen. Microbiol.* **89**, 310–318.

Francis, D. (1998). High frequency recombination during the sexual cycle of *Dictyostelium discoideum. Genetics* **148**, 1829–1832.

Francis, D., and Eisenberg, R. (1993). Genetic structure of a natural population of *Dictyostelium discoideum*, a cellular slime mold. *Mol. Ecol.* **2**, 385–392.

Frank, S. A. (1995). Mutual policing and repression of competition in the evolution of cooperative groups. *Nature* **377**, 520–522.

Franke, J., and Kessin, R. (1977). A defined minimal medium for axenic strains of *Dictyostelium discoideum. Proc. Natl. Acad. Sci. USA* **74**, 2157–2161.

Franke, J., and Kessin, R. H. (1981). The cyclic nucleotide phosphodiesterase inhibitory protein of *Dictyostelium discoideum*. Purification and characterization. *J. Biol. Chem.* **256**, 7628–7637.

Franke, J., and Sussman, M. (1973). Accumulation of UDPG-pyrophosphorylase in *Dictyostelium discoideum* via preferential synthesis. *J. Mol. Biol.* **81**, 173–185.

Franke, J., Faure, M., Wu, L., Hall, A. L., Podgorski, G. J., and Kessin, R. H. (1991). Cyclic nucleotide phosphodiesterase of *Dictyostelium discoideum* and its glycoprotein inhibitor – structure and expression of their genes. *Dev. Genet.* **12**, 104–112.

Freeland, T. M., Guyer, R. B., Ling, A. Z., and Deering, R. A. (1996). Apurinic/apyrimidinic (AP) endonuclease from *Dictyostelium discoideum*: Cloning, nucleotide sequence and induction by sublethal levels of DNA damaging agents. *Nucleic Acids Res.* **24**, 1950–1953.

Freeze, H. H. (1997). Post-translational modification and sorting of lysosomal enzymes in *Dictyostelium*: a perspective. In *Dictyostelium – A model system for cell and developmental biology.* (Y. Maeda, K. Inouye, and I. Takeuchi, Eds.), pp. 93–107. Universal Academy Press, Tokyo, Japan.

Freeze, H., and Loomis, W. (1977). The isolation and characterization of a component of the surface sheath of *Dictyostelium discoideum*. *J. Biol. Chem.* **252**, 820–824.

Freeze, H. H., and Wolgast, D. (1986). Biosynthesis of methylphosphomannosyl residues in the oligosaccharides of *Dictyostelium discoideum* glycoproteins. *J. Biol. Chem.* **261**, 135–141.

Freeze, H. H., Koza-Taylor, P., Saunders, A., and Cardelli, J. A. (1989). The effects of altered N-linked oligosaccharide structures on maturation and targeting of lysosomal enzymes in *Dictyostelium discoideum*. *J. Biol. Chem.* **264**, 19278–19286.

Fuchs, M., Jones, M. K., and Williams, K. L. (1993). Characterization of an epithelium-like layer of cells in the multicellular *Dictyostelium discoideum* slug. *J. Cell Sci.* **105**, 243–253.

Fukuhara, H. (1982). Restriction map and gene organization of the mitochondrial DNA from *Dictyostelium discoideum*. *Biol. Cell* **46**, 321–324.

Fukui, Y. (1993). Toward a new concept of cell motility: cytoskeletal dynamics in amoeboid movement and cell division. *Int. Rev. Cytol.* **144**, 85–127.

Fukui, Y., and Imamoto, N. (1983). Phagoskeleton – characterization of a new cortical cytoskeletal element involved in the phagocytosis of *Dictyostelium*. *Cell Struct. Funct.* **8**, 459.

Fukui, Y., and Inoue, S. (1997). Amoeboid movement anchored by eupodia, new actin-rich knobby feet in *Dictyostelium*. *Cell Motil. Cytoskel.* **36**, 339–354.

Fukui, Y., Yumura, S., and Yumura, T. K. (1987). Agar-overlay immunofluorescence: high-resolution studies of cytoskeletal components and their changes during chemotaxis. *Methods Cell Biol.* **28**, 347–356.

Funamoto, S., and Ochiai, H. (1996). Antisense RNA inactivation of gene expression of a cell-cell adhesion protein (gp64) in the cellular slime mold *Polysphondylium pallidum*. *J. Cell Sci.* **109**, 1009–1016.

Furukawa, R., and Fechheimer, M. (1997). The structure, function, and assembly of actin filament bundles. *Int. Rev. Cytol.* **175**, 29–90.

Furukawa, R., Butz, S., Fleischmann, E., and Fechheimer, M. (1992). The *Dictyostelium discoideum* 30,000-dalton protein contributes to phagocytosis. *Protoplasma* **169**, 18–27.

Futrelle, R. P., Traut, J., and McKee, W. G. (1982). Cell behavior in *Dictyostelium discoideum* preaggregation response to localized cAMP pulses. *J. Cell Biol.* **92**, 807–821.

Gadagkar, R., and Bonner, J. T. (1994). Social insects and social amoebae. *J. Biosci.* **19**, 219–245.

Gao, E. N., Shier, P., and Siu, C. H. (1992). Purification and partial characterization of a cell adhesion molecule (gp150) involved in postaggregation stage cell-cell binding in *Dictyostelium discoideum*. *J. Biol. Chem.* **267**, 9409–9415.

Garside, K., and MacLean, N. (1987). The histones of *Dictyostelium discoideum*. *Experientia* **43**, 147–151.

Gaskell, M. J., Jermyn, K. A., Watts, D. J., Treffry, T., and Williams, J. G. (1992). Immunolocalization and separation of multiple prestalk cell types in *Dictyostelium*. *Differentiation* **51**, 171–176.

Gaskins, C., Clark, A. M., Aubry, L., Segall, J. E., and Firtel, R. A. (1996). The *Dictyostelium* MAP kinase ERK2 regulates multiple, independent developmental pathways. *Genes Dev.* **10**, 118–128.

Gee, K., Russell, F., and Gross, J. D. (1994). Ammonia hypersensitivity of slugger mutants of *D. discoideum*. *J. Cell Sci.* **107**, 701–708.

Geier, A., Horn, J., Dingermann, T., and Winckler, T. (1996). Nuclear protein factor binds specifically to the 3′-regulatory module of the long-interspersed-nuclear-element-like *Dictyostelium* repetitive element. *Eur. J. Biochem.* **241**, 70–76.

George, R. P., Albrecht, R. M., Raper, K. B., Sachs, I. B., and MacKenzie, A. P. (1972). Scanning electron microscopy of spore germination in *Dictyostelium discoideum*. *J. Bacteriol.* **112**, 1383–1386.

Gerald, N., Dai, J. W., Ting-Beall, H. P., and De Lozanne, A. (1998). A role for *Dictyostelium* RacE in cortical tension and cleavage furrow progression. *J. Cell Biol.* **141**, 483–492.

Gerisch, G. (1961). Zellkontaktbildung vegetativer und aggregationsreifer Zellen von *Dictyostelium discoideum*. *Naturwissenschaften* **48**, 436–437.

Gerisch, G. (1968). Cell aggregation and differentiation in *Dictyostelium*. In *Current Topics in Developmental Biology*. (A. A. Moscona, and A. Monroy, Eds.), pp. 157–197. Academic Press, New York.

Gerisch, G. (1980). Univalent antibody fragments as tools for the analysis of cell interactions in *Dictyostelium*. In *Current Topics in Developmental Biology*. (M. Friedlander, Ed.), pp. 243–269. Academic Press, New York.

Gerisch, G., and Hess, B. (1974). Cyclic-AMP-controlled oscillations in suspended *Dictyostelium* cells: Their relation to morphogenetic cell interactions. *Proc. Natl. Acad. Sci. USA* **71**, 2118–2122.

Gerisch, G., and Malchow, D. (1976). Cyclic AMP receptors and the control of cell aggregation in *Dictyostelium*. *Adv. Cyclic Nucleotide Res.* **7**, 49–68.

Gerisch, G., and Wick, U. (1975). Intracellular oscillations and release of cyclic AMP from *Dictyostelium* cells. *Biochem. Biophys. Res. Commun.* **65**, 364–370.

Gerisch, G., Malchow, D., Riedel, V., Müller, E., and Every, M. (1972). Cyclic AMP phosphodiesterase and its inhibitor in slime mould development. *Nature New Biol.* **235**, 90–92.

Gerisch, G., Hülser, D., Malchow, D., and Wick, U. (1975). Cell communication by periodic cyclic-AMP pulses. *Phil. Trans. R. Soc. London. B* **272**, 181–192.

Gerisch, G., Albrecht, R., Heizer, C., Hodgkinson, S., and Maniak, M. (1995). Chemoattractant-controlled accumulation of coronin at the leading edge of *Dictyostelium* cells monitored using green fluorescent protein-coronin fusion protein. *Curr. Biol.* **5**, 1280–1285.

Gezelius, K. (1974). Inorganic polyphosphates and enzymes of polyphosphate metabolism in the cellular slime mold *Dictyostelium discoideum*. *Arch. Microbiol.* **98**, 311–329.

Giglione, C., and Gross, J. D. (1995). Anion effects on vesicle acidification in *Dictyostelium*. *Biochem. Mol. Biol. Int.* **36**, 1057–1065.

Ginsburg, G. T., and Kimmel, A. R. (1997). Autonomous and nonautonomous regulation of axis formation by antagonistic signaling via 7-span cAMP receptors and GSK3 in *Dictyostelium. Genes Dev.* **11**, 2112–2123.

Giorda, R., and Ennis, H. L. (1987). Structure of two developmentally regulated *Dictyostelium discoideum* ubiquitin genes. *Mol. Cell. Biol.* **6**, 2097–2103.

Giorda, R., Ohmachi, T., and Ennis, H. L. (1989). Organization of a gene family developmentally regulated during *Dictyostelium discoideum* spore germination. *J. Mol. Biol.* **205**, 63–69.

Giorda, R., Ohmachi, T., Shaw, D. R., and Ennis, H. L. (1990). A shared internal threonine-glutamic acid-threonine-proline repeat defines a family of *Dictyostelium discoideum* spore germination specific proteins. *Biochemistry* **29**, 7264–7269.

Giri, J. G., and Ennis, H. L. (1977). Protein and RNA synthesis during spore germination in the cellular slime mold *Dictyostelium discoideum. Biochem. Biophys. Res. Commun.* **77**, 282–289.

Giri, J. G., and Ennis, H. L. (1978). Developmental changes in RNA and protein synthesis during germination of *Dictyostelium discoideum* spores. *Dev. Biol.* **67**, 189–201.

Glazer, P. M., and Newell, P. C. (1981). Initiation of aggregation by *Dictyostelium discoideum* in mutant populations lacking pulsatile signaling. *J. Gen. Microbiol.* **125**, 221–232.

Goldbeter, A. (1996). *Biochemical oscillations and cellular rhythms. The molecular bases of periodic and chaotic behaviour.* Cambridge University Press, Cambridge.

Gollop, R., and Kimmel, A. R. (1997). Control of cell-type specific gene expression in *Dictyostelium* by the general transcription factor GBF. *Development* **124**, 3395–3405.

Gomer, R. H., and Ammann, R. R. (1996). A cell-cycle phase-associated cell-type choice mechanism monitors the cell cycle rather than using an independent timer. *Dev. Biol.* **174**, 82–91.

Gomer, R. H., and Firtel, R. A. (1987). Cell-autonomous determination of cell-type choice in *Dictyostelium* development by cell-cycle phase. *Science* **237**, 758–762.

Gomer, R. H., Datta, S., and Firtel, R. A. (1986). Cellular and subcellular distribution of a cAMP-regulated prestalk protein and prespore protein in *Dictyostelium discoideum*: A study on the ontogeny of prestalk and prespore cells. *J. Cell Biol.* **103**, 1999–2015.

Gomer, R. H., Yuen, I. S., and Firtel, R. A. (1991). A secreted 80×10^3 Mr protein mediates sensing of cell density and the onset of development in *Dictyostelium. Development* **112**, 269–278.

Gonzales, C. M., Spencer, T. D., Pendley, S. S., and Welker, D. L. (1999). Dgp1 and Dfp1 are closely related plasmids in the *Dictyostelium* Ddp2 plasmid family. *Plasmid* **41**, 89–96.

Goodloe-Holland, C. M., and Luna, E. J. (1987). Purification and characterization of *Dictyostelium discoideum* plasma membranes. *Methods Cell Biol.* **28**, 103–128.

Gottwald, U., Brokamp, R., Karakesisoglou, I., Schleicher, M., and Noegel, A. A. (1996). Identification of a cyclase-associated protein (CAP) homologue in *Dictyostelium discoideum* and characterization of its interaction with actin. *Mol. Biol. Cell* **7**, 261–272.

Gräf, R., Euteneuer, U., Ueda, M., and Schliwa, M. (1998). Isolation of nucleation-competent centrosomes from *Dictyostelium discoideum. Eur. J. Cell Biol.* **76**, 167–175.

Grant, W. N., and Williams, K. L. (1983). Monoclonal antibody characterization of slime sheath: the extracellular matrix of *Dictyostelium discoideum. EMBO J.* **2**, 935–940.

Green, A. A., and Newell, P. C. (1975). Evidence for the existence of two types of cAMP binding sites in aggregating cells of *Dictyostelium discoideum. Cell* **6**, 129–136.

Gregg, J. H., Hackney, A. L., and Krivanek, J. O. (1954). Nitrogen metabolism of the slime mold *Dictyostelium discoideum* during growth and morphogenesis. *Biol. Bull.* **107**, 226–235.

Gregg, K., Carrin, I., and Cox, E. C. (1996). Positional information and whorl morphogenesis in *Polysphondylium. Dev. Biol.* **180**, 511–518.

Grimson, M. J., Haigler, C. H., and Blanton, R. L. (1996). Cellulase microfibrils, cell motility, and plasma membrane protein organization change in parallel during culmination in *Dictyostelium discoideum. J. Cell Sci.* **109**, 3079–3087.

Gross, J. D. (1994). Developmental decisions in *Dictyostelium discoideum. Microbiol. Rev.* **58**, 330–351.

Gross, J. D., Peacey, M. J., and Trevan, D. J. (1976). Signal emission and signal propagation during early aggregation in *Dictyostelium discoideum. J. Cell Sci.* **22**, 645–656.

Gross, J. D., Peacey, M. J., and Pogge von Strandmann, R. (1988). Plasma membrane proton pump inhibition and stalk cell differentiation in *Dictyostelium discoideum. Differentiation* **38**, 91–98.

Guialis, A., and Deering, R. A. (1976). Repair of DNA in ultraviolet light-sensitive and -resistant *Dictyostelium discoideum* strains. *J. Bacteriol.* **127**, 59–66.

Gustafson, G., Kong, W. Y., and Wright, B. (1973). Analysis of UDPG-pyrophosphorylase synthesis during differentiation in *Dictyostelium discoideum. J. Biol. Chem.* **248**, 5188–5196.

Guyer, R. B., Skantar, A. M., and Deering, R. A. (1985). Acid DNAse activity from *Dictyostelium discoideum. Biochim. Biophys. Acta* **826**, 151–153.

Guyer, R. B., Nonnemaker, J. M., and Deering, R. A. (1986). Uracil-DNA glycosylase activity from *Dictyostelium discoideum. Biochim. Biophys. Acta* **868**, 262–264.

Haberstroh, L., and Firtel, R. A. (1990). A spatial gradient of expresssion of a cAMP-regulated prespore cell type-specific gene in *Dictyostelium. Genes Dev.* **4**, 596–612.

Haberstroh, L., Galindo, J., and Firtel, R. A. (1991). Developmental and spatial regulation of a *Dictyostelium* prespore gene – Cis-acting elements and a cAMP-induced, developmentally regulated DNA binding activity. *Development* **113**, 947–958.

Habura, A., Tikhonenko, I., Chisholm, R. L., and Koonce, M. P. (1999). Interaction mapping of a dynein heavy chain – Identification of dimerization and intermediate-chain binding domains. *J. Biol. Chem.* **274**, 15447–15453.

Hacker, U., Albrecht, R., and Maniak, M. (1997). Fluid-phase uptake by macropinocytosis in *Dictyostelium. J. Cell Sci.* **110**, 105–112.

Häder, D. P., and Burkart, U. (1983). Optical properties of *Dictyostelium discoideum* pseudoplasmodia responsible for phototactic orientation. *Exp. Mycol.* **7**, 1–8.

Hadwiger, J. A., and Firtel, R. A. (1992). Analysis of Gα4, a G-protein subunit required for multicellular development in *Dictyostelium. Genes Dev.* **6**, 38–49.

Hadwiger, J. A., Lee, S., and Firtel, R. A. (1994). The Gα subunit Gα4 couples to pterin receptors and identifies a signaling pathway that is essential for multicellular development in *Dictyostelium. Proc. Natl. Acad. Sci. USA* **91**, 10566–10570.

Hadwiger, J. A., Natarajan, K., and Firtel, R. A. (1996). Mutations in the *Dictyostelium* heterotrimeric G protein α subunit Gα5 alter the kinetics of tip morphogenesis. *Development* **122**, 1215–1224.

Hagiwara, H. (1989). *The taxonomic study of Japanese Dictyostelid cellular slime molds.* National Science Museum, Tokyo, Japan.

Haigler, C. H., and Blanton, R. L. (1996). New hope for old dreams: evidence that plant cellulose synthase genes have finally been identified. *Proc. Natl. Acad. Sci. USA* **93**, 12082–12085.

Hall, A. L., Schlein, A., and Condeelis, J. (1988). Relationship of pseudopod extension to chemotactic hormone-induced actin polymerization in amoeboid cells. *J. Cell. Biochem.* **37**, 285–299.

Hall, A. L., Franke, J., Faure, M., and Kessin, R. H. (1993). The role of the cyclic nucleotide phosphodiesterase of *Dictyostelium discoideum* during growth, aggregation, and morphogenesis: overexpression and localization studies with the separate promoters of the pde. *Dev. Biol.* **157**, 73–84.

Hamilton, W. D. (1964). The genetical evolution of social behaviour. *J. Theor. Biol.* **7**, 1–16.

Han, Z., and Firtel, R. A. (1998). The homeobox-containing gene *Wariai* regulates anterior-posterior patterning and cell-type homeostasis in *Dictyostelium*. *Development* **125**, 313–325.

Hanakam, F., Gerisch, G., Lotz, S., Alt, T., and Seelig, A. (1996). Binding of hisactophilin I and II to lipid membranes is controlled by a pH-dependent myristoyl-histidine switch. *Biochemistry* **35**, 11036–11044.

Harloff, C., Gerisch, G., and Noegel, A. A. (1989). Selective elimination of the contact site A protein of *Dictyostelium discoideum* by gene disruption. *Genes Dev.* **3**, 2011–2019.

Harper, R. A. (1926). Morphogenesis in *Dictyostelium*. *Bull. Torrey Bot. Club* **53**, 229–268.

Harper, R. A. (1929). Morphogenesis in *Polysphondylium*. *Bull. Torrey Bot. Club* **56**, 227–258.

Harper, R. A. (1932). Organization and light relations in *Polysphondylium*. *Bull. Torrey Bot. Club* **59**, 49–84.

Harwood, A. J., Early, A. E., Jermyn, K. A., and Williams, J. (1991). Unexpected localization of cells expressing a prespore marker of *Dictyostelium discoideum*. *Differentiation* **46**, 7–13.

Harwood, A. J., Hopper, N. A., Simon, M. N., Bouzid, S., Véron, M., and Williams, J. G. (1992a). Multiple roles for cAMP-dependent protein kinase during *Dictyostelium* development. *Dev. Biol.* **149**, 90–99.

Harwood, A. J., Hopper, N. A., Simon, M. N., Driscoll, D. M., Véron, M., and Williams, J. G. (1992b). Culmination in *Dictyostelium* is regulated by the cAMP-dependent protein kinase. *Cell* **69**, 615–624.

Harwood, A. J., Early, A., and Williams, J. G. (1993). A repressor controls the timing and spatial localization of stalk cell-specific gene expression in *Dictyostelium*. *Development* **118**, 1041–1048.

Harwood, A. J., Plyte, S. E., Woodgett, J., Strutt, H., and Kay, R. R. (1995). Glycogen synthase kinase 3 regulates cell fate in *Dictyostelium*. *Cell* **80**, 139–148.

Haser, H., and Häder, D. P. (1992). Orientation and phototaxis in pseudoplasmodia of an axenic strain of the cellular slime mold, *Dictyostelium discoideum*. *Exp. Mycol.* **16**, 119–131.

Haugwitz, M., Noegel, A. A., Rieger, D., Lottspeich, F., and Schleicher, M. (1991). *Dictyostelium discoideum* contains two profilin isoforms that differ in structure and function. *J. Cell Sci.* **100**, 481–489.

Haugwitz, M., Noegel, A. A., Karakesisoglou, J., and Schleicher, M. (1994). *Dictyostelium* amoebae that lack G-actin-sequestering profilins show defects in F-actin content, cytokinesis, and development. *Cell* **79**, 303–314.

Haus, U., Hartmann, H., Trommler, P., Noegel, A. A., and Schleicher, M. (1991). F-actin capping by Cap32/34 requires heterodimeric conformation and can be inhibited with PIP$_2$. *Biochem. Biophys. Res. Commun.* **181**, 833–839.

Haus, U., Trommler, P., Fisher, P. R., Hartmann, H., Lottspeich, F., Noegel, A. A., and Schleicher, M. (1993). The heat shock cognate protein from *Dictyostelium* affects actin polymerization through interaction with the actin-binding protein cap32/34. *EMBO J.* **12**, 3763–3771.

Hauser, L. J., Dhar, M. S., and Olins, D. E. (1995). *Dictyostelium discoideum* contains a single-copy gene encoding a unique subtype of histone H1. *Gene* **154**, 119–122.

Hayashi, H., and Suga, T. (1978). Some characteristics of peroxisomes in the slime mold, *Dictyostelium discoideum*. *J. Biochem.* **84**, 513–521.

Heads, R. J., Carpenter, B. G., Rickenberg, H. V., and Chambers, T. C. (1992). The lysine-rich H1-histones from the slime moulds, *Physarum polycephalum* and *Dictyostelium discoideum*, lack phosphorylation sites recognised by cyclic AMP-dependent protein kinase *in vitro*. *FEBS Lett.* **306**, 66–70.

Hemmes, D. E., Kojima-Buddenhagen, E. S., and Hohl, H. R. (1972). Structural and enzymatic analysis of the spore wall layer in *Dictyostelium discoideum*. *J. Ultrastruct. Res.* **41**, 406–417.

Henderson, E. J. (1975). The cyclic adenosine 3′,5′-monophosphate receptor of *Dictyostelium discoideum*. Binding characteristics of aggregation competent cells and variation of binding levels during the life cycle. *J. Biol. Chem.* **250**, 4730–4736.

Hereld, D., and Devreotes, P. N. (1992). The cAMP receptor family of *Dictyostelium*. *Int. Rev. Cytol.* **137B**, 35–47.

Hereld, D., Vaughan, R., Kim, J. Y., Borleis, J., and Devreotes, P. (1994). Localization of ligand-induced phosphorylation sites to serine clusters in the C-terminal domain of the *Dictyostelium* cAMP receptor, cAR1. *J. Biol. Chem.* **269**, 7036–7044.

Heuser, J., Zhu, Q. L., and Clarke, M. (1993). Proton pumps populate the contractile vacuoles of *Dictyostelium* amoebae. *J. Cell Biol.* **121**, 1311–1327.

Higuchi, I., Kanemura, Y., Shimizu, H., and Urushihara, H. (1995). Self- and non-self-recognition in bisexual mating of *Dictyostelium discoideum*. *Devel. Growth Differ.* **37**, 311–317.

Hildebrandt, M., and Nellen, W. (1992). Differential antisense transcription from the *Dictyostelium* EB4 gene locus – Implications on antisense-mediated regulation of messenger RNA stability. *Cell* **69**, 197–204.

Hildebrandt, M., Humbel, B. M., and Nellen, W. (1991). The *Dictyostelium discoideum* EB4 gene product and a truncated mutant form of the protein are localized in prespore vesicles but absent from mature spores. *Dev. Biol.* **144**, 212–214.

Hirth, K. P., Edwards, C. A., and Firtel, R. A. (1982). A DNA-mediated transformation system for *Dictyostelium discoideum*. *Proc. Natl. Acad. Sci. USA* **79**, 7356–7360.

Hitt, A. L., Hartwig, J. H., and Luna, E. J. (1994). Ponticulin is the major high affinity link between the plasma membrane and the cortical actin network in *Dictyostelium*. *J. Cell Biol.* **126**, 1433–1444.

Hjorth, A. L., Khanna, N. C., and Firtel, R. A. (1989). A trans-acting factor required for cAMP-induced gene expression in *Dictyostelium* is regulated developmentally and induced by cAMP. *Genes Dev.* **3**, 747–759.

Hjorth, A. L., Pears, C., Williams, J. G., and Firtel, R. A. (1990). A developmentally regulated trans-acting factor recognizes dissimilar G/C-rich elements controlling a class of cAMP-inducible *Dictyostelium* genes. *Genes Dev.* **4**, 419–432.

Hofmann, A., Noegel, A. A., Bomblies, L., Lottspeich, F., and Schleicher, M. (1993). The 100 kDa F-actin capping protein of *Dictyostelium* amoebae is a villin prototype (protovillin). *FEBS Lett.* **328**, 71–76.

Hohl, H. R., and Hamamoto, S. T. (1969). Ultrastructure of spore differentiation in *Dictyostelium*: The prespore vacuole. *J. Ultrastruct. Res.* **26**, 442–453.

Hohl, H. R., and Jehli, J. (1973). The presence of cellulose microfibrils in the proteinaceous slime track of *Dictyostelium discoideum*. *Arch. Mikrobiol.* **92**, 179–187.

Horn, F., and Gross, J. (1996). A role for calcineurin in *Dictyostelium discoideum* development. *Differentiation* **60**, 269–275.

Horvitz, H. R., and Sternberg, P. W. (1991). Multiple intercellular signaling systems control the development of the *Caenorhabditis elegans* vulva. *Nature* **351**, 535–541.

Hughes, J. E., and Welker, D. L. (1989). Copy number control and compatibility of nuclear plasmids in *Dictyostelium discoideum*. *Plasmid* **22**, 215–223.

Hughes, J. E., Ashktorab, H., and Welker, D. L. (1988). Nuclear plasmids in the *Dictyostelium* slime molds. *Dev. Genet.* **9**, 495–504.

Hughes, J. E., Kiyosawa, H., and Welker, D. L. (1994). Plasmid maintenance functions encoded on *Dictyostelium discoideum* nuclear plasmid Ddp1. *Mol. Cell. Biol.* **14**, 6117–6124.

Huss, M. J. (1989). Dispersal of cellular slime moulds by two soil invertebrates. *Mycologia* **81**, 677–682.

Ihara, M., Taya, Y., and Nishimura, S. (1980). Developmental regulation of cytokinin, spore germination inhibitor discadenine and related enzymes in *Dictyostelium discoideum*. *Exp. Cell Res.* **126**, 273–278.

Ihara, M., Tanaka, Y., Yanagisawa, K., Taya, Y., and Nishimura, S. (1986). Purification and some properties of discadenine synthase from *Dictyostelium discoideum*. *Biochim. Biophys. Acta* **881**, 135–140.

Iijima, N., Takagi, T., and Maeda, Y. (1995). A proteinous factor mediating intercellular communication during the transition of *Dictyostelium* cells from growth to differentiation. *Zool. Sci.* **12**, 61–69.

Inouye, K. (1989). Control of cell type proportions by a secreted factor in *Dictyostelium discoideum*. *Development* **107**, 605–610.

Inouye, K. (1992). Patterning in the cellular slime moulds. *J. Biosci.* **17**, 115–128.

Inouye, K., and Gross, J. (1993). *In vitro* stalk cell differentiation in wild-type and slugger mutants of *Dictyostelium discoideum*. *Development* **118**, 523–526.

Inouye, K., and Takeuchi, I. (1982). Correlations between prestalk-prespore tendencies and cAMP-related activities in *Dictyostelium discoideum*. *Exp. Cell Res.* **138**, 311–318.

Insall, R. (1995). Glycogen synthase kinase and *Dictyostelium* development: Old pathways pointing in new directions? *Trends Genet.* **11**, 37–39.

Insall, R. H. (1996). Osmoregulation: Cyclic GMP and the big squeeze. *Curr. Biol.* **6**, 516–518.

Insall, R., and Kay, R. R. (1990). A specific DIF binding protein in *Dictyostelium*. *EMBO J.* **9**, 3323–3328.

Insall, R., Nayler, O., and Kay, R. R. (1992). DIF-1 induces its own breakdown in *Dictyostelium. EMBO J.* **11**, 2849–2854.

Insall, R., Kuspa, A., Lilly, P. J., Shaulsky, G., Levin, L. R., Loomis, W. F., and Devreotes, P. (1994). CRAC, a cytosolic protein containing a pleckstrin homology domain, is required for receptor and G protein-mediated activation of adenylyl cyclase in *Dictyostelium. J. Cell Biol.* **126**, 1537–1545.

Insall, R. H., Borleis, J., and Devreotes, P. N. (1996). The aimless RasGEF is required for processing of chemotactic signals through G-protein-coupled receptors in *Dictyostelium. Curr. Biol.* **6**, 719–729.

Ivatt, R. L., Das, O. P., Henderson, E. J., and Robbins, P. W. (1984). Glycoprotein biosynthesis in *Dictyostelium discoideum*: developmental regulation of the protein-linked glycans. *Cell* **38**, 561–567.

Jacquet, M., Part, D., and Felenbok, B. (1981). Changes in the polyadenylated mRNA population during development of *Dictyostelium discoideum. Dev. Biol.* **81**, 155–166.

Jacquet, M., Guilbaud, R., and Garreau, H. (1988). Sequence analysis of the DdPYR5-6 gene coding for UMP synthase in *Dictyostelium discoideum* and comparison with orotate phosphoribosyl transferases and OMP decarboxylases. *Mol. Gen. Genet.* **211**, 441–445.

Jaffe, L. F. (1997). The roles of calcium in pattern formation. In *Dictyostelium – A model system for cell and developmental biology.* (Y. Maeda, K. Inouye, and I. Takeuchi, Eds.), pp. 267—277. Universal Academy Press, Tokyo, Japan.

Jain, R., Yuen, I. S., Taphouse, C. R., Gomer, R. H. (1992). A density-sensing factor controls development in *Dictyostelium. Genes Dev.* **6**, 390–400.

Jain, R., Brazill, D. T., Cardelli, J. A., Bush, J., and Gomer, R. H. (1997). Autocrine factors controlling early development. In *Dictyostelium – A model system for cell and developmental biology.* (Y. Maeda, K. Inouye, and I. Takeuchi, Eds.), pp. 219–234. Universal Academy Press, Tokyo, Japan.

Jensen, C. G., Bollard, S. M., and Roos, U. P. (1991). Analysis of microtubule arms and bridges in mitotic and interphase amebae of *Dictyostelium discoideum. Eur. J. Cell Biol.* **54**, 121–131.

Jentsch, S., and Ulrich, H. D. (1998). Protein breakdown. Ubiquitous déjà vu [news and views]. *Nature* **395**, 321, 323.

Jermyn, K. A., and Williams, J. G. (1991). An analysis of culmination in *Dictyostelium* using prestalk and stalk-specific cell autonomous markers. *Development* **111**, 779–787.

Jermyn, K., and Williams, J. (1995). Comparison of the *Dictyostelium rasD* and *ecmA* genes reveals two distinct mechanisms whereby an mRNA may become enriched in prestalk cells. *Differentiation* **58**, 261–267.

Jermyn, K. A., Berks, M., Kay, R. R., and Williams, J. G. (1987). Two distinct classes of prestalk-enriched mRNA sequences in *Dictyostelium discoideum. Development* **100**, 745–755.

Jermyn, K. A., Duffy, K. T., and Williams, J. G. (1989). A new anatomy of the prestalk zone in *Dictyostelium. Nature* **340**, 144–146.

Jermyn, K., Traynor, D., and Williams, J. (1996). The initiation of basal disc formation in *Dictyostelium discoideum* is an early event in culmination. *Development* **122**, 753–760.

Jin, T., Soede, R. D. M., Liu, J. C., Kimmel, A. R., Devreotes, P. N., and Schaap, P. (1998). Temperature-sensitive Gβ mutants discriminate between G protein-dependent and -independent signaling mediated by serpentine receptors. *EMBO J.* **17**, 5076–5084.

Johnson, R. L., Vaughan, R. A., Caterina, M. J., van Haastert, P. J. M., and Devreotes, P. N. (1991). Overexpression of the cAMP receptor 1 in growing *Dictyostelium* cells. *Biochemistry* **30**, 6982–6986.

Johnson, R. L., van Haastert, P. J. M., Kimmel, A. R., Saxe III, C. L., Jastorff, B., and Devreotes, P. N. (1992). The cyclic nucleotide specificity of three cAMP receptors in *Dictyostelium*. *J. Biol. Chem.* **267**, 4600–4607.

Johnson, J. D., Edman, J. C., and Rutter, W. J. (1993a). A receptor tyrosine kinase found in breast carcinoma cells has an extracellular discoidin I-like domain. *Proc. Natl. Acad. Sci. USA* **90**, 5677–5681.

Johnson, R. L., Saxe III, C. L., Gollop, R., Kimmel, A. R., and Devreotes, P. N. (1993b). Identification and targeted gene disruption of cAR3, a cAMP receptor subtype expressed during multicellular stages of *Dictyostelium* development. *Genes Dev.* **7**, 273–282.

Juliani, M. H., and Klein, C. (1981). Photoaffinity labeling of the cell surface adenosine 3′,5′-monophosphate receptor of *Dictyostelium discoideum* and its modification in down-regulated cells. *J. Biol. Chem.* **256**, 613–619.

Jung, G., Fukui, Y., Martin, B., and Hammer III, J. A. (1993). Sequence, expression pattern, intracellular localization, and targeted disruption of the *Dictyostelium* myosin ID heavy chain isoform. *J. Biol. Chem.* **268**, 14981–14990.

Jung, G., Wu, X. F., and Hammer, J. A. (1996). *Dictyostelium* mutants lacking multiple classic myosin I isoforms reveal combinations of shared and distinct functions. *J. Cell Biol.* **133**, 305–323.

Kaiser, D. (1986). Control of multicellular development: *Dictyostelium* and *Myxococcus*. *Annu. Rev. Genet.* **20**, 539–566.

Kaiser, D. (1993). Roland Thaxter's legacy and the origins of multicellular development. *Genetics* **135**, 249–254.

Kakutani, T., and Takeuchi, I. (1986). Characterization of anterior-like cells in *Dictyostelium* as analyzed by their movement. *Dev. Biol.* **115**, 439–445.

Kalt, A., and Schliwa, M. (1996). A novel structural component of the *Dictyostelium* centrosome. *J. Cell Sci.* **109**, 3103–3112.

Kamboj, R. K., Gariepy, J., and Siu, C. H. (1989). Identification of an octapeptide involved in homophilic interaction of the cell adhesion molecule gp80 of *Dictyostelium discoideum*. *Cell* **59**, 615–625.

Karakesisoglou, I., Janssen, K. P., Eichinger, L., Noegel, A. A., and Schleicher, M. (1999). Identification of a suppressor of the *Dictyostelium* profilin-minus phenotype as a CD36/LIMP-II homologue. *J. Cell Biol.* **145**, 167–181.

Kasbekar, D. P., Madigan, S., and Katz, E. R. (1983). Use of nystatin-resistant mutations in parasexual genetic analysis in *Dictyostelium discoideum*. *Genetics* **104**, 271–277.

Kasbekar, D. P., Madigan, S., Campbell, E., and Katz, E. R. (1988). Rapid identification of non-allelic nystatin resistance mutations in *Dictyostelium discoideum*. *J. Genet.* **67**, 23–28.

Katz, E. R., and Kao, V. (1974). Evidence for mitotic recombination in the cellular slime mold *Dictyostelium discoideum*. *Proc. Natl. Acad. Sci. USA* **71**, 4025–4026.

Katz, E. R., and Sussman, M. (1972). Parasexual recombination in *Dictyostelium discoideum*: selection of stable diploid heterozygotes and stable haploid segregants. *Proc. Natl. Acad. Sci. USA* **69**, 495–498.

Kawata, T., Early, A., and Williams, J. (1996). Evidence that a combined activator-repressor protein regulates *Dictyostelium* stalk cell differentiation. *EMBO J.* **15**, 3085–3092.

Kawata, T., Shevchenko, A., Fukuzawa, M., Jermyn, K. A., Totty, N. F., Zhukovskaya, N. V., Sterling, A. E., Mann, M., and Williams, J. G. (1997). SH2 signaling in a lower eukaryote: a STAT protein that regulates stalk cell differentiation in *Dictyostelium*. *Cell* **89**, 909–916.

Kay, R. R. (1979). Gene expression in *Dictyostelium discoideum*: mutually antagonistic roles of cyclic AMP and ammonia. *J. Embryol. Exp. Morphol.* **52**, 171–182.

Kay, R. R. (1987). Cell differentiation in monolayers and the investigation of slime mold morphogens. *Methods Cell Biol.* **28**, 433–448.

Kay, R. R. (1989). Evidence that elevated intracellular cyclic AMP triggers spore maturation in *Dictyostelium*. *Development* **105**, 753–759.

Kay, R. R. (1992). Cell differentiation and patterning in *Dictyostelium*. *Curr. Opin. Cell Biol.* **4**, 934–938.

Kay, R. R. (1997). DIF signalling. In *Dictyostelium – A model system for cell and developmental biology*. (Y. Maeda, K. Inouye, and I. Takeuchi, Eds.), pp. 279–292. Universal Academy Press, Tokyo, Japan.

Kay, R. R. (1998). The biosynthesis of differentiation-inducing factor, a chlorinated signal molecule regulating *Dictyostelium* development. *J. Biol. Chem.* **273**, 2669–2675.

Kay, R. R., and Jermyn, K. A. (1983). A possible morphogen controlling differentiation in *Dictyostelium*. *Nature* **303**, 242–244.

Kay, R. R., Dhokia, B., and Jermyn, K. A. (1983). Purification of stalk-cell-inducing morphogens from *Dictyostelium discoideum*. *Eur. J. Biochem.* **136**, 51–56.

Kay, R. R., Berks, M., Traynor, D., Taylor, G. W., Masento, M. S., and Morris, H. R. (1988). Signals controlling cell differentiation and pattern formation in *Dictyostelium*. *Dev. Genet.* **9**, 579–587.

Kay, R. R., Taylor, G. W., Jermyn, K. A., and Traynor, D. (1992). Chlorine-containing compounds produced during *Dictyostelium* development – Detection by labeling with Cl-36. *Biochem. J.* **281**, 155–161.

Kay, R. R., Large, S., Traynor, D., and Nayler, O. (1993). A localized differentiation-inducing-factor sink in the front of the *Dictyostelium* slug. *Proc. Natl. Acad. Sci. USA* **90**, 487–491.

Kelly, L. J., Kelly, R., and Ennis, H. L. (1983). Characterization of cDNA clones specific for sequences developmentally regulated during *Dictyostelium discoideum* spore germination. *Mol. Cell. Biol.* **3**, 1943–1948.

Kelly, R., Kelly, L. J., and Ennis, H. L. (1985). *Dictyostelium discoideum* mRNAs developmentally regulated during spore germination have short half-lives. *Mol. Cell. Biol.* **5**, 133–139.

Ken, R., and Singleton, C. K. (1994). Redundant regulatory elements account for the developmental control of a ribosomal protein gene of *Dictyostelium discoideum*. *Differentiation* **55**, 97–103.

Kesbeke, F., Snaar-Jagalska, B. E., and van Haastert, P. J. M. (1988). Signal transduction in *Dictyostelium fgdA* mutants with a defective interaction between surface cAMP receptors and a GTP-binding regulatory protein. *J. Cell Biol.* **107**, 521–528.

Kesbeke, F., van Haastert, P. J. M., De Wit, R. J. W., and Snaar-Jagalska, B. E. (1990). Chemotaxis to cyclic AMP and folic acid is mediated by different G-proteins in *Dictyostelium discoideum*. *J. Cell Sci.* **96**, 669–673.

Kessin, R. H. (1973). RNA metabolism during vegetative growth and morphogenesis of the cellular slime mold *Dictyostelium discoideum*. *Dev. Biol.* **31**, 242–251.

Kessin, R. (1988). Genetics of early *Dictyostelium discoideum* development. *Microbiol. Rev.* **52**, 29–49.

Kessin, R. H., and van Lookeren Campagne, M. M. (1992). The development of a social amoeba. *Am. Scientist* **80**, 556–565.

Kessin, R. H., Williams, K. L., and Newell, P. C. (1974). Linkage analysis in *Dictyostelium discoideum* using temperature-sensitive growth mutants selected with bromodeoxyuridine. *J. Bacteriol.* **119**, 776–783.

Kessin, R. H., Orlow, S. J., Shapiro, R. I., and Franke, J. (1979). Binding of inhibitor alters kinetic and physical properties of extracellular cyclic AMP phosphodiesterase from *Dictyostelium discoideum*. *Proc. Natl. Acad. Sci. USA* **76**, 5450–5454.

Kessin, R. H., Gundersen, G. G., Zaydfudim, V., Grimson, M., and Blanton, R. L. (1996). How cellular slime molds evade nematodes. *Proc. Natl. Acad. Sci. USA* **93**, 4857–4861.

Killick, K. A. (1979). Trehalose-6-phosphate synthase from *Dictyostelium discoideum*: partial purification and characterization of the enzyme from young sorocarps. *Arch. Biochem. Biophys.* **196**, 121–133.

Killick, K. A. (1983). Trehalase from the cellular slime mold *Dictyostelium discoideum*: purification and characterization of the homogeneous enzyme from myxamoebae. *Arch. Biochem. Biophys.* **222**, 561–573.

Kim, H. J., Chang, W. T., Meima, M., Gross, J. D., and Schaap, P. (1998). A novel adenylyl cyclase detected in rapidly developing mutants of *Dictyostelium*. *J. Biol. Chem.* **273**, 30859–30862.

Kim, J. Y., and Devreotes, P. N. (1994). Random chimeragenesis of G-protein-coupled receptors – Mapping the affinity of the cAMP chemoattractant receptors in *Dictyostelium*. *J. Biol. Chem.* **269**, 28724–28731.

Kim, J. Y., van Haastert, P., and Devreotes, P. N. (1996). Social senses: G-protein-coupled receptor signaling pathways in *Dictyostelium discoideum*. *Chem. Biol.* **3**, 239–243.

Kim, J. Y., Caterina, M. J., Milne, J. L. S., Lin, K. C., Borleis, J. A., and Devreotes, P. N. (1997a). Random mutagenesis of the cAMP chemoattractant receptor, cAR1, of *Dictyostelium* – Mutant classes that cause discrete shifts in agonist affinity and lock the receptor in a novel activational intermediate. *J. Biol. Chem.* **272**, 2060–2068.

Kim, J. Y., Soede, R. D. M., Schaap, P., Valkema, R., Borleis, J. A., van Haastert, P. J. M., Devreotes, P. N., and Hereld, D. (1997b). Phosphorylation of chemoattractant receptors is not essential for chemotaxis or termination of G-protein-mediated responses. *J. Biol. Chem.* **272**, 27313–27318.

Kimble, M., Khodjakov, A. L., and Kuriyama, R. (1992). Identification of ubiquitous high-molecular-mass, heat-stable microtubule-associated proteins (MAPs) that are related to the *Drosophila* 205-kDa MAP but are not related to the mammalian MAP-4. *Proc. Natl. Acad. Sci. USA* **89**, 7693–7697.

Kimmel, A. R., and Firtel, R. A. (1982). The organization and expression of the *Dictyostelium* genome. In *The development of Dictyostelium discoideum* (W. F. Loomis, Ed.), pp. 234–324. Academic Press, New York.

Kimmel, A. R., and Firtel, R. A. (1985). Sequence organization and developmental expression of an interspersed, repetitive element and associated single-copy DNA sequences in *Dictyostelium discoideum*. *Mol. Cell. Biol.* **5**, 2123–2130.

Kiyosawa, H., Hughes, J. E., Podgorski, G. J., and Welker, D. L. (1993). Small circular plasmids of the eukaryote *Dictyostelium purpureum* define two novel plasmid families. *Plasmid* **30**, 106–118.

Kiyosawa, H., Hughes, J. E., and Welker, D. L. (1994). Compatible *Dictyostelium mucoroides* nuclear plasmids dmp1 and dmp2 both belong to the ddp1 plasmid family. *Plasmid* **31**, 121–130.

Kiyosawa, H., Hughes, J. E., and Welker, D. L. (1995). The replication origin position and its relationship to a negative trans-acting transcription regulator encoded by *Dictyostelium discoideum* nuclear plasmid Ddp1. *Curr. Genet.* **27**, 479–485.

Klaus, M., and George, R. P. (1974). Microdissection of the developmental stages of the cellular slime mold, *Dictyostelium discoideum*, using a ruby laser. *Dev. Biol.* **39**, 183–188.

Klein, P., Vaughan, R., Borleis, J., and Devreotes, P. (1987). The surface cyclic AMP receptor in *Dictyostelium* – Levels of ligand-induced phosphorylation, solubilization, identification of primary transcript, and developmental regulation of expression. *J. Biol. Chem.* **262**, 358–364.

Klein, P. S., Sun, T. J., Saxe III, C. L., Kimmel, A. R., Johnson, R. L., and Devreotes, P. N. (1988). A chemoattractant receptor controls development in *Dictyostelium discoideum*. *Science* **241**, 1467–1472.

Knecht, D. A., and Loomis, W. F. (1987). Antisense RNA inactivation of myosin heavy chain gene expression in *Dictyostelium discoideum*. *Science* **236**, 1081–1085.

Knecht, D. A., Fuller, D. L., and Loomis, W. F. (1987). Surface glycoprotein, gp24, involved in early adhesion of *Dictyostelium discoideum*. *Dev. Biol.* **121**, 277–283.

Komori, K., Kuroe, K., Yanagisawa, K., and Tanaka, Y. (1997a). Cloning and characterization of the gene encoding a mitochondrially localized DNA topoisomerase II in *Dictyostelium discoideum* – Western blot analysis. *Biochim. Biophys. Acta* **1352**, 63–72.

Komori, K., Maruo, F., Morio, T., Urushihara, H., and Tanaka, Y. (1997b). Localization of a DNA topoisomerase-II to mitochondria in *Dictyostelium discoideum* – Deletion mutant analysis and mitochondrial targeting signal presequence. *J. Plant Res.* **110**, 65–75.

Kon, T., Adachi, H., and Sutoh, K. (2000). *amiB*, a novel gene required for the growth/ differentiation transition in *Dictyostelium*. *Genes to Cells* **5**, 43–55.

Konijn, T. M. (1968). Chemotaxis in the cellular slime molds. 2. The effect of density. *Biol. Bull.* **134**, 298–304.

Konijn, T. M., van de Meene, J. G. C., Bonner, J. T., and Barkley, D. S. (1967). The acrasin activity of adenosine-3′,5′-cyclic phosphate. *Proc. Natl. Acad. Sci. USA* **58**, 1152–1154.

Konijn, T. M., Barkley, D. S., Chang, Y. Y., and Bonner, J. T. (1968). Cyclic AMP: a naturally occurring acrasin in the cellular slime molds. *Am. Naturalist* **102**, 225–233.

Koonce, M. P., and Samso, M. (1996). Overexpression of cytoplasmic dynein's globular head causes a collapse of the interphase microtubule network in *Dictyostelium*. *Mol. Biol. Cell* **7**, 935–948.

Koonce, M. P., and Knecht, D. A. (1998). Cytoplasmic dynein heavy chain is an essential gene product in *Dictyostelium*. *Cell Motil. Cytoskel.* **39**, 63–72.

Koonce, M. P., Grissom, P. M., Lyon, M., Pope, T., and McIntosh, J. R. (1994). Molecular characterization of a cytoplasmic dynein from *Dictyostelium*. *J. Euk. Microbiol.* **41**, 645–651.

Kopachik, W., Oochata, W., Dhokia, B., Brookman, J. J., and Kay, R. R. (1983). *Dictyostelium* mutants lacking DIF, a putative morphogen. *Cell* **33**, 397–403.

Kopachik, W. J., Dhokia, B., and Kay, R. R. (1985). Selective induction of stalk-cell-specific proteins in *Dictyostelium*. *Differentiation* **28**, 209–216.

Kraft, B., Chandrasekhar, A., Rotman, M., Klein, C., and Soll, D. R. (1989). *Dictyostelium* erasure mutant Hi4 abnormally retains development-specific mRNAs during dedifferentiation. *Dev. Biol.* **136**, 363–371.

Kreitmeier, M., Gerisch, G., Heizer, C., and Müller-Taubenberger, A. (1995). A talin homologue of *Dictyostelium* rapidly assembles at the leading edge of cells in response to chemoattractant. *J. Cell Biol.* **129**, 179–188.

Kubohara, Y., Maeda, M., and Okamoto, K. (1993). Analysis of the maturation process of prestalk cells in *Dictyostelium discoideum*. *Exp. Cell Res.* **207**, 107–114.

Kuma, K., Nikoh, N., Iwabe, N., and Miyata, T. (1995). Phylogenetic position of *Dictyostelium* inferred from multiple protein data sets. *J. Mol. Evol.* **41**, 238–246.

Kumagai, A., Hadwiger, J. A., Pupillo, M., and Firtel, R. A. (1991). Molecular genetic analysis of two Gα protein subunits in *Dictyostelium*. *J. Biol. Chem.* **266**, 1220–1228.

Kuspa, A., and Loomis, W. F. (1992). Tagging developmental genes in *Dictyostelium* by restriction enzyme-mediated integration of plasmid DNA. *Proc. Natl. Acad. Sci. USA* **89**, 8803–8807.

Kuspa, A., and Loomis, W. F. (1994a). REMI-RFLP mapping in the *Dictyostelium* genome. *Genetics* **138**, 665–674.

Kuspa, A., and Loomis, W. F. (1994b). Transformation of *Dictyostelium* – Gene disruptions, insertional mutagenesis, and promoter traps. *Methods Mol. Genet.* **3**, 3–21.

Kuspa, A., and Loomis, W. F. (1996). Ordered yeast artificial chromosome clones representing the *Dictyostelium discoideum* genome. *Proc. Natl. Acad. Sci. USA* **93**, 5562–5566.

Kuspa, A., Maghakian, D., Bergesch, P., and Loomis, W. F. (1992). Physical mapping of genes to specific chromosomes in *Dictyostelium discoideum*. *Genomics* **13**, 49–61.

Kuwayama, H., and van Haastert, P. J. M. (1996). Regulation of guanylyl cyclase by a cGMP-binding protein during chemotaxis in *Dictyostelium discoideum*. *J. Biol. Chem.* **271**, 23718–23724.

Kuwayama, H., and van Haastert, P. J. M. (1998a). Chemotactic and osmotic signals share a cGMP transduction pathway in *Dictyostelium discoideum*. *FEBS Lett.* **424**, 248–252.

Kuwayama, H., and van Haastert, P. J. M. (1998b). Chemotactic and osmotic signals share a cGMP transduction pathway in *Dictyostelium discoideum*. *FEBS Lett.* **424**, 248–252.

Kuwayama, H., Ishida, S., and van Haastert, P. J. M. (1993). Non-chemotactic *Dictyostelium discoideum* mutants with altered cGMP signal transduction. *J. Cell Biol.* **123**, 1453–1462.

Kuwayama, H., Viel, G. T., Ishida, S., and van Haastert, P. J. M. (1995). Aberrant cGMP-binding activity in non-chemotactic *Dictyostelium discoideum* mutants. *Biochim. Biophys. Acta* **1268**, 214–220.

Kuwayama, H., Ecke, M., Gerisch, G., and van Haastert, P. J. M. (1996). Protection against osmotic stress by cGMP-mediated myosin phosphorylation. *Science* **271**, 207–209.

Labrousse, A., and Satre, M. (1997). A new collection of thermosensitive endocytosis mutants in the cellular slime mold *Dictyostelium discoideum*. *J. Euk. Microbiol.* **44**, 620–625.

Lacombe, M. L., Podgorski, G. J., Franke, J., and Kessin, R. H. (1986). Molecular cloning and developmental expression of the cyclic nucleotide phosphodiesterase gene of *Dictyostelium discoideum*. *J. Biol. Chem.* **261**, 16811–16817.

Landfear, S. M., and Lodish, H. F. (1980). A role for cAMP in expression of developmentally regulated genes in *Dictyostelium discoideum*. *Proc. Natl. Acad. Sci. USA* **77**, 1044–1048.

Larochelle, D. A., Vithalani, K. K., and De Lozanne, A. (1996). A novel member of the rho family of small GTP-binding proteins is specifically required for cytokinesis. *J. Cell Biol.* **133**, 1321–1329.

Larochelle, D. A., Vithalani, K. K., and De Lozanne, A. (1997). Role of *Dictyostelium* racE in cytokinesis: Mutational analysis and localization studies by use of green fluorescent protein. *Mol. Biol. Cell* **8**, 935–944.

Larson, M. A., Kelly, D. L., and Weber, A. T. (1994). A mutant of *Dictyostelium mucoroides* lacking the sorocarp sheath and primary macrocyst wall. *Trans. Amer. Microsc. Soc.* **113**, 200–210.

Lauzeral, J., Halloy, J., and Goldbeter, A. (1997). Desynchronization of cells on the developmental path triggers the formation of spiral waves of cAMP during *Dictyostelium* aggregation. *Proc. Natl. Acad. Sci. USA* **94**, 9153–9158.

Leach, C. K., Ashworth, J. M., and Garrod, D. R. (1973). Cell sorting out during the differentiation of mixtures of metabolically distinct populations of *Dictyostelium discoideum*. *J. Embryol. Exp. Morphol.* **29**, 647–661.

Lee, K. L., Cox, E. C., and Goldstein, R. E. (1996). Competing patterns of signaling activity in *Dictyostelium discoideum*. *Physiol. Rev. Lett.* **76**, 1174–1177.

Lee, S., Escalante, R., and Firtel, R. A. (1997a). A Ras GAP is essential for cytokinesis and spatial patterning in *Dictyostelium*. *Development* **124**, 983–996.

Lee, S. K., Yu, S. L., Garcia, M. X., Alexander, H., and Alexander, S. (1997b). Differential developmental expression of the *repB* and *repD* Xeroderma pigmentosum related DNA helicase genes from *Dictyostelium discoideum*. *Nucleic Acids Res.* **25**, 2365–2374.

Lee, S. K., Yu, S. L., Alexander, H., and Alexander, S. (1998). A mutation in *repB*, the *Dictyostelium* homolog of the human xeroderma pigmentosum B gene, has increased sensitivity to UV-light but normal morphogenesis. *Biochim. Biophys. Acta* **1399**, 161–172.

Leiting, B., and Noegel, A. (1988). Construction of an extrachromosomally replicating transformation vector for *Dictyostelium discoideum*. *Plasmid* **20**, 241–248.

Leiting, B., Lindner, I. J., and Noegel, A. A. (1990). The extrachromosomal replication of *Dictyostelium* plasmid Ddp2 requires a cis-acting element and a plasmid-encoded transacting factor. *Mol. Cell. Biol.* **10**, 3727–3736.

Leng, P., Klatte, D. H., Schumann, G., Boeke, J. D., and Steck, T. L. (1998). Skipper, an LTR retrotransposon of *Dictyostelium*. *Nucleic Acids Res.* **26**, 2008–2015.

Levine, H., Aranson, I., Tsimring, L., and Truong, T. V. (1996). Positive genetic feedback governs cAMP spiral wave formation in *Dictyostelium*. *Proc. Natl. Acad. Sci. USA* **93**, 6382–6386.

Lewis, K. E., and O'Day, D. H. (1977). Sex hormone of *Dictyostelium discoideum* is volatile. *Nature* **268**, 730–731.

Li, G., Alexander, H., Schneider, N., and Alexander, S. (2000). Molecular basis for resistance to the anticancer drug cisplatin in *Dictyostelium*. *Microbiology*, in press.

Lilly, P. J., and Devreotes, P. N. (1994). Identification of CRAC, a cytosolic regulator required for guanine nucleotide stimulation of adenylyl cyclase in *Dictyostelium*. *J. Biol. Chem.* **269**, 14123–14129.

Lilly, P. J., and Devreotes, P. N. (1995). Chemoattractant and GTPγS-mediated stimulation of adenylyl cyclase in *Dictyostelium* requires translocation of CRAC to membranes. *J. Cell Biol.* **129**, 1659–1665.

Lilly, P., Wu, L. J., Welker, D. L., and Devreotes, P. N. (1993). A G-protein β-subunit is essential for *Dictyostelium* development. *Genes Dev.* **7**, 986–995.

Liu, G., and Newell, P. C. (1991). Evidence that cyclic GMP may regulate the association of myosin-II heavy chain with the cytoskeleton by inhibiting its phosphorylation. *J. Cell Sci.* **98**, 483–490.

Liu, G., and Newell, P. C. (1994). Regulation of myosin regulatory light chain phosphorylation via cyclic GMP during chemotaxis of *Dictyostelium*. *J. Cell Sci.* **107**, 1737–1743.

Liu, G., Kuwayama, H., Ishida, S., and Newell, P. C. (1993). The role of cyclic GMP in regulating myosin during chemotaxis of *Dictyostelium* – Evidence from a mutant lacking the normal cyclic GMP response to cyclic AMP. *J. Cell Sci.* **106**, 591–596.

Liu, G., Edmonds, B. T., and Condeelis, J. (1996). pH, EF-1α and the cytoskeleton. *Trends Cell Biol.* **6**, 168–171.

Liu, T. Y., and Clarke, M. (1996). The vacuolar proton pump of *Dictyostelium discoideum*: molecular cloning and analysis of the 100 kDa subunit. *J. Cell Sci.* **109**, 1041–1051.

Lonski, J. (1976). The effect of ammonia on fruiting body size and microcyst formation in the cellular slime mold. *Dev. Biol.* **51**, 158–165.

Loomis, W. F. (1969). Temperature-sensitive mutants of *Dictyostelium discoideum*. *J. Bacteriol.* **99**, 65–69.

Loomis, W. F. (1988). Cell–cell adhesion in *Dictyostelium discoideum*. *Dev. Genet.* **9**, 549–559.

Loomis, W. F. (1993). Lateral inhibition and pattern formation in *Dictyostelium*. *Curr. Topics Dev. Biol.* **28**, 1–46.

Loomis, W. F. (1998). Role of PKA in the timing of developmental events in *Dictyostelium* cells. *Microbiol. Mol. Biol. Rev.* **62**, 684.

Loomis, W. F., and Ashworth, J. M. (1968). Plaque-size mutants of the cellular slime mould *Dictyostelium discoideum*. *J. Gen. Microbiol.* **53**, 181–186.

Loomis, W. F., and Kuspa, A. (1997). The genome of *Dictyostelium discoideum*. In *Dictyostelium – A model system for cell and developmental biology*. (Y. Maeda, K. Inouye, and I. Takeuchi, Eds.), pp. 15–30. Universal Academy Press, Tokyo, Japan.

Loomis, W. F., and Smith, D. W. (1995). Consensus phylogeny of *Dictyostelium*. *Experientia* **51**, 1110–1115.

Loomis, W. F., Welker, D., Hughes, J., Maghakian, D., and Kuspa, A. (1995). Integrated maps of the chromosomes in *Dictyostelium discoideum*. *Genetics* **141**, 147–157.

Loomis, W. F., Kuspa, A., and Shaulsky, G. (1998). Two-component signal transduction systems in eukaryotic microorganisms. *Curr. Opin. Microbiol.* **1**, 643–648.

Louis, J. M., Saxe III, C. L., and Kimmel, A. R. (1993). Two transmembrane signaling mechanisms control expression of the cAMP receptor gene CAR1 during *Dictyostelium* development. *Proc. Natl. Acad. Sci. USA* **90**, 5969–5973.

Louis, J. M., Ginsburg, G. T., and Kimmel, A. R. (1994). The cAMP receptor CAR4 regulates axial patterning and cellular differentiation during late development of *Dictyostelium*. *Genes Dev.* **8**, 2086–2096.

Louis, S. A., Spiegelman, G. B., and Weeks, G. (1997). Expression of an activated rasD gene changes cell fate decisions during *Dictyostelium* development. *Mol. Biol. Cell* **8**, 303–312.

Luna, E. J., Hitt, A. L., Shutt, D., Wessels, D., Soll, D., Jay, P., Hug, C., Elson, E. L. Vesley, A., Downey, G. P., Wang, M., Block, S. M., Sigurdson, W., and Sachs, F. (1998). Role of ponticulin in pseudopod dynamics, cell–cell adhesion, and mechanical stability of an amoeboid membrane skeleton. *Biol. Bull.* **194**, 345–346.

Lydan, M. A., and Cotter, D. A. (1994). Spore swelling in *Dictyostelium* is a dynamic process mediated by calmodulin. *FEMS Microbiol. Lett.* **115**, 137–142.

Lydan, M. A., and Cotter, D. A. (1995). The role of Ca^{2+} during spore germination in *Dictyostelium*: autoactivation is mediated by the mobilization of Ca^{2+} while amoebal emergence requires entry of external Ca^{2+}. *J. Cell Sci.* **108**, 1921–1930.

Lyttle, T. W. (1993). Cheaters sometimes prosper: distortion of mendelian segregation by meiotic drive. *Trends Genet.* **9**, 205–210.

Ma, G. C. L., and Firtel, R. A. (1978). Regulation of the synthesis of two carbohydrate-binding proteins in *Dictyostelium discoideum*. *J. Biol. Chem.* **253**, 3924–3932.

Ma, H., Gamper, M., Parent, C., and Firtel, R. A. (1997). The *Dictyostelium* MAP kinase kinase DdMEK1 regulates chemotaxis and is essential for chemoattractant-mediated activation of guanylyl cyclase. *EMBO J.* **16**, 4317–4332.

MacInnes, M. A., and Francis, D. (1974). Meiosis in *Dictyostelium mucoroides*. *Nature* **251**, 321–324.

MacWilliams, H. K. (1991). Models of pattern formation in *Hydra* and *Dictyostelium*. *Semin. Dev. Biol.* **2**, 119–128.

MacWilliams, H. K., and Bonner, J. T. (1979). The prestalk-prespore pattern in cellular slime molds. *Differentiation* **14**, 1–22.

MacWilliams, H. K., and David, C. N. (1984). Pattern formation in *Dictyostelium*. In *Microbial Development Monograph Series*. (R. Losick, and L. Shapiro, Eds.), pp. 255–274. Cold Spring Harbor Laboratories, Cold Spring Harbor, USA.

Maeda, M. (1988). Dual effects of cAMP on the stability of prespore vesicles and 8-bromo-cAMP enhanced maturation of spore and stalk cells of *Dictyostelium discoideum*. *Devel. Growth Differ.* **30**, 573–588.

Maeda, M. (1992). Efficient induction of sporulation of *Dictyostelium* prespore cells by 8-bromocyclic AMP under both submerged- and shaken-culture conditions and involvement of protein kinase(s) in its action. *Devel. Growth Differ.* **34**, 263–275.

Maeda, Y., and Gerisch, G. (1977). Vesicle formation in *Dictyostelium discoideum* cells during oscillations of cAMP synthesis and release. *Exp. Cell Res.* **110**, 119–126.

Maeda, Y., and Takeuchi, I. (1969). Cell differentiation and fine structures in the development of the cellular slime molds. *Devel. Growth Differ.* **11**, 232–245.

Maeda, Y., Ohmori, T., Abe, T., Abe, F., and Amagai, A. (1989). Transition of starving *Dictyostelium* cells to differentiation phase at a particular position of the cell cycle. *Differentiation* **41**, 169–175.

Malchow, D., and Gerisch, G. (1974). Short-term binding and hydrolysis of cyclic $3':5'$-adenosine monophosphate by aggregating *Dictyostelium* cells. *Proc. Natl. Acad. Sci. USA* **71**, 2423–2427.

Malchow, D., Nägele, B., Schwartz, H., and Gerisch, G. (1972). Membrane-bound cyclic AMP phosphodiesterase in chemotactically responding cells of *Dictyostelium discoideum*. *Eur. J. Biochem.* **28**, 136–142.

Malchow, D., Fuchila, J., and Nanjundiah, V. (1975). A plausible role for a membrane-bound cyclic AMP phosphodiesterase in cellular slime mold chemotaxis. *Biochim. Biophys. Acta* **385**, 421–428.

Malchow, D., Nanjundiah, V., and Gerisch, G. (1978a). pH oscillations in cell suspensions of *Dictyostelium discoideum*: their relation to cyclic-AMP signals. *J. Cell Sci.* **30**, 319–330.

Malchow, D., Nanjundiah, V., Wurster, B., Eckstein, F., and Gerisch, G. (1978b). Cyclic AMP-induced pH changes in *Dictyostelium discoideum* and their control by calcium. *Biochim. Biophys. Acta* **538**, 473–480.

Malchow, D., Mutzel, R., and Schlatterer, C. (1996a). On the role of calcium during chemotactic signaling and differentiation of the cellular slime mould *Dictyostelium discoideum*. *Int. J. Dev. Biol.* **40**, 135–139.

Malchow, D., Schaloske, R., and Schlatterer, C. (1996b). An increase in cytosolic Ca^{2+} delays cAMP oscillations in *Dictyostelium* cells. *Biochem. J.* **319**, 323–327.

Manabe, R., Saito, T., Kumazaki, T., Sakaitani, T., Nakata, N., and Ochiai, H. (1994). Molecular cloning and the COOH-terminal processing of gp64, a putative cell–cell adhesion protein of the cellular slime mold *Polysphondylium pallidum*. *J. Biol. Chem.* **269**, 528–535.

Maniak, M., Rauchenberger, R., Albrecht, R., Murphy, J., and Gerisch, G. (1995). Coronin involved in phagocytosis: dynamics of particle-induced relocalization visualized by a green fluorescent protein tag. *Cell* **83**, 915–924.

Mann, S. K., and Firtel, R. A. (1989). Two-phase regulatory pathway controls cAMP receptor-mediated expression of early genes in *Dictyostelium*. *Proc. Natl. Acad. Sci. USA* **86**, 1924–1928.

Mann, S. K. O., and Firtel, R. A. (1991). A developmentally regulated, putative serine/threonine protein kinase is essential for development in *Dictyostelium*. *Mech. Devel.* **35**, 89–101.

Mann, S. K. O., Yonemoto, W. M., Taylor, S. S., and Firtel, R. A. (1992). DdPK3, which plays essential roles during *Dictyostelium* development, encodes the catalytic subunit of cAMP-dependent protein kinase. *Proc. Natl. Acad. Sci. USA* **89**, 10701–10705.

Mann, S. K. O., Brown, J. M., Briscoe, C., Parent, C., Pitt, G., Devreotes, P. N., and Firtel, R. A. (1997). Role of cAMP-dependent protein kinase in controlling aggregation and postaggregative development in *Dictyostelium*. *Dev. Biol.* **183**, 208–221.

Manstein, D. J., Titus, M. A., De Lozanne, A., and Spudich, J. A. (1989). Gene replacement in *Dictyostelium*: generation of myosin null mutants. *EMBO J.* **8**, 923–932.

Margolskee, J. P., and Lodish, H. F. (1980). The regulation of the synthesis of actin and two other proteins induced early in *Dictyostelium discoideum* development. *Dev. Biol.* **74**, 50–64.

Margolskee, J. P., Froshauer, S., Skrinska, R., and Lodish, H. F. (1980). The effects of cell density and starvation on early developmental events in *Dictyostelium discoideum*. *Dev. Biol.* **74**, 409–421.

Marin, F. (1976). The role of amino acid starvation and cell contact in the regulation of early development in *Dictyostelium discoideum*. *J. Cell Biol.* **70**, 115a.

Marschalek, R., Brechner, T., Amon-Böhm, E., and Dingermann, T. (1989). Transfer RNA genes: landmarks for integration of mobile genetic elements in *Dictyostelium discoideum*. *Science* **244**, 1493–1496.

Marschalek, R., Borschet, G., and Dingermann, T. (1990). Genomic organization of the transposable element Tdd-3 from *Dictyostelium discoideum*. *Nucleic Acids Res.* **18**, 5751–5757.

Marschalek, R., Hofmann, J., Schumann, G., and Dingermann, T. (1992a). Two distinct subforms of the retrotransposable DRE element in NC4 strains of *Dictyostelium discoideum*. *Nucleic Acids Res.* **20**, 6247–6252.

Marschalek, R., Hofmann, J., Schumann, G., Gosseringer, R., and Dingermann, T. (1992b). Structure of DRE, a retrotransposable element which integrates with position specificity upstream of *Dictyostelium discoideum* tRNA genes. *Mol. Cell. Biol.* **12**, 229–239.

Marshak, D. R., Clarke, M., Roberts, D. M., and Watterson, D. M. (1984). Structural and functional properties of calmodulin from the eukaryotic microorganism *Dictyostelium discoideum*. *Biochemistry* **23**, 2891–2899.

Martiel, J. L., and Goldbeter, A. (1984). Oscillations et relais des signaux d'AMP cyclique chez *Dictyostelium discoideum*: analyse d'un modèle fonde sur la modification du récepteur pour l'AMP cyclique. *C.R. Acad. Sc. Paris, Série C* **298**, 549–552.

Masento, M. S., Morris, H. R., Taylor, G. W., Johnson, S. J., Skapski, A. C., and Kay, R. R. (1988). Differentiation-inducing factor from the slime mould *Dictyostelium discoideum* and its analogues. *Biochem. J.* **256**, 23–28.

Mato, J. M., Losada, A., Nanjundiah, V., and Konijn, T. M. (1975). Signal input for a chemotactic response in the cellular slime mold *Dictyostelium discoideum*. *Proc. Natl. Acad. Sci. USA* **72**, 4991–4993.

Mato, J. M., Krens, F. A., van Haastert, P. J. M., and Konijn, J. M. (1977a). Unified control of chemotaxis and cAMP mediated cGMP accumulation by cAMP in *Dictyostelium discoideum*. *Biochem. Biophys. Res. Commun.* **77**, 399–402.

Mato, J. M., van Haastert, P. J. M., Krens, F. A., Rhijnsburger, E. H., Dobbe, F. C. P. M., and Konijn, T. M. (1977b). Cyclic AMP and folic acid mediated cyclic GMP accumulation in *Dictyostelium discoideum*. *FEBS Lett.* **79**, 331–336.

McCaffrey, G., and Vale, R. D. (1989). Identification of a kinesin-like microtubule based motor protein in *Dictyostelium discoideum*. *EMBO J.* **8**, 3229–3234.

McCarroll, R., Olson, G. J., Stahl, X. D., Woese, C. R., and Sogin, M. L. (1983). Nucleotide sequence of the *Dictyostelium discoideum* small-subunit ribosomal ribonucleic acid inferred from the gene sequence: evolutionary implications. *Biochemistry* **22**, 5858–5868.

McDonald, S. A., and Durston, A. J. (1984). The cell cycle and sorting behaviour in *Dictyostelium discoideum*. *J. Cell Sci.* **66**, 195–204.

McGuire, V., and Alexander, S. (1996). PsB multiprotein complex of *Dictyostelium discoideum* – Demonstration of cellulose binding activity and order of protein subunit assembly. *J. Biol. Chem.* **271**, 14596–14603.

McIntosh, J. R., Roos, U. P., Neighbors, B., and McDonald, K. L. (1985). Architecture of the microtubule component of mitotic spindles from *Dictyostelium discoideum*. *J. Cell Sci.* **75**, 93–129.

McKeown, M., Taylor, W. C., Kindle, K. L., Firtel, R. A., Bender, W., and Davidson, N. (1978). Multiple, heterogeneous actin genes in *Dictyostelium*. *Cell* **15**, 789–800.

McNally, J. G., and Cox, E. C. (1989). Spots and stripes: the patterning spectrum in the cellular slime mould *Polysphondylium pallidum*. *Development* **105**, 323–333.

McPherson, C. E., and Singleton, C. K. (1992). V4, a gene required for the transition from growth to development in *Dictyostelium discoideum*. *Dev. Biol.* **150**, 231–242.

McPherson, C. E., and Singleton, C. K. (1993). Nutrient-responsive promoter elements of the V4 gene of *Dictyostelium discoideum*. *J. Mol. Biol.* **232**, 386–396.

McRobbie, S. J., and Newell, P. C. (1983). Changes in actin associated with cytoskeleton following chemotactic stimulation of *Dictyostelium discoideum*. *Biochem. Biophys. Res. Commun.* **115**, 351–359.

McRobbie, S. J., and Newell, P. C. (1985a). Cytoskeletal accumulation of a specific iso-actin during chemotaxis of *Dictyostelium*. *FEBS Lett.* **181**, 100–102.

McRobbie, S. J., and Newell, P. C. (1985b). Effects of cytochalasin B on cell movements and chemoattractant elicited actin changes in *Dictyostelium*. *Exp. Cell Res.* **160**, 275–286.

McRobbie, S. J., Jermyn, K. A., Duffy, K., Blight, K., and Williams, J. G. (1988). Two DIF-inducible, prestalk-specific mRNAs of *Dictyostelium* encode extracellular matrix protein of the slug. *Development* **104**, 275–284.

Mehdy, M. C., and Firtel, R. A. (1985). A secreted factor and cyclic AMP jointly regulate cell-type-specific gene expression in *Dictyostelium discoideum*. *Mol. Cell. Biol.* **5**, 705–713.

Meili, R., Ellsworth, C., Lee, S., Reddy, T. B., Ma, H., and Firtel, R. A. (1999). Chemoattractant-mediated transient activation and membrane localization of Akt/PKB is required for efficient chemotaxis to cAMP in *Dictyostelium*. *EMBO J* **18**, 2092–2105.

Menz, S., Bumann, J., Jaworski, E., and Malchow, D. (1991). Mutant analysis suggests that cyclic GMP mediates the cyclic AMP-induced Ca^{2+} uptake in *Dictyostelium*. *J. Cell Sci.* **99**, 187–191.

Mermall, V., Post, P. L., and Mooseker, M. S. (1998). Unconventional myosins in cell movement, membrane traffic, and signal transduction. *Science* **279**, 527–533.

Miller, C., McDonald, J., and Francis, D. (1996). Evolution of promoter sequences: Elements of a canonical promoter for prespore genes of *Dictyostelium*. *J. Mol. Evol.* **43**, 185–193.

Milne, J. L., and Coukell, M. B. (1991). A Ca^{2+} transport system associated with the plasma membrane of *Dictyostelium discoideum* is activated by different chemoattractant receptors. *J. Cell Biol.* **112**, 103–110.

Milne, J. L., and Devreotes, P. N. (1993). The surface cyclic AMP receptors, cAR1, cAR2, and cAR3, promote Ca^{2+} influx in *Dictyostelium discoideum* by a Gα2-independent mechanism. *Mol. Biol. Cell* **4**, 283–292.

Milne, J. L. S., Wu, L. J., Caterina, M. J., and Devreotes, P. N. (1995). Seven helix cAMP receptors stimulate Ca^{2+} entry in the absence of functional G proteins in *Dictyostelium*. *J. Biol. Chem.* **270**, 5926–5931.

Milne, J. L. S., Caterina, M. J., and Devreotes, P. N. (1997). Random mutagenesis of the cAMP chemoattractant receptor, cAR1, of *Dictyostelium* – Evidence for multiple states of activation. *J. Biol. Chem.* **272**, 2069–2076.

Mirfakhrai, M., Tanaka, Y., and Yanagisawa, K. (1990). Evidence for mitochondrial DNA polymorphism and uniparental inheritance in the cellular slime mold *Polysphondylium pallidum*: Effect of intraspecies mating on mitochondrial DNA transmission. *Genetics* **124**, 607–613.

Miura, K., and Siegert, F. (2000). Light affects cAMP signaling and cell movement activity in *Dictyostelium discoideum*. *Proc. Natl. Acad. Sci. USA* **97**, 2111–2116.

Mizutani, A., and Yanagisawa, K. (1990). Cell-division inhibitor produced by a killer strain of cellular slime mold *Polysphondylium pallidum*. *Devel. Growth Differ.* **32**, 397–402.

Moens, P. B. (1976). Spindle and kinetochore of *Dictyostelium discoideum*. *J. Cell Biol.* **68**, 113–122.

Moniakis, J., Coukell, M. B., and Forer, A. (1995). Molecular cloning of an intracellular P-type ATPase from *Dictyostelium* that is up-regulated in calcium-adapted cells. *J. Biol. Chem.* **270**, 28276–28281.

Monnat, J., Hacker, U., Geissler, H., Rauchenberger, R., Neuhaus, E. M., Maniak, M., and Soldati, T. (1997). *Dictyostelium discoideum* protein disulfide isomerase, an endoplasmic reticulum resident enzyme lacking a KDEL-type retrieval signal. *FEBS Lett.* **418**, 357–362.

Moores, S. L., and Spudich, J. A. (1998). Conditional loss-of-myosin-II-function mutants reveal a position in the tail that is critical for filament nucleation. *Mol. Cell* **1**, 1043–1050.

Moores, S. L., Sabry, J. H., and Spudich, J. A. (1996). Myosin dynamics in live *Dictyostelium* cells. *Proc. Natl. Acad. Sci. USA* **93**, 443–446.

Morita, Y. S., Jung, G., Hammer III, J. A., and Fukui, Y. (1996). Localization of *Dictyostelium* MyoB and MyoD to filopodia and cell–cell contact sites using isoform-specific antibodies. *Eur. J. Cell Biol.* **71**, 371–379.

Morris, H. R., Taylor, G. W., Masento, M. S., Jermyn, K. A., and Kay, R. R. (1987). Chemical structure of the morphogen differentiation inducing factor from *Dictyostelium discoideum*. *Nature* **328**, 811–814.

Morrison, A., Blanton, R. L., Grimson, M., Fuchs, M., Williams, K., and Williams, J. (1994). Disruption of the gene encoding the EcmA, extracellular matrix protein of *Dictyostelium* alters slug morphology. *Dev. Biol.* **163**, 457–466.

Morrissey, J. H., and Loomis, W. F. (1981). Parasexual genetic analysis of cell proportioning mutants of *Dictyostelium discoideum*. *Genetics* **99**, 183–196.

Morrissey, J. H., Farnsworth, P. A., and Loomis, W. F. (1981). Pattern formation in *Dictyostelium discoideum*: an analysis of mutants altered in cell proportioning. *Dev. Biol.* **83**, 1–8.

Mu, X. Q., Lee, B., Louis, J. M., and Kimmel, A. R. (1998). Sequence-specific protein interaction with a transcriptional enhancer involved in the autoregulated expression of cAMP receptor 1 in *Dictyostelium*. *Development* **125**, 3689–3698.

Müller, K., and Gerisch, G. (1978). A specific glycoprotein as the target site of adhesion blocking Fab in aggregating *Dictyostelium* cells. *Nature* **274**, 445–449.

Müller, U., and Hartung, K. (1990). Properties of three different ion channels in the plasma membrane of the slime mold *Dictyostelium discoideum*. *Biochim. Biophys. Acta* **1026**, 204–212.

Müller, U., Malchow, D., and Hartung, K. (1986). Single ion channels in the slime mold *Dictyostelium discoideum*. *Biochim. Biophys. Acta* **857**, 287–290.

Murgia, I., MacIver, S. K., and Morandini, P. (1995). An actin-related protein from *Dictyostelium discoideum* is developmentally regulated and associated with mitochondria. *FEBS Lett.* **360**, 235–241.

Murray, B. A. (1982). Membranes. In *The development of Dictyostelium discoideum* (W. F. Loomis, Ed.), pp. 71–116. Academic Press, New York.

Mutzel, R., Lacombe, M. L., Simon, M. N., De Gunzburg, J., and Véron, M. (1987). Cloning and cDNA sequence of the regulatory subunit of cAMP-dependent protein kinase from *Dictyostelium discoideum*. *Proc. Natl. Acad. Sci. USA* **84**, 6–10.

Nadson, G. A. (1899/1900). Des cultures du *Dictyostelium mucoroides* Bref. et des cultures pures des amides en général. *Scripta Bot. Horti Univ. Imp. Petropolitanae* **15**, 188–190.

Nakao, H., Yamamoto, A., Takeuchi, I., and Tasaka, M. (1994). *Dictyostelium* pre-spore-specific gene dp87 encodes a sorus matrix protein. *J. Cell Sci.* **107**, 397–403.

Nanjundiah, V. (1997). Models for pattern formation in the Dictyostelid slime molds. In *Dictyostelium – A model system for cell and developmental biology.* (Y. Maeda, K. Inouye, and I. Takeuchi, Eds.), pp. 305–322. Universal Academy Press, Tokyo, Japan.

Nanjundiah, V., and Malchow, D. (1976). A theoretical study of the effects of cyclic AMP phosphodiesterases during aggregation in *Dictyostelium. J. Cell Sci.* **22**, 49–58.

Nayler, O., Insall, R., and Kay, R. R. (1992). Differentiation-inducing-factor dechlorinase, a novel cytosolic dechlorinating enzyme from *Dictyostelium discoideum. Eur. J. Biochem.* **208**, 531–536.

Nebl, T., and Fisher, P. R. (1997). Intracellular Ca^{2+} signals in *Dictyostelium* chemotaxis are mediated exclusively by Ca^{2+} influx. *J. Cell Sci.* **110**, 2845–2853.

Nellen, W., Silan, C., and Firtel, R. A. (1984). DNA-mediated transformation in *Dictyostelium.* In *Molecular Biology of Development.* (E. H. Davidson, and R. A. Firtel, Eds.), pp. 633–645. A.R. Liss, New York.

Nes, W. D., Norton, R. A., Crumley, F. G., Madigan, S. J., and Katz, E. R. (1990). Sterol phylogenesis and algal evolution. *Proc. Natl. Acad. Sci. USA* **87**, 7565–7569.

Nestle, M., and Sussman, M. (1972). The effect of cyclic AMP on morphogenesis and enzyme accumulation in *Dictyostelium discoideum. Dev. Biol.* **28**, 545–554.

Newell, P. C. (1982). Cell surface binding of adenosine to *Dictyostelium* and inhibition of pulsatile signaling. *FEMS Microbiol. Lett.* **13**, 417–421.

Newell, P. C. (1995a). Calcium, cyclic GMP and the control of myosin II during chemotactic signal transduction of *Dictyostelium. J. Biosci.* **20**, 289–310.

Newell, P. C. (1995b). Signal transduction and motility of *Dictyostelium. Biosci. Reports* **15**, 445–462.

Newell, P. C., and Liu, G. (1992). Streamer F mutants and chemotaxis of *Dictyostelium. BioEssays* **14**, 473–479.

Newell, P. C., and Ross, F. M. (1982a). Genetic analysis of the slug stage of *Dictyostelium discoideum. J. Gen. Microbiol.* **128**, 1639–1652.

Newell, P. C., and Ross, F. M. (1982b). Inhibition by adenosine of aggregation centre initiation and cyclic AMP binding in *Dictyostelium. J. Gen. Microbiol.* **128**, 2715–2724.

Newell, P. C., Telser, A., and Sussman, M. (1969). Alternative developmental pathways determined by environmental conditions in the cellular slime mold *Dictyostelium discoideum. J. Bacteriol.* **100**, 763–768.

Newell, P. C., Franke, J., and Sussman, M. (1972). Regulation of four functionally related enzymes during shifts in the developmental program of *Dictyostelium discoideum. J. Mol. Biol.* **63**, 373–382.

Newell, P. C., Malchow, D., and Gross, J. D. (1995). The role of calcium in aggregation and development of *Dictyostelium. Experientia* **51**, 1155–1165.

Newth, C. K., Arnal, C., Goulart, R. A., and Hanna, M. H. (1987). Founder cell differentiation and acrasin production in an aggregateless mutant of *Polysphondylium violaceum. Devel. Growth Differ.* **29**, 479–488.

Niewohner, J., Weber, I., Maniak, M., Müller-Taubenberger, A., and Gerisch, G. (1997). Talin-null cells of *Dictyostelium* are strongly defective in adhesion to particle and substrate surfaces and slightly impaired in cytokinesis. *J. Cell Biol.* **138**, 349–361.

Niswonger, M. L., and O'Halloran, T. J. (1997). A novel role for clathrin in cytokinesis. *Proc. Natl. Acad. Sci. USA* **94**, 8575–8578.

Noegel, A. A., and Luna, J. E. (1995). The *Dictyostelium* cytoskeleton. *Experientia* **51**, 1135–1143.

Noegel, A., and Schleicher, M. (2000). The actin cytoskeleton of *Dictyostelium*: a story told by mutants. *J. Cell Sci.* **113**, 759–766.

Noegel, A. A., and Witke, W. (1988). Inactivation of the α-actinin gene in *Dictyostelium*. *Dev. Genet.* **9**, 531–538.

Noegel, A., Welker, D. L., Metz, B. A., and Williams, K. L. (1985). Presence of nuclear associated plasmids in the lower eukaryote *Dictyostelium discoideum*. *J. Mol. Biol.* **185**, 447–450.

Noegel, A., Gerisch, G., Stadler, J., and Westphal, M. (1986). Complete sequence and transcript regulation of a cell adhesion protein from aggregating *Dictyostelium* cells. *EMBO J.* **5**, 1473–1476.

Noegel, A. A., Koppel, B., Gottwald, U., Witke, W., Albrecht, R., and Schleicher, M. (1995). Actin and actin-binding proteins in the motility of *Dictyostelium*. In *The Cytoskeleton (45. Colloquium – Mosbach 1994)*. (B. M. Jockusch, E. Mandelkow, and K. Weber, Eds), pp. 117–126. Springer-Verlag, Berlin.

Noegel, A. A., Rivero, F., Fucini, P., Bracco, E., Janssen, K. P., and Schleicher, M. (1997). Actin binding proteins: role and regulation. In *Dictyostelium – A model system for cell and developmental biology*. (Y. Maeda, K. Inouye, and I. Takeuchi, Eds), pp. 33–42. Universal Academy Press, Tokyo, Japan.

Nolta, K. V., and Steck, T. L. (1994). Isolation and initial characterization of the bipartite contractile vacuole complex from *Dictyostelium discoideum*. *J. Biol. Chem.* **269**, 2225–2233.

Nolta, K. V., Rodriguez-Paris, J. M., and Steck, T. L. (1994). Analysis of successive endocytic compartments isolated from *Dictyostelium discoideum* by magnetic fractionation. *Biochim. Biophys. Acta* **1224**, 237–246.

North, M. J., and Cotter, D. A. (1991). Regulation of cysteine proteinases during different pathways of differentiation in cellular slime molds. *Dev. Genet.* **12**, 154–162.

Novak, K. D., and Titus, M. A. (1997). Myosin I overexpression impairs cell migration. *J. Cell Biol.* **136**, 633–647.

Novak, K. D., Peterson, M. D., Reedy, M. C., and Titus, M. A. (1995). *Dictyostelium* myosin I double mutants exhibit conditional defects in pinocytosis. *J. Cell Biol.* **131**, 1205–1221.

Ochiai, H., Hata, K., Saito, T., Funamoto, S., and Nakata, N. (1996). Dimer formation of a cell–cell adhesion protein, gp64 of the cellular slime mold, *Polysphondylium pallidum*. *Plant Cell Physiol.* **37**, 135–139.

Ochiai, H., Saito, T., and Funamoto, S. (1997). Cell–cell adhesion protein gp64 of *Polysphondylium pallidum*. In *Dictyostelium – A model system for cell and developmental biology*. (Y. Maeda, K. Inouye, and I. Takeuchi, Eds), pp. 123–136. Universal Academy Press, Tokyo, Japan.

O'Day, D. H. (1973). α-Mannosidase and microcyst differentiation in the cellular mold *Polysphondylium pallidum*. *J. Bacteriol.* **113**, 192–197.

Ogawa, S., Naito, K., Angata, K., Morio, T., Urushihara, H., and Tanaka, Y. (1997). A site-specific DNA endonuclease specified by one of two ORFs encoded by a group I intron in *Dictyostelium discoideum* mitochondrial DNA. *Gene* **191**, 115–121.

O'Halloran, T. J., and Anderson, R. G. W. (1992). Clathrin heavy chain is required for pinocytosis, the presence of large vacuoles, and development in *Dictyostelium*. *J. Cell Biol.* **118**, 1371–1377.

Ohmori, R., and Maeda, Y. (1987). The developmental fate of *Dictyostelium discoideum* cells depends greatly on the cell-cycle position at the onset of starvation. *Cell Differ.* **22**, 11–18.

Okaichi, K., Cubitt, A. B., Pitt, G. S., and Firtel, R. A. (1992). Amino acid substitutions in the *Dictyostelium* Gα subunit Gα2 produce dominant negative phenotypes and inhibit the activation of adenylyl cyclase, guanylyl cyclase, and phospholipase C. *Mol. Biol. Cell* **3**, 735–747.

Olie, R. A., Durrieu, F., Cornillon, S., Loughran, G., Gross, J., Earnshaw, W. C., and Golstein, P. (1998). Apparent caspase independence of programmed cell death in *Dictyostelium. Curr. Biol.* **8**, 955–958.

Olive, E. W. (1901). A preliminary enumeration of the Sorophoreae. *Proc. Am. Acad. Arts Sci.* **37**, 333–344.

Olive, E. W. (1902). Monograph of the Acrasieae. *Proc. Boston Soc. Natur. Hist.* **30**, 451–513.

Omura, F., and Fukui, Y. (1985). *Dictyostelium* microtubule organizing center: structure and linkage to the nucleus. *Protoplasma* **127**, 212–221.

Orlow, S. J., Shapiro, I., Franke, J., and Kessin, R. H. (1981). The extracellular cyclic nucleotide phosphodiesterase of *Dictyostelium discoideum*. Purification and characterization. *J. Biol. Chem.* **256**, 7620–7627.

Orlowski, M., and Loomis, W. F. (1979). Plasma membrane proteins of *Dictyostelium*: the spore coat proteins. *Dev. Biol.* **71**, 297–307.

Othmer, H. G., and Schaap, P. (1998). Oscillatory cAMP signaling in the development of *Dictyostelium discoideum. Comments Theor. Biol.* **5**, 175–282.

Otsuka, H., and van Haastert, P. J. M. (1998). A novel Myb homolog initiates *Dictyostelium* development by induction of adenylyl cyclase expression. *Genes Dev.* **12**, 1738–1748.

Ozaki, T., Hoshikawa, Y., Iida, Y., and Iwabuchi, M. (1984). Sequence analysis of the transcribed and 5′non-transcribed regions of the ribosomal RNA gene in *Dictyostelium discoideum. Nucleic Acids Res.* **12**, 4171–4178.

Padh, H., Ha, J. H., Lavasa, M., and Steck, T. L. (1993). A post-lysosomal compartment in *Dictyostelium discoideum. J. Biol. Chem.* **268**, 6742–6747.

Palsson, E., and Cox, E. C. (1996). Origin and evolution of circular waves and spirals in *Dictyostelium discoideum* territories. *Proc. Natl. Acad. Sci. USA* **93**, 1151–1155.

Palsson, E., Lee, K. J., Goldstein, R. E., Franke, J., Kessin, R. H., and Cox, E. C. (1997). Selection for spiral waves in the social amoebae *Dictyostelium. Proc. Natl. Acad. Sci. USA* **94**, 13719–13723.

Pan, P., Hall, E. M., and Bonner, J. T. (1972). Folic acid as second chemotactic substance in the cellular slime moulds. *Nature New Biol.* **237**, 181–182.

Parent, C. A., and Devreotes, P. N. (1995). Isolation of inactive and G protein-resistant adenylyl cyclase mutants using random mutagenesis. *J. Biol. Chem.* **270**, 22693–22696.

Parent, C. A., and Devreotes, P. N. (1996a). Constitutively active adenylyl cyclase mutant requires neither G proteins nor cytosolic regulators. *J. Biol. Chem.* **271**, 18333–18336.

Parent, C. A., and Devreotes, P. N. (1996b). Molecular genetics of signal transduction in *Dictyostelium. Annu. Rev. Biochem.* **65**, 411–440.

Parent, C. A., and Devreotes, P. N. (1999). A cell's sense of direction. *Science* **284**, 765–770.

Parent, C. A., Blacklock, B. J., Froehlich, W. M., Murphy, D. B., and Devreotes, P. N. (1998). G protein signaling events are activated at the leading edge of chemotactic cells. *Cell* **95**, 81–91.

Parish, R. W. (1975). Mitochondria and peroxisomes from the cellular slime mold *Dictyostelium discoideum*. Isolation techniques and urate oxidase association with peroxisomes. *Eur. J. Biochem.* **58**, 523–531.

Parissenti, A. M., and Coukell, M. B. (1990). Effects of DNA and synthetic oligodeoxyribonucleotides on the binding properties of a cGMP-binding protein from *Dictyostelium discoideum*. *Biochim. Biophys. Acta* **1040**, 294–300.

Patterson, B., and Spudich, J. A. (1995). A novel positive selection for identifying cold-sensitive myosin II mutants in *Dictyostelium*. *Genetics* **140**, 505–515.

Peskin, C. S., Odell, G. M., and Oster, G. F. (1993). Cellular motions and thermal fluctuations: the Brownian ratchet. *Biophys. J.* **65**, 316–324.

Pinter, K., and Gross, J. (1995). Calcium and cell-type-specific gene expression in *Dictyostelium*. *Differentiation* **59**, 201–206.

Pitt, G. S., Gundersen, R. E., and Devreotes, P. N. (1990). Mechanisms of excitation and adaptation in *Dictyostelium*. *Semin. Cell Biol.* **1**, 99–104.

Pitt, G. S., Milona, N., Borleis, J., Lin, K. C., Reed, R. R., and Devreotes, P. N. (1992). Structurally distinct and stage-specific adenylyl cyclase genes play different roles in *Dictyostelium* development. *Cell* **69**, 305–315.

Plyte, S. E., O'Donovan, E., Woodgett, J. R., and Harwood, A. J. (1999). Glycogen synthase kinase-3 (GSK-3) is regulated during *Dictyostelium* development via the serpentine receptor cAR3. *Development* **126**, 325–333.

Podgorski, G. J., Franke, J., Faure, M., and Kessin, R. H. (1989). The cyclic nucleotide phosphodiesterase gene of *Dictyostelium discoideum* utilizes alternate promotors and splicing for the synthesis of multiple mRNAs. *Mol. Cell. Biol.* **9**, 3938–3950.

Podolski, J. L., and Steck, T. L. (1990). Length distribution of F-actin in *Dictyostelium discoideum*. *J. Biol. Chem.* **265**, 1312–1318.

Poff, K. L., and Butler, W. L. (1974). Spectral characteristics of the photoreceptor pigment of phototaxis in *Dictyostelium discoideum*. *Photochem. Photobiol.* **20**, 241–244.

Poff, K. L., and Häder, D. P. (1984). An action spectrum for phototaxis by pseudoplasmodia of *Dictyostelium discoideum*. *Photochem. Photobiol.* **39**, 433–436.

Poff, K. L., and Loomis, W. F. (1973). Control of phototactic migration in *Dictyostelium discoideum*. *Exp. Cell Res.* **82**, 236–240.

Poff, K. L., and Skokut, M. (1977). Thermotaxis by pseudoplasmodia of *Dictyostelium discoideum*. *Proc. Natl. Acad. Sci. USA* **74**, 2007–2010.

Poff, K. L., Loomis, W. F., and Butler, W. L. (1974). Isolation and purification of the photoreceptor pigment associated with phototaxis in *Dictyostelium discoideum*. *J. Biol. Chem.* **249**, 2164–2167.

Pogge-von Strandmann, R., Kay, R. R., and Dufour, J.-P. (1984). An electrogenic proton pump in plasma membranes from the cellular slime mould *Dictyostelium discoideum*. *FEBS Lett.* **175**, 422–427.

Ponte, E., Bracco, E., Faix, J., and Bozzaro, S. (1998). Detection of subtle phenotypes: the case of the cell adhesion molecule csA in *Dictyostelium*. *Proc. Natl. Acad. Sci. USA* **95**, 9360–9365.

Ponte, E., Rivero, F, Fechheimer, M., Noegel, A., and Bozzaro, S. (2000). Severe developmental defects in *Dictyostelium* null mutants for actin-binding proteins. *Mech. Dev.* **91**, 153–161.

Poole, S. J., and Firtel, R. A. (1984). Genomic instability and mobile genetic elements in regions surrounding two discoidin I genes of *Dictyostelium discoideum*. *Mol. Cell. Biol.* **4**, 671–680.

Poole, S., Firtel, R. A., Lamar, E., and Rowekamp, W. (1981). Sequence and expression of the discoidin I gene family in *Dictyostelium discoideum*. *J. Mol. Biol.* **153**, 273–289.

Potts, G. (1902). Zur Physiologie des *Dictyostelium mucoroides*. *Flora (Jena)* **91**, 281–347.

Powell-Coffman, J. A., and Firtel, R. A. (1994). Characterization of a novel *Dictyostelium discoideum* prespore-specific gene, PspB, reveals conserved regulatory sequences. *Development* **120**, 1601–1611.

Powell-Coffman, J. A., Schnitzler, G. R., and Firtel, R. A. (1994). A GBF-binding site and a novel AT element define the minimal sequences sufficient to direct prespore-specific expression in *Dictyostelium discoideum*. *Mol. Cell. Biol.* **14**, 5840–5849.

Primpke, G., Iassonidou, V., Nellen, W., and Wetterauer, B. (2000). Role of cAMP-dependent protein kinase during growth and early development of *Dictyostelium discoideum*. *Dev. Biol.* **222**, 101–111.

Pukatzki, S., Tordilla, N., Franke, J., and Kessin, R. H. (1998). A novel component involved in ubiquitination is required for development of *Dictyostelium discoideum*. *J. Biol. Chem.* **273**, 24131–24138.

Pupillo, M., Insall, R., Pitt, G. S., and Devreotes, P. N. (1992). Multiple cyclic AMP receptors are linked to adenylyl cyclase in *Dictyostelium*. *Mol. Biol. Cell.* **3**, 1229–1234.

Raff, R., A. (1996). *The Shape of Life: Genes, Development and the Evolution of Animal Form*. The University of Chicago Press: Chicago and London. 520 pp.

Rall, T. W., Sutherland, E. W., and Berthet, J. (1957). The relationship of epinephrine and glucagon to liver phosphorylase. *J. Biol. Chem.* **224**, 453–475.

Ramalingam, R., and Ennis, H. L. (1997). Characterization of the *Dictyostelium discoideum* cellulose-binding protein CelB and regulation of gene expression. *J. Biol. Chem.* **272**, 26166–26172.

Ramalingam, R., Blume, J. E., Ganguly, K., and Ennis, H. L. (1995). AT-rich upstream sequence elements regulate spore germination-specific expression of the *Dictyostelium discoideum celA* gene. *Nucleic Acids Res.* **23**, 3018–3025.

Raman, R. K. (1976). Analysis of the chemotactic response during aggregation in *Dictyostelium minutum*. *J. Cell Sci.* **20**, 497–512.

Raman, R. K., Hashimoto, Y., Cohen, M. H., and Robertson, A. (1976). Differentiation for aggregation in the cellular slime mold. The emergence of autonomously signaling cells in *Dictyostelium discoideum*. *J. Cell Sci.* **21**, 243–259.

Rand, K. D., and Sussman, M. (1983). The morphogenetic sequence followed by migrating slugs of *Dictyostelium discoideum* during reentry into the fruiting body. *Differentiation* **24**, 88–96.

Raper, K. B. (1935). *Dictyostelium discoideum*, a new species of slime mold from decaying forest leaves. *J. Agr. Res.* **50**, 135–147.

Raper, K. B. (1937). Growth and development of *Dictyostelium discoideum* with different bacterial associates. *J. Agr. Res.* **55**, 289–316.

Raper, K. B. (1940a). The communal nature of the fruiting process in the Acrasieae. *Am. J. Bot.* **27**, 436–448.

Raper, K. B. (1940b). Pseudoplasmodium formation and organization in *Dictyostelium discoideum*. *J. Elisha Mitchell Sci. Soc.* **56**, 241–282.

Raper, K. B. (1984). *The Dictyostelids*. Princeton University Press, Princeton, NJ.

Raper, K. B., and Fennell, D. I. (1952). Stalk formation in *Dictyostelium. Bull. Torrey Bot. Club* **79**, 25–51.

Raper, K. B., and Fennell, D. I. (1967). The crampon-based Dictyostelia. *Am. J. Bot.* **54**, 515–528.

Raper, K. B., and Smith, N. R. (1939). The growth of *Dictyostelium discoideum* on pathogenic bacteria. *J. Bacteriol.* **38**, 431–444.

Raper, K. B., and Thom, C. (1941). Interspecific mixtures in the Dictyosteliaceae. *Am. J. Bot.* **28**, 69–78.

Rathi, A., and Clarke, M. (1992). Expression of early developmental genes in *Dictyostelium discoideum* is initiated during exponential growth by an autocrine-dependent mechanism. *Mech. Devel.* **36**, 173–182.

Rathi, A., Kayman, S. C., and Clarke, M. (1991). Induction of gene expression in *Dictyostelium* by prestarvation factor, a factor secreted by growing cells. *Dev. Genet.* **12**, 82–87.

Rayment, I., Smith, C., and Yount, R. G. (1996). The active site of myosin. *Annu. Rev. Physiol.* **58**, 671–702.

Reymond, C. D., Gomer, R. H., Mehdy, M. C., and Firtel, R. A. (1984). Developmental regulation of a *Dictyostelium* gene encoding a protein homologous to mammalian Ras protein. *Cell* **39**, 141–148.

Reymond, C. D., Nellen, W., Gomer, R. H., and Firtel, R. A. (1986). Regulation of the *Dictyostelium* ras gene during development and in transformants. In *Progress in Developmental Biology. Series: Progress in Clinical and Biological Research.* (H. C. Slavkin, Ed.), pp. 17–21. A.R. Liss, New York.

Reymond, C. D., Ludérus, M. E. E., Europe-Finner, G. N., Thompson, N. A., Burki, E., van Driel, R., and Newell, P. C. (1989). Analysis of the *ras* gene function in *Dictyostelium discoideum*. In *The guanine-nucleotide binding proteins: Common structural and functional properties.* (L. Bosch, B. Kraal, and A. Parmeggiani, Eds.), pp. 265–272. Plenum Press, New York.

Richardson, D. L., and Loomis, W. F. (1992). Disruption of the sporulation-specific gene *spiA* in *Dictyostelium discoideum* leads to spore instability. *Genes Dev.* **6**, 1058–1070.

Richardson, D. L., Loomis, W. F., and Kimmel, A. R. (1994). Progression of an inductive signal activates sporulation in *Dictyostelium discoideum*. *Development* **120**, 2891–2900.

Rieben, W. K., Gonzales, C. M, Gonzales, S. T., Pilkington, K. J., Kiyosawa, H., Hughes, J. E., and Welker, D. L. (1998). *Dictyostelium discoideum* nuclear plasmid Ddp5 is a chimera related to the Ddp1 and Ddp2 plasmid families. *Genetics* **148**, 1117–1125.

Rietdorf, J., Siegert, F., and Weijer, C. J. (1996). Analysis of optical density wave propagation and cell movement during mound formation in *Dictyostelium discoideum*. *Dev. Biol.* **177**, 427–438.

Rietdorf, J., Siegert, F., Dharmawardhane, S., Firtel, R. A., and Weijer, C. J. (1997). Analysis of cell movement and signaling during ring formation in an activated Gα1 mutant of *Dictyostelium discoideum* that is defective in prestalk zone formation. *Dev. Biol.* **181**, 79–90.

Rivero, F., Furukawa, R., Noegel, A. A., and Fechheimer, M. (1996a). *Dictyostelium discoideum* cells lacking the 34,000-Dalton actin-binding protein can grow, locomote, and develop, but exhibit defects in regulation of cell structure and movement: A case of partial redundancy. *J. Cell Biol.* **135**, 965–980.

Rivero, F., Koppel, B., Peracino, B., Bozzaro, S., Siegert, F., Weijer, C. J., Schleicher, M., Albrecht, R., and Noegel, A. A. (1996b). The role of the cortical cytoskeleton: F-actin crosslinking proteins protect against osmotic stress, ensure cell size, cell shape and motility, and contribute to phagocytosis and development. *J. Cell Sci.* **109**, 2679–2691.

Rivero, F., Kuspa, A., Brokamp, R., Matzner, M., and Noegel, A. A. (1998). Interaptin, an actin-binding protein of the alpha-actinin superfamily in *Dictyostelium discoideum*, is developmentally and cAMP-regulated and associates with intracellular membrane compartments. *J. Cell Biol.* **142**, 735–750.

Robinson, V., and Williams, J. (1997). A marker of terminal stalk cell terminal differentiation in *Dictyostelium*. *Differentiation* **61**, 223–228.

Robson, G. E., and Williams, K. L. (1977). The mitotic chromosomes of the cellular slime mould *Dictyostelium discoideum*: A karyotype based on giemsa banding. *J. Gen. Microbiol.* **99**, 191–200.

Robson, G. E., and Williams, K. L. (1979). Vegetative incompatibility and the mating-type locus in the cellular slime mold *Dictyostelium discoideum*. *Genetics* **93**, 861–875.

Rodriguez-Paris, J. M., Nolta, K. V., and Steck, T. L. (1993). Characterization of lysosomes isolated from *Dictyostelium discoideum* by magnetic fractionation. *J. Biol. Chem.* **268**, 9110–9116.

Rogers, K. C., Ginsburg, G. T., Mu, X., Gollop, R., Balint-Kurti, P., Louis, J. M., and Kimmel, A. R. (1997). The cAMP receptor gene family of *Dictyostelium discoideum*: expression, regulation, function. In *Dictyostelium – A model system for cell and developmental biology*. (Y. Maeda, K. Inouye, and I. Takeuchi, Eds), pp. 163–172. Universal Academy Press, Tokyo, Japan.

Rogers, M. J., Xiong, X. J., Brown, R. J., Watts, D. J., Russell, R. G. G., Bayless, A. V., and Ebetino, F. H. (1995). Structure–activity relationships of new heterocycle-containing bisphosphonates as inhibitors of bone resorption and as inhibitors of growth of *Dictyostelium discoideum* amoebae. *Mol. Pharmacol.* **47**, 398–402.

Romans, P., and Firtel, R. A. (1985). Organization of the actin multigene family of *Dictyostelium discoideum* and analysis of variability in the protein coding regions. *J. Mol. Biol.* **186**, 321–335.

Roos, U. P., De Brabander, M., and Nuydens, R. (1987). Movements of intracellular particles in undifferentiated amebae of *Dictyostelium discoideum*. *Cell Motil. Cytoskel.* **7**, 258–271.

Rosen, E., Sivertsen, A., and Firtel, R. A. (1983). An unusual transposon encoding heat shock inducible and developmentally regulated transcripts in *Dictyostelium*. *Cell* **35**, 243–251.

Ross, F. M., and Newell, P. C. (1981). Streamers: chemotactic mutants of *Dictyostelium discoideum* with altered cyclic GMP metabolism. *J. Gen. Microbiol.* **127**, 339–350.

Rubin, J., and Robertson, A. (1975). The tip of the *Dictyostelium discoideum* pseudoplasmodium as an organizer. *J. Embryol. Exp. Morphol.* **33**, 227–241.

Rubino, S., Fighetti, M., Unger, E., and Cappuccinelli, P. (1984). Location of actin, myosin and microtubular structures during directed locomotion of *Dictyostelium* amebae. *J. Cell Biol.* **98**, 382–390.

Runyon, E. H. (1942). Aggregation of separate cells of *Dictyostelium* to form a multicellular body. *The Collecting Net (Woods Hole, Mass.)* **17**, 88.

Ruppel, K. M., and Spudich, J. A. (1995). Myosin motor function: structural and mutagenic approaches. *Curr. Opin. Cell Biol.* **7**, 89–93.

Ruppel, K. M., and Spudich, J. A. (1996a). Structure–function analysis of the motor domain of myosin. *Annu. Rev. Cell. Dev. Biol.* **12**, 543–573.

Ruppel, K. M., and Spudich, J. A. (1996b). Structure–function studies of the myosin motor domain: importance of the 50-kDa cleft. *Mol. Biol. Cell* **7**, 1123–1136.

Rutherford, C. L., and Jefferson, B. L. (1976). Trehalose accumulation in stalk and spore cells of *Dictyostelium discoideum*. *Dev. Biol.* **52**, 52–60.

Rutherford, C. L., McCaffery, I., and Favis, R. (1997). Regulation of glycogen phosphorylase and synthase genes during cell differentiation of *Dictyostelium*. In *Dictyostelium – A model system for cell and developmental biology*. (Y. Maeda, K. Inouye, and I. Takeuchi, Eds.), pp. 363–377. Universal Academy Press, Tokyo, Japan.

Sadeghi, H., da Silva, A. M., and Klein, C. (1988). Evidence that a glycolipid tail anchors antigen 117 to the plasma membrane of *Dictyostelium discoideum* cells. *Proc. Natl. Acad. Sci. USA* **85**, 5512–5515.

Sakai, Y. (1973). Cell type conversion in isolated prestalk and prespore fragments of the cellular slime mold *Dictyostelium discoideum*. *Devel. Growth Differ.* **15**, 11–19.

Sakai, Y., and Takeuchi, I. (1971). Changes of the prespore specific structure during dedifferentiation and cell type conversion of a slime mold cell. *Devel. Growth Differ.* **13**, 231–240.

Sameshima, M., Imai, Y., and Hashimoto, Y. (1988). The position of the microtubule-organizing center relative to the nucleus is independent of the direction of cell migration in *Dictyostelium discoideum*. *Cell Motil. Cytoskel.* **9**, 111–116.

Sandona, D., Gastaldello, S., Rizzuto, R., and Bisson, R. (1995). Expression of cytochrome c oxidase during growth and development of *Dictyostelium*. *J. Biol. Chem.* **270**, 5587–5593.

Sarkar, S., and Steck, T. L. (1997). An NADH-dependent disulfide reductase activity in the endoplasmic reticulum of *Dictyostelium discoideum*. *Biochem. Biophys. Res. Commun.* **234**, 313–315.

Saxe III, C. L., Johnson, R., Devreotes, P. N., and Kimmel, A. R. (1991a). Multiple genes for cell surface cAMP receptors in *Dictyostelium discoideum*. *Dev. Genet.* **12**, 6–13.

Saxe III, C. L., Johnson, R. L., Devreotes, P. N., and Kimmel, A. R. (1991b). Expression of a cAMP receptor gene of *Dictyostelium* and evidence for a multigene family. *Genes Dev.* **5**, 1–8.

Saxe III, C. L., Ginsburg, G. T., Louis, J. M., Johnson, R., Devreotes, P. N., and Kimmel, A. R. (1993). CAR2, a prestalk cAMP receptor required for normal tip formation and late development of *Dictyostelium discoideum*. *Genes Dev.* **7**, 262–272.

Saxe III, C. L., Yu, Y. M., Jones, C., Bauman, A., and Haynes, C. (1996). The cAMP receptor subtype cAR2 is restricted to a subset of prestalk cells during *Dictyostelium* development and displays unexpected DIF-1 responsiveness. *Dev. Biol.* **174**, 202–213.

Saxena, I. M., Kudlicka, K., Okuda, K., Brown, R. M., Jr. (1994). Characterization of genes in the cellulose-synthesizing operon (acs operon) of *Acetobacter xylinum*: implications for cellulose crystallization. *J. Bacteriol.* **176**, 5735–5752.

Schaap, P., and Wang, M. (1986). Interactions between adenosine and oscillatory cAMP signaling regulate size and pattern in *Dictyostelium*. *Cell* **45**, 137–144.

Schaap, P., Nebl, T., and Fisher, P. R. (1996a). A slow sustained increase in cytosolic Ca^{2+} levels mediates stalk gene induction by differentiation inducing factor in *Dictyostelium*. *EMBO J.* **15**, 5177–5183.

Schaap, P., Tang, Y. H., and Othmer, H. G. (1996b). A model for pattern formation in *Dictyostelium discoideum*. *Differentiation* **60**, 1–16.

Schafer, D. A., and Cooper, J. A. (1995). Control of actin assembly at filament ends. *Annu. Rev. Cell. Dev. Biol.* **11**, 497–518.

Schaloske, R., and Malchow, D. (1997). Mechanism of cAMP-induced Ca^{2+} influx in *Dictyostelium*: role of phospholipase A_2. *Biochem. J.* **327**, 233–238.

Schatzle, J., Bush, J., and Cardelli, J. (1992). Molecular cloning and characterization of the structural gene coding for the developmentally regulated lysosomal enzyme, α-mannosidase, in *Dictyostelium discoideum*. *J. Biol. Chem.* **267**, 4000–4007.

Scheel, J., Ziegelbauer, K., Kupke, T., Humbel, B. M., Noegel, A. A., Gerisch, G., and Schleicher, M. (1989). Hisactophilin, a histidine-rich actin-binding protein from *Dictyostelium discoideum*. *J. Biol. Chem.* **264**, 2832–2839.

Schindler, J., and Sussman, M. (1977). Ammonia determines the choice of morphogenetic pathways in *Dictyostelium discoideum*. *J. Mol. Biol.* **116**, 161–169.

Schindler, J., and Sussman, M. (1979). Inhibition by ammonia of intracellular cAMP accumulation in *Dictyostelium discoideum*: its significance for the regulation of morphogenesis. *Dev. Genet.* **1**, 13–20.

Schleicher, M., and Noegel, A. A. (1992). Dynamics of the *Dictyostelium* cytoskeleton during chemotaxis. *New Biol.* **4**, 461–472.

Schlenkrich, T., Fleischmann, P., and Häder, D. P. (1995). Biochemical and spectroscopic characterization of the putative photoreceptor for phototaxis in amoebae of the cellular slime mould *Dictyostelium discoideum*. *J. Photochem. Photobiol. B-Biol.* **30**, 139–143.

Schnitzler, G. R., Fischer, W. H., and Firtel, R. A. (1994). Cloning and characterization of the G-box binding factor, an essential component of the developmental switch between early and late development in *Dictyostelium*. *Genes Dev.* **8**, 502–514.

Schnitzler, G. R., Briscoe, C., Brown, J. M., and Firtel, R. A. (1995). Serpentine cAMP receptors may act through a G protein-independent pathway to induce postaggregative development in *Dictyostelium*. *Cell* **81**, 737–745.

Schoen, C. D., Schulkes, C. C. G. M., Arents, J. C., and van Driel, R. (1996). Guanylate cyclase activity in permeabilized *Dictyostelium discoideum* cells. *J. Cell. Biochem.* **60**, 411–423.

Schuster, S. C., Noegel, A. A., Oehme, F., Gerisch, G., and Simon, M. I. (1996). The hybrid histidine kinase DokA is part of the osmotic response system of *Dictyostelium*. *EMBO J.* **15**, 3880–3889.

Schwartz, M. A., and Luna, E. J. (1988). How actin binds and assembles onto plasma membranes from *Dictyostelium discoideum*. *J. Cell Biol.* **107**, 201–209.

Seastone, D. J., Lee, E., Bush, J., Knecht, D., and Cardelli, J. (1998). Overexpression of a novel Rho family GTPase, RacC, induces unusual actin-based structures and positively affects phagocytosis in *Dictyostelium discoideum*. *Mol. Biol. Cell* **9**, 2891–2904.

Segall, J. E. (1992). Behavioral responses of streamer F mutants of *Dictyostelium discoideum*: effects of cyclic GMP on cell motility. *J. Cell Sci.* **101**, 589–597.

Segall, J. E., Kuspa, A., Shaulsky, G., Ecke, M., Maeda, M., Gaskins, C., and Firtel, R. A. (1995). A MAP kinase necessary for receptor-mediated activation of adenylyl cyclase in *Dictyostelium*. *J. Cell Biol.* **128**, 405–413.

Segel, L. A., Goldbeter, A., Devreotes, P. N., and Knox, B. E. (1986). A mechanism for exact sensory adaptation based on receptor modification. *J. Theor. Biol.* **120**, 151–179.

Sesaki, H., Wong, E. F. S., and Siu, C. H. (1997). The cell adhesion molecule DdCAD-1 in *Dictyostelium* is targeted to the cell surface by a nonclassical transport pathway involving contractile vacuoles. *J. Cell Biol.* **138**, 939–951.

Shaffer, B. M. (1956). Acrasin, the chemotactic agent in cellular slime moulds. *J. Exp. Biol.* **33**, 645–657.

Shaffer, B. M. (1961). Aggregation in the Dictyosteliaceae. *Recent advances in botany. (Proc. 9th Int. Bot Congress; 1959; Montreal)* **2**, 355.

Shaffer, B. M. (1962). The Acrasina. *Adv. Morphogenesis* **2**, 109–182.

Shaffer, B. M. (1963). Inhibition by existing aggregations of founder differentiation in the cellular slime mould *Polysphondylium violaceum*. *Exp. Cell Res.* **31**, 432–435.

Shammat, I. M., and Welker, D. L. (1999). Mechanism of action of the Rep protein from the *Dictyostelium* Ddp2 plasmid family. *Plasmid* **41**, 248–259.

Shammat, I. M., Gonzales, C. M., and Welker, D. L. (1998). *Dictyostelium discoideum* nuclear plasmid Ddp6 is a new member of the Ddp2 plasmid family. *Curr. Genet.* **33**, 77–82.

Shariff, A., and Luna, E. J. (1990). *Dictyostelium discoideum* plasma membranes contain an actin-nucleating activity that requires ponticulin, an integral membrane glycoprotein. *J. Cell Biol.* **110**, 681–692.

Shaulsky, G., and Loomis, W. F. (1995). Mitochondrial DNA replication but no nuclear DNA replication during development of *Dictyostelium*. *Proc. Natl. Acad. Sci. USA* **92**, 5660–5663.

Shaulsky, G., and Loomis, W. F. (1996). Initial cell type divergence in *Dictyostelium* is independent of DIF-1. *Dev. Biol.* **174**, 214–220.

Shaulsky, G., Kuspa, A., and Loomis, W. F. (1995). A multidrug resistance transporter serine protease gene is required for prestalk specialization in *Dictyostelium*. *Genes Dev.* **9**, 1111–1122.

Shaulsky, G., Escalante, R., and Loomis, W. F. (1996). Developmental signal transduction pathways uncovered by genetic suppressors. *Proc. Natl. Acad. Sci. USA* **93**, 15260–15265.

Shaulsky, G., Fuller, D., and Loomis, W. F. (1998). A cAMP-phosphodiesterase controls PKA-dependent differentiation. *Development* **125**, 691–699.

Shaw, D. R., Richter, H., Giorda, R., Ohmachi, T., and Ennis, H. L. (1989). Nucleotide sequences of *Dictyostelium discoideum* developmentally regulated cDNAs rich in (AAC) imply proteins that contain clusters of asparagine, glutamine, or threonine. *Mol. Gen. Genet.* **218**, 453–459.

Shimizu, H., Julius, M. A., Giarre, M., Zheng, Z., Brown, A. M., and Kitajewski, J. (1997). Transformation by Wnt family proteins correlates with regulation of β-catenin. *Cell. Growth Differ.* **8**, 1349–1358.

Shimomura, O., Suthers, H. L. B., and Bonner, J. T. (1982). Chemical identity of the acrasin of the cellular slime mold *Polysphondylium violacium*. *Proc. Natl. Acad. Sci. USA* **79**, 7376–7379.

Shutt, D. C., Wessels, D., Wagenknecht, K., Chandrasekhar, A., Hitt, A. L., Luna, E. J., and Soll, D. R. (1995). Ponticulin plays a role in the positional stabilization of pseudopods. *J. Cell Biol.* **131**, 1495–1506.

Siegert, F., and Weijer, C. (1989). Digital image processing of optical density wave propagation in *Dictyostelium discoideum* and analysis of the effects of caffeine and ammonia. *J. Cell Sci.* **93**, 325–335.

Siegert, F., and Weijer, C. J. (1991). Analysis of optical density wave propagation and cell movement in the cellular slime mold *Dictyostelium discoideum*. *Physica D* **49**, 224–232.

Siegert, F., and Weijer, C. J. (1992). Three-dimensional scroll waves organize *Dictyostelium* slugs. *Proc. Natl. Acad. Sci. USA* **89**, 6433–6437.

Siegert, F., and Weijer, C. J. (1993). The role of periodic signals in the morphogenesis of *Dictyostelium discoideum*. In *Oscillations and Morphogenesis*. (L. Rensing, Ed.), pp. 133–152. M. Dekker, New York.

Siegert, F., and Weijer, C. J. (1995). Spiral and concentric waves organize multicellular *Dictyostelium* mounds. *Curr. Biol.* **5**, 937–943.

Siegert, F., and Weijer, C. (1997). Control of cell movement during multicellular morphogenesis. In *Dictyostelium – A model system for cell and developmental biology.* (Y. Maeda, K. Inouye, and I. Takeuchi, Eds.), pp. 425–436. Universal Academy Press, Tokyo, Japan.

Silveira, L. A., Smith, J. L., Tan, J. L., and Spudich, J. A. (1998). MLCK-A, an unconventional myosin light chain kinase from *Dictyostelium*, is activated by a cGMP-dependent pathway. *Proc. Natl. Acad. Sci. USA* **95**, 13000–13005.

Simmonds, A. C., Ellison, J. E., and Garrod, D. R. (1987). Cytoskeleton-attached membrane protein of *Dictyostelium discoideum* is absent from phagocytosis mutant. *Biochem. Soc. Trans.* **15**, 850.

Simon, M. N., Driscoll, D., Mutzel, R., Part, D., Williams, J., and Véron, M. (1989). Overproduction of the regulatory subunit of the cAMP-dependent protein kinase blocks the differentiation of *Dictyostelium discoideum*. *EMBO J.* **8**, 2039–2044.

Simon, M. N., Pelegrini, O., Véron, M., and Kay, R. R. (1992). Mutation of protein kinase-A causes heterochronic development of *Dictyostelium*. *Nature* **356**, 171–172.

Singleton, C. K., Gregoli, P. A., Manning, S. S., and Northington, S. J. (1988). Characterization of genes which are transiently expressed during the preaggregative phase of development of *Dictyostelium discoideum*. *Dev. Biol.* **129**, 140–146.

Singleton, C. K., Zinda, M. J., Mykytka, B., and Yang, P. (1998). The histidine kinase *dhkC* regulates the choice between migrating slugs and terminal differentiation in *Dictyostelium discoideum*. *Dev. Biol.* **203**, 345–357.

Siu, C. H., and Kamboj, R. K. (1990). Cell–cell adhesion and morphogenesis in *Dictyostelium discoideum*. *Dev. Genet.* **11**, 377–387.

Siu, C. H., Harris, T. J. C., Wong, E. F. S., Yang, C., Sesaki, H., and Wang, J. (1997). Cell adhesion molecules in *Dictyostelium*. In *Dictyostelium – A model system for cell and developmental biology.* (Y. Maeda, K. Inouye, and I. Takeuchi, Eds.), pp. 111–121. Universal Academy Press, Tokyo, Japan.

Slade, M. B., Chang, A. C. M., and Williams, K. L. (1990). The sequence and organization of Ddp2, a high-copy-number nuclear plasmid of *Dictyostelium discoideum*. *Plasmid* **24**, 195–207.

Smith, S. S., and Ratner, D. I. (1991). Lack of 5-methylcytosine in *Dictyostelium discoideum* DNA. *Biochem. J.* **277**, 273–275.

Snaar-Jagalska, B. E., Jakobs, K. H., and van Haastert, P. J. M. (1988). Agonist-stimulated high-affinity GTPase in *Dictyostelium* membranes. *FEBS Lett.* **236**, 139–144.

Soll, D. R., and Waddell, D. R. (1975). Morphogenesis in the slime mold *Dictyostelium discoideum*. 1. The accumulation and erasure of "morphogenetic information". *Dev. Biol.* **47**, 292–302.

Soll, D. R., and Wessels, D. (1998). *Motion analysis of living cells*. Wiley-Liss, New York.

Souza, G. M., Hirai, J., Mehta, D. P., and Freeze, H. H. (1995). Identification of two novel *Dictyostelium discoideum* cysteine proteinases that carry *N*-acetylglucosamine-1-P modification. *J. Biol. Chem.* **270**, 28938–28945.

Souza, G. M., Mehta, D. P., Lammertz, M., Rodriguez-Paris, J., Wu, R. R., Cardelli, J. A., and Freeze, H. H. (1997). *Dictyostelium* lysosomal proteins with different sugar modifications sort to functionally distinct compartments. *J. Cell Sci.* **110**, 2239–2248.

Souza, G. M., Lu, S. J., and Kuspa, A. (1998). YakA, a protein kinase required for the transition from growth to development in *Dictyostelium*. *Development* **125**, 2291–2302.

Souza, G. M., da Silva, A. M., and Kuspa, A. (1999). Starvation promotes *Dictyostelium* development by relieving PufA inhibition of PKA translation through the YakA kinase pathway. *Development* **126**, 3263–3274.

Spann, T. P., Brock, D. A., Lindsey, D. F., Wood, S. A., and Gomer, R. H. (1996). Mutagenesis and gene identification in *Dictyostelium* by shotgun antisense. *Proc. Natl. Acad. Sci. USA* **93**, 5003–5007.

Springer, W. R., Cooper, D. N. W., and Barondes, S. H. (1984). Discoidin I is implicated in cell-substratum attachment and ordered cell migration of *Dictyostelium discoideum* and resembles fibronectin. *Cell* **39**, 557–564.

Spudich, J. A. (1989). In pursuit of myosin function. *Cell Regulation* **1**, 1–11.

Spudich, J. A., Finer, J., Simmons, B., Ruppel, K., Patterson, B., and Uyeda, T. (1995). Myosin structure and function. *Cold Spring Harbor Symp. Quant. Biol.* **60**, 783–791.

Srinivasan, S., Alexander, H., and Alexander, S. (1999). The prespore vesicles of *Dictyostelium discoideum* – Purification, characterization, and developmental regulation. *J. Biol. Chem.* **274**, 35823–35831.

Srinivasan, S., Alexander, H., and Alexander, S. (2000). Crossing the finish line of development: regulated secretion of *Dictyostelium* proteins. *Trends Cell Biol.* **10**, 215–219.

Stadler, J., Keenan, T. W., Bauer, G., and Gerisch, G. (1989). The contact site A glycoprotein of *Dictyostelium discoideum* carries a phospholipid anchor of a novel type. *EMBO J.* **8**, 371–377.

Staples, S. O., and Gregg, J. H. (1967). Carotenoid pigments in the cellular slime mold, *Dictyostelium discoideum*. *Biol. Bull.* **132**, 413–422.

Steck, T. L., and Lavasa, M. (1994). A general method for plasma membrane isolation by colloidal gold density shift. *Anal. Biochem.* **223**, 47–50.

Steck, T. L., Chiaraviglio, L., and Meredith, S. (1997). Osmotic homeostasis in *Dictyostelium discoideum*: excretion of amino acids and ingested solutes. *J. Euk. Microbiol.* **44**, 503–510.

Stege, J. T., Shaulsky, G., and Loomis, W. F. (1997). Sorting of the initial cell types in *Dictyostelium* is dependent on the *tipA* gene. *Dev. Biol.* **185**, 34–41.

Sternfeld, J. (1992). A study of pstB cells during *Dictyostelium* migration and culmination reveals a unidirectional cell type conversion process. *W.R. Arch. Dev. Biol.* **201**, 354–363.

Sternfeld, J. (1998). The anterior-like cells in *Dictyostelium* are required for the elevation of the spores during culmination. *Dev. Genes Evol.* **208**, 487–494.

Sternfeld, J., and David, C. N. (1981). Cell sorting during pattern formation in *Dictyostelium*. *Differentiation* **20**, 10–21.

Sternfeld, J., and David, C. N. (1982). Fate and regulation of anterior-like cells in *Dictyostelium* slugs. *Dev. Biol.* **93**, 111–118.

Stites, J., Wessels, D., Uhl, A., Egelhoff, T., Shutt, D., and Soll, D. R. (1998). Phosphorylation of the *Dictyostelium* myosin II heavy chain is necessary for maintaining cellular polarity and suppressing turning during chemotaxis. *Cell. Motil. Cytoskel.* **39**, 31–51.

Stoeckelhuber, M., Noegel, A. A., Eckerskorn, C., Köhler, J., Rieger, D., and Schleicher, M. (1996). Structure/function studies on the pH-dependent actin-binding protein hisactophilin in *Dictyostelium* mutants. *J. Cell Sci.* **109**, 1825–1835.

Sucgang, R., Weijer, C. J., Siegert, F., Franke, J., and Kessin, R. H. (1997). Null mutations of the *Dictyostelium* cyclic nucleotide phosphodiesterase gene block chemotactic cell movement in developing aggregates. *Dev. Biol.* **192**, 181–192.

Sukumaran, S., Brown, J. M., Firtel, R. A., and McNally, J. G. (1998). *lagC*-null and *gbf*-null cells define key steps in the morphogenesis of *Dictyostelium* mounds. *Dev. Biol.* **200**, 16–26.

Sun, T. J., and Devreotes, P. N. (1991). Gene targeting of the aggregation stage cAMP receptor cAR1 in *Dictyostelium. Genes Dev.* **5**, 572–582.

Sussman, M. (1955). 'Fruity' and other mutants of the cellular slime mould, *Dictyostelium discoideum*: a study of developmental aberrations. *J. Gen. Microbiol.* **13**, 295–309.

Sussman, M., and Lovgren, N. (1965). Preferential release of the enzyme UDP-galactose polysaccharide transferase during cellular differentiation in the slime mold, *Dictyostelium discoideum. Exp. Cell Res.* **38**, 97–105.

Sussman, M., and Osborn, M. J. (1964). UDP-galactose polysaccharide transferase in the cellular slime mold *Dictyostelium discoideum*: appearance and disappearance of activity during cell differentiation. *Proc. Natl. Acad. Sci. USA* **52**, 81–87.

Sussman, M., Schindler, J., and Kim, H. (1978). "Sluggers", a new class of morphogenetic mutants of *D. discoideum. Exp. Cell Res.* **116**, 217–227.

Sussman, R. R. (1961). A method for staining the chromosomes of *Dictyostelium discoideum* myxamoebae in the vegetative stage. *Exp. Cell Res.* **24**, 154–155.

Sussman, R., and Rayner, E. P. (1971). Physical characterization of deoxyribonucleic acids in *Dictyostelium discoideum. Arch. Biochem. Biophys.* **144**, 127–137.

Sussman, R. R., and Sussman, M. (1953). Cellular differentiation in Dictyosteliaceae: heritable modifications of the developmental pattern. *Ann. N. Y. Acad. Sci.* **56**, 949–960.

Sussman, R., and Sussman, M. (1967). Cultivation of *Dictyostelium discoideum* in axenic culture. *Biochem. Biophys. Res. Commun.* **29**, 53–55.

Suthers, H. B. (1985). Ground-feeding migratory songbirds as cellular slime mold distribution vectors. *Oecologia* **65**, 526–530.

Swanson, J. A., and Taylor, D. L. (1982). Local and spatially coordinated movements in *Dictyostelium discoideum* amoebae during chemotaxis. *Cell* **28**, 225–232.

Takeuchi, I. (1963). Immunochemical and immunohistochemical studies on the development of the cellular slime mold *Dictyostelium mucoroides. Dev. Biol.* **8**, 1–26.

Takeuchi, I., and Sakai, Y. (1971). Dedifferentiation of the disaggregated slug cell of the cellular slime mold *Dictyostelium discoideum. Devel. Growth Differ.* **13**, 201–210.

Takeuchi, I., Tasaka, M., Oyama, M., Yamamoto, A., and Amagai, A. (1982). Pattern formation in the development of *Dictyostelium discoideum*. In *Embryonic Development. Part B. Cellular aspects.* (M. M. Burger, and R. Weber, Eds.), pp. 283–294. A.R. Liss, New York.

Tan, J. L., and Spudich, J. A. (1990). Developmentally regulated protein-tyrosine kinase genes in *Dictyostelium discoideum. Mol. Cell. Biol.* **10**, 3578–3583.

Tang, Y. H., and Othmer, H. G. (1995). Excitation, oscillations and wave propagation in a G-protein-based model of signal transduction in *Dictyostelium discoideum. Phil. Trans. R. Soc. Lond. B* **349**, 179–195.

Tasaka, M., Hasegawa, M., Ozaki, T., Iwabuchi, M., and Takeuchi, I. (1990). Isolation and characterization of spore coat protein (sp96) gene of *Dictyostelium discoideum*. *Cell Differ. Devel.* **31**, 1–10.

Tejedor, F., Zhu, X. R., Kaltenbach, E., Ackermann, A., Baumann, A., Canal, I., Heisenberg, M., Fischbach, K. F., and Pongs, O. (1995). Minibrain: a new protein kinase family involved in postembryonic neurogenesis in *Drosophila*. *Neuron* **14**, 287–301.

Temesvari, L. A., and Cotter, D. A. (1997). Trehalase of *Dictyostelium discoideum*: inhibition by amino-containing analogs of trehalose and affinity purification. *Biochimie* **79**, 229–239.

Temesvari, L., Rodriguez-Paris, J., Bush, J., Steck, T. L., and Cardelli, J. (1994). Characterization of lysosomal membrane proteins of *Dictyostelium discoideum* – A complex population of acidic integral membrane glycoproteins, Rab GTP-binding proteins, and vacuolar ATPase subunits. *J. Biol. Chem.* **269**, 25719–25727.

Theibert, A., and Devreotes, P. N. (1983). Cyclic $3',5'$-AMP relay in *Dictyostelium discoideum*: adaptation is independent of activation of adenylate cyclase. *J. Cell Biol.* **97**, 173–177.

Theibert, A., and Devreotes, P. (1984). Adenosine and its derivatives inhibit the cAMP signaling response in *Dictyostelium discoideum*. *Dev. Biol.* **106**, 166–173.

Theibert, A., and Devreotes, P. (1986). Surface receptor-mediated activation of adenylate cyclase in *Dictyostelium*. Regulation by guanine nucleotides in wild-type cells and aggregation deficient mutants. *J. Biol. Chem.* **261**, 15121–15125.

Theibert, A., Klein, P., and Devreotes, P. N. (1984). Specific photo affinity labeling of the cAMP surface receptor in *Dictyostelium discoideum*. *J. Biol. Chem.* **259**, 12318–12321.

Thilo, L., and Vogel, G. (1980). Kinetics of membrane internalization and recycling during pinocytosis in *Dictyostelium discoideum*. *Proc. Natl. Acad. Sci. USA* **77**, 1015–1019.

Thomason, P. A., Traynor, D., Cavet, G., Chang, W. T., Harwood, A. J., and Kay, R. R. (1998). An intersection of the cAMP/PKA and two-component signal transduction systems in *Dictyostelium*. *EMBO J.* **17**, 2838–2845.

Thomason, P., Traynor, D., and Kay, R. (1999). Taking the plunge terminal differentiation in *Dictyostelium*. *Trends Genet.* **15**, 15–19.

Titus, M. A., Wessels, D., Spudich, J. A., and Soll, D. (1993). The unconventional myosin encoded by the *myoA* gene plays a role in *Dictyostelium* motility. *Mol. Biol. Cell* **4**, 233–246.

Tomchik, K. J., and Devreotes, P. N. (1981). Adenosine $3',5'$-monophosphate waves in *Dictyostelium discoideum*: a demonstration by isotope dilution-fluorography technique. *Science* **212**, 443–446.

Town, C., and Gross, J. (1978). The role of cyclic nucleotides and cell agglomeration in postaggregative enzyme synthesis in *Dictyostelium discoideum*. *Dev. Biol.* **63**, 412–420.

Town, C., and Stanford, E. (1979). An oligosaccharide-containing factor that induces cell differentiation in *Dictyostelium discoideum*. *Proc. Natl. Acad. Sci. USA* **76**, 308–312.

Town, C. D., Gross, J. D., and Kay, R. R. (1976). Cell differentiation without morphogenesis in *Dictyostelium discoideum*. *Nature* **262**, 717–719.

Traynor, D., Kessin, R. H., and Williams, J. G. (1992). Chemotactic sorting to cAMP in the multicellular stages of *Dictyostelium* development. *Proc. Natl. Acad. Sci. USA* **89**, 8303–8307.

Trivinos-Lagos, L., Ohmachi, T., Albrightson, C., Burns, R. G., Ennis, H. L., and Chisholm, R. L. (1993). The highly divergent α-tubulins and β-tubulins from *Dictyostelium discoideum* are encoded by single genes. *J. Cell Sci.* **105**, 903–911.

Tsujioka, M., Machesky, L. M., Cole, S. L., Yahata, K., and Inouye, K. (1999). A unique talin homologue with a villin headpiece-like domain is required for multicellular morphogenesis in *Dictyostelium. Curr. Biol.* **9**, 389–392.

Tuxworth, R. I., Cheetham, J. L., Machesky, L. M., Spiegelmann, G. B., Weeks, G., and Insall, R. H. (1997). *Dictyostelium* RasG is required for normal motility and cytokinesis, but not growth. *J. Cell Biol.* **138**, 605–614.

Tyson, J. J., Alexander, K. A., Manoranjan, V. S., and Murray, J. D. (1989). Spiral waves of cyclic AMP in a model of slime mold aggregation. *Physica D* **34**, 193–207.

Ueda, M., Gräf, R., MacWilliams, H. K., Schliwa, M., and Euteneuer, U. (1997). Centrosome positioning and directionality of cell movements. *Proc. Natl. Acad. Sci. USA* **94**, 9674–9678.

Ueda, M., Schliwa, M., and Euteneuer, U. (1999). Unusual centrosome cycle in *Dictyostelium*: correlation of dynamic behavior and structural changes. *Mol. Biol. Cell* **10**, 151–160.

Urushihara, H. (1997). Cell recognition in the sexual development of *Dictyostelium discoideum*. In *Dictyostelium – A model system for cell and developmental biology*. (Y. Maeda, K. Inouye, and I. Takeuchi, Eds.), pp. 137–142. Universal Academy Press, Tokyo, Japan.

Uyeda, T. Q., and Titus, M. A. (1997). The myosins of *Dictyostelium*. In *Dictyostelium – A model system for cell and developmental biology*. (Y. Maeda, K. Inouye, and I. Takeuchi, Eds.), pp. 43–64. Universal Academy Press, Tokyo, Japan.

Uyeda, T. Q. P., Abramson, P. D., and Spudich, J. A. (1996). The neck region of the myosin motor domain acts as a lever arm to generate movement. *Proc. Natl. Acad. Sci. USA* **93**, 4459–4464.

Valkema, R., and van Haastert, P. J. M. (1994). A model for cAMP-mediated cGMP response in *Dictyostelium discoideum. Mol. Biol. Cell* **5**, 575–585.

van Duijn, B., and Vogelzang, S. A. (1989). The membrane potential of the cellular slime mold *Dictyostelium discoideum* is mainly generated by an electrogenic proton pump. *Biochim. Biophys. Acta* **983**, 186–192.

van Es, S., Hodgkinson, S., Schaap, P., and Kay, R. R. (1994). Metabolic pathways for differentiation-inducing factor-1 and their regulation are conserved between closely related *Dictyostelium* species, but not between distant members of the family. *Differentiation* **58**, 95–100.

van Es, S., Virdy, K. J., Pitt, G. S., Meima, M., Sands, T. W., Devreotes, P. N., Cotter, D. A., and Schaap, P. (1996). Adenylyl cyclase G, an osmosensor controlling germination of *Dictyostelium* spores. *J. Biol. Chem.* **271**, 23623–23625.

van Haastert, P. J. M. (1983). Binding of cAMP and adenosine derivatives to *Dictyostelium discoideum* cells. *J. Biol. Chem.* **258**, 9643–9648.

van Haastert, P. J. M. (1987). Differential effects of temperature on cAMP-induced excitation, adaptation, and deadaptation of adenylate and guanylate cyclase in *Dictyostelium discoideum. J. Cell Biol.* **105**, 2301–2306.

van Haastert, P. J. M. (1995). Transduction of the chemotactic cAMP signal across the plasma membrane of *Dictyostelium* cells. *Experientia* **51**, 1144–1154.

van Haastert, P. J. M. (1997). Transduction of the chemotactic cAMP signal across the plasma membrane. In *Dictyostelium – A model system for cell and developmental biology*. (Y. Maeda, K. Inouye, and I. Takeuchi, Eds.), pp. 173–191. Universal Academy Press, Tokyo, Japan.

van Haastert, P. J. M., and Devreotes, P. N. (1993). Biochemistry and genetics of sensory transduction in *Dictyostelium*. In *Signal transduction: prokaryotic and simple eukaryotic systems*. (J. Kurjan, and B. L. Taylor, Eds.), pp. 329–352. Academic Press, San Diego.

van Haastert, P. J. M., and de Wit, R. J. W. (1984). Demonstration of receptor heterogeneity and affinity modulation by nonequilibrium binding experiments. *J. Biol. Chem.* **259**, 13321–13328.

van Haastert, P. J. M., and Kuwayama, H. (1997). cGMP as second messenger during *Dictyostelium* chemotaxis. *FEBS Lett.* **410**, 25–28.

van Haastert, P. J. M., and van Dijken, P. (1997). Biochemistry and genetics of inositol phosphate metabolism in *Dictyostelium*. *FEBS Lett.* **410**, 39–43.

van Haastert, P. J. M., de Wit, R. J. W., Grijpma, Y., and Konijn, T. M. (1982). Identification of a pterin as the acrasin of the cellular slime mold *Dictyostelium lacteum*. *Proc. Natl. Acad. Sci. USA* **79**, 6270–6274.

van Haastert, P. J. M., van Lookeren Campagne, M. M., and Kesbeke, F. (1983). Multiple degradation pathways of chemoattractant mediated cGMP accumulation in *Dictyostelium*. *Biochim. Biophys. Acta* **756**, 67–71.

van Haastert, P. J. M., Bishop, J. D., and Gomer, R. H. (1996). The cell density factor CMF regulates the chemoattractant receptor cAR1 in *Dictyostelium*. *J. Cell Biol.* **134**, 1543–1549.

van Lookeren Campagne, M. M., Schaap, P., and van Haastert, P. J. M. (1986). Specificity of adenosine inhibition of cAMP-induced responses in *Dictyostelium* resembles that of the P site of higher organisms. *Dev. Biol.* **117**, 245–251.

van Lookeren Campagne, M. M., Franke, J., and Kessin, R. H. (1991). Functional cloning of a *Dictyostelium discoideum* cDNA encoding GMP synthetase. *J. Biol. Chem.* **266**, 16448–16452.

van Tieghem, M. P. H. (1880). Sur quelques Myxomycètes a plasmode agrégé. *Bull. Soc. Bot. Fr.* **27**, 317–322.

Vardy, P. H., Fisher, L. R., Smith, E., and Williams, K. L. (1986). Traction proteins in the extracellular matrix of *Dictyostelium discoideum* slugs. *Nature* **320**, 526–529.

Varnum, B., and Soll, D. R. (1984). Effects of cAMP on single cell motility in *Dictyostelium*. *J. Cell Biol.* **99**, 1151–1155.

Vasiev, B., and Weijer, C. J. (1999). Modeling chemotactic cell sorting during *Dictyostelium discoideum* mound formation. *Biophys. J.* **76**, 595–605.

Vasiev, B., Siegert, F., and Weijer, C. J. (1997). A hydrodynamic model for *Dictyostelium discoideum* mound formation. *J. Theor. Biol.* **184**, 441.

Vasieva, O. O., Vasiev, B. N., Karpov, V. A., and Zaikin, A. N. (1994). A model of *Dictyostelium discoideum* aggregation. *J. Theor. Biol.* **171**, 361–368.

Vauti, F., Morandini, P., Blusch, J., Sachse, A., and Nellen, W. (1990). Regulation of the discoidin-Iγ-gene in *Dictyostelium discoideum* – identification of individual promoter elements mediating induction of transcription and repression by cyclic AMP. *Mol. Cell. Biol.* **10**, 4080–4088.

Verkerke-van Wijk, I., and Schaap, P. (1997). cAMP, a signal for survival. In *Dictyostelium – A model system for cell and developmental biology*. (Y. Maeda, K. Inouye, and I. Takeuchi, Eds.), pp. 145–162. Universal Academy Press, Tokyo, Japan.

Verkerke-van Wijk, I., Brandt, R., Bosman, L., and Schaap, P. (1998). Two distinct signaling pathways mediate DIF induction of prestalk gene expression in *Dictyostelium*. *Exp. Cell Res.* **245**, 179–185.

Vicker, M. G. (1994). The regulation of chemotaxis and chemokinesis in *Dictyostelium* amoebae by temporal signals and spatial gradients of cyclic AMP. *J. Cell Sci.* **107**, 659–667.

Vogel, G., Thilo, L., Schwarz, H., and Steinhart, R. (1980). Mechanism of phagocytosis in *Dictyostelium discoideum*: phagocytosis is mediated by different recognition sites as disclosed by mutants with altered phagocytotic properties. *J. Cell Biol.* **86**, 456–465.

Vogel, W., Gish, G. D., Alves, F., and Pawson, T. (1997). The discoidin domain receptor tyrosine kinases are activated by collagen. *Mol. Cell* **1**, 13–23.

Vuillemin, P. (1903). Une Acrasiée bactériophage. *C.R. Acad. Sci. Paris* **137**, 387–389.

Waddell, D. (1986). Self non-self recognition in the cellular slime mold, *Dictyostelium caveatum*. *Am. Zool.* **26**, A 19.

Waddell, D. R. (1982). A predatory slime mould. *Nature* **298**, 464–466.

Waddell, D. R., and Soll, D. R. (1977). A characterization of the erasure phenomenon in *Dictyostelium*. *Dev. Biol.* **60**, 83–92.

Waddell, D. R., and Vogel, G. (1985). Phagocytic behavior of the predatory slime mold, *Dictyostelium caveatum*. Cell nibbling. *Exp. Cell Res.* **159**, 323–334.

Wallace, J. S., and Newell, P. C. (1982). Genetic analysis by mitotic recombination in *Dictyostelium discoideum* of growth and developmental loci on linkage group VII. *J. Gen. Microbiol.* **128**, 953–964.

Wallace, M. A., and Raper, K. B. (1979). Genetic exchanges in the macrocysts of *Dictyostelium discoideum*. *J. Gen. Microbiol.* **113**, 327–337.

Wang, B., and Kuspa, A. (1997). *Dictyostelium* development in the absence of cAMP. *Science* **277**, 251–254.

Wang, M., and Schaap, P. (1985). Correlations between tip dominance, prestalk/prespore pattern, and cAMP-relay efficiency in slugs of *Dictyostelium discoideum*. *Differentiation* **30**, 7–14.

Wang, M., Aerts, R. J., Spek, W., and Schaap, P. (1988a). Cell cycle phase in *Dictyostelium discoideum* is correlated with the expression of cyclic AMP production, detection, and degradation. *Dev. Biol.* **125**, 410–416.

Wang, M., van Driel, R., and Schaap, P. (1988b). Cyclic AMP-phosphodiesterase induces dedifferentiation of prespore cells in *Dictyostelium discoideum* slugs: evidence that cyclic AMP is the morphogenetic signal for prespore differentiation. *Development* **103**, 611–618.

Wang, N., Shaulsky, G., Escalante, R., and Loomis, W. F. (1996). A two-component histidine kinase gene that functions in *Dictyostelium* development. *EMBO J.* **15**, 3890–3898.

Wang, Y., Liu, J., and Segall, J. E. (1998). MAP kinase function in amoeboid cells. *J. Cell Sci.* **111**, 373–383.

Wanner, R., and Wurster, B. (1990). Cyclic GMP-activated protein kinase from *Dictyostelium discoideum*. *Biochim. Biophys. Acta* **1053**, 179–184.

Warren, J., Warren, W., and Cox, E. (1975). Genetic complexity of aggregation in the cellular slime mold *Polysphondylium violaceum*. *Proc. Natl. Acad. Sci. USA* **72**, 1041–1042.

Warrick, H. M., and Spudich, J. A. (1987). Myosin structure and function in cell motility. *Annu. Rev. Cell Biol.* **3**, 379–421.

Watson, N., Williams, K. L., and Alexander, S. (1993). A developmentally regulated glycoprotein complex from *Dictyostelium discoideum*. *J. Biol. Chem.* **268**, 22634–22641.

Watson, N., McGuire, V., and Alexander, S. (1994). The PsB glycoprotein complex is secreted as a preassembled precursor of the spore coat in *Dictyostelium discoideum*. *J. Cell Sci.* **107**, 2567–2579.

Watts, D. J., and Ashworth, J. M. (1970). Growth of myxamoebae of the cellular slime mould *Dictyostelium discoideum* in axenic culture. *Biochem. J.* **119**, 171–174.

Weeks, G., and Herring, F. G. (1980). The lipid composition and membrane fluidity of *Dictyostelium discoideum* plasma membranes at various stages during differentiation. *J. Lipid. Res.* **21**, 681–686.

Weeks, G., and Weijer, C. J. (1994). The *Dictyostelium* cell cycle and its relationship to differentiation. (Minireview). *FEMS Microbiol. Lett.* **124**, 123–130.

Weijer, C. J., Duschl, G., and David, C. N. (1984a). Dependence of cell-type proportioning and sorting on cell cycle phase in *Dictyostelium discoideum*. *J. Cell Sci.* **70**, 133–145.

Weijer, C. J., Duschl, G., and David, C. N. (1984b). A revision of the *Dictyostelium discoideum* cell cycle. *J. Cell Sci.* **70**, 111–131.

Weiner, O. H., Murphy, J., Griffiths, G., Schleicher, M., and Noegel, A. A. (1993). The actin-binding protein comitin (p24) is a component of the golgi apparatus. *J. Cell Biol.* **123**, 23–34.

Welker, D. L., and Deering, R. A. (1978). Genetics of radiation sensitivity in the slime mould *Dictyostelium discoideum*. *J. Gen. Microbiol.* **109**, 11–23.

Welker, D. L., and Williams, K. L. (1982). A genetic map of *Dictyostelium discoideum* based on mitotic recombination. *Genetics* **102**, 691–710.

Welker, D. L., Hirth, K. P., Romans, P., Noegel, A., Firtel, R. A., and Williams, K. L. (1986). The use of restriction fragment length polymorphisms and DNA duplications to study the organization of the actin multigene family in *Dictyostelium discoideum*. *Genetics* **112**, 27–42.

Wessels, D., Murray, J., Jung, G., Hammer III, J. A., and Soll, D. R. (1991). Myosin IB null mutants of *Dictyostelium* exhibit abnormalities in motility. *Cell Motil. Cytoskel.* **20**, 301–315.

Wessels, D., Murray, J., and Soll, D. R. (1992). Behavior of *Dictyostelium* amoebae is regulated primarily by the temporal dynamics of the natural cAMP wave. *Cell Motil. Cytoskel.* **23**, 145–156.

Wessels, D., Titus, M., and Soll, D. R. (1996). A *Dictyostelium* myosin I plays a crucial role in regulating the frequency of pseudopods formed on the substratum. *Cell Motil. Cytoskel.* **33**, 64–79.

Wessels, D., Voss, E., Von Bergen, N., Burns, R., Stites, J., and Soll, D. R. (1998). A computer-assisted system for reconstructing and interpreting the dynamic three-dimensional relationships of the outer surface, nucleus and pseudopods of crawling cells. *Cell Motil. Cytoskel.* **41**, 225–246.

West, C. M., and Erdos, G. W. (1990). Formation of the *Dictyostelium* spore coat. *Dev. Genet.* **11**, 492–506.

West, C. M., and Erdos, G. W. (1992). Incorporation of protein into spore coats is not cell autonomous in *Dictyostelium*. *J. Cell Biol.* **116**, 1291–1300.

West, C. M., Mao, J., van der Wel, H., Erdos, G. W., and Zhang, Y. Y. (1996). SP75 is encoded by the DP87 gene and belongs to a family of modular *Dictyostelium discoideum* outer layer spore coat proteins. *Microbiology* **142**, 2227–2243.

Westphal, M., Jungbluth, A., Heidecker, M., Mühlbauer, B., Heizer, C., Schwartz, J. M., Marriott, G., and Gerisch, G. (1997). Microfilament dynamics during cell movement and chemotaxis monitored using a GFP-actin fusion protein. *Curr. Biol.* **7**, 176–183.

Wharton, R. P., Sonoda, J., Lee, T., Patterson, M., and Murata, Y. (1998). The Pumilio RNA-binding domain is also a translational regulator. *Mol. Cell.* **1**, 863–872.

White, E., Tolbert, E. M., and Katz, E. R. (1983). Identification of tubulin in *Dictyostelium discoideum*: characterization of some unique properties. *J. Cell. Biol.* **97**, 1011–1019.

White, G. J., and Sussman, M. (1961). Metabolism of major cell components during slime mold morphogenesis. *Biochim. Biophys. Acta* **53**, 285–293.

Wier, P. W. (1977). Cyclic AMP, cyclic AMP phosphodiesterase, and the duration of the interphase in *Dictyostelium discoideum*. *Differentiation* **9**, 183–191.

Wilczynska, Z., and Fisher, P. R. (1994). Analysis of a complex plasmid insertion in a phototaxis-deficient transformant of *Dictyostelium discoideum* selected on a *Micrococcus luteus* lawn. *Plasmid* **32**, 182–194.

Wilczynska, Z., Barth, C., and Fisher, P. R. (1997). Mitochondrial mutations impair signal transduction in *Dictyostelium discoideum* slugs. *Biochem. Biophys. Res. Commun.* **234**, 39–43.

Wilkins, M. R., and Williams, K. L. (1995). The extracellular matrix of the *Dictyostelium discoideum* slug. *Experientia* **51**, 1189–1196.

Williams, J. (1995). Morphogenesis in *Dictyostelium*: New twists to a not-so-old tale. *Curr. Opin. Genet. Devel.* **5**, 426–431.

Williams, J. (1997). Prestalk and stalk heterogeneity in *Dictyostelium*. In *Dictyostelium – A model system for cell and developmental biology*. (Y. Maeda, K. Inouye, and I. Takeuchi, Eds.), pp. 293–304. Universal Academy Press, Tokyo, Japan.

Williams, J., and Morrison, A. (1994). Prestalk cell-differentiation and movement during the morphogenesis of *Dictyostelium discoideum*. *Prog. Nucleic Acid Res. Mol. Biol.* **47**, 1–27.

Williams, J. G., and Jermyn, K. A. (1991). Cell sorting and positional differentiation during *Dictyostelium* morphogenesis. In *Cell–Cell Interactions in Early Development* (J. Gerhart, Ed.), pp. 261–272. Wiley-Liss, New York.

Williams, J. G., Ceccarelli, A., McRobbie, S., Mahbubani, H., Kay, R. R., Farly, A., Berks, M., and Jermyn, K. A. (1987). Direct induction of *Dictyostelium* prestalk gene expression by DIF provides evidence that DIF is a morphogen. *Cell* **49**, 185–192.

Williams, J. G., Duffy, K. T., Lane, D. P., McRobbie, S. J., Harwood, A. J., Traynor, D., Kay, R. R., and Jermyn, K. A. (1989). Origins of the prestalk-prespore pattern in *Dictyostelium* development. *Cell* **59**, 1157–1163.

Williams, K. L., and Newell, P. C. (1976). A genetic study of aggregation in the cellular slime mould *Dictyostelium discoideum* using complementation analysis. *Genetics* **82**, 287–307.

Williams, K. L., Kessin, R. H., and Newell, P. C. (1974a). Genetics of growth in axenic medium of the cellular slime mould *Dictyostelium discoideum*. *Nature* **247**, 142–143.

Williams, K. L., Kessin, R. H., and Newell, P. C. (1974b). Parasexual genetics in *Dictyostelium discoideum*: mitotic analysis of acriflavin resistance and growth in axenic medium. *J. Gen. Microbiol.* **84**, 59–69.

Wilson, A. K., Pollenz, R. S., Chisholm, R. L., and de Lanerolle, P. (1992). The role of myosin-I and myosin-II in cell motility. *Cancer Metastasis Rev.* **11**, 79–91.

Wilson, J. B., and Rutherford, C. L. (1978). ATP, trehalose, glucose and ammonium ion localization in the two cell types of *Dictyostelium discoideum*. *J. Cell. Physiol.* **94**, 37–48.

Wolf, W. A., Chew, T. L., and Chisholm, R. L. (1999). Regulation of cytokinesis. *Cell. Mol. Life Sci.* **55**, 108–120.

Wong, E. F. S., Brar, S. K., Sesaki, H., Yang, C. Z., and Siu, C. H. (1996). Molecular cloning and characterization of DdCAD-1, a Ca^{2+}-dependent cell–cell adhesion molecule, in *Dictyostelium discoideum*. *J. Biol. Chem.* **271**, 16399–16408.

Wong, L. M., and Siu, C.-H. (1986). Cloning of cDNA for the contact site A glycoprotein of *Dictyostelium discoideum*. *Proc. Natl. Acad. Sci. USA* **83**, 4248–4252.

Wood, S. A., Ammann, R. R., Brock, D. A., Li, L., Spann, T., and Gomer, R. H. (1996). RtoA links initial cell type choice to the cell cycle in *Dictyostelium*. *Development* **122**, 3677–3685.

Woychik, N. A., and Dimond, R. L. (1987). A single mutation prevents the normal intracellular transport of multiple lysosomal proteins from the rough endoplasmic reticulum. *J. Biol. Chem.* **262**, 10008–10014.

Wright, B. E. (1991). Construction of kinetic models to understand metabolism *in vivo*. *J. Chromatogr.* **566**, 309–326.

Wright, B. E., and Albe, K. R. (1994). Carbohydrate metabolism in *Dictyostelium discoideum*. 1. Model construction. *J. Theor. Biol.* **169**, 231–241.

Wright, B. E., and Gustafson, G. L. (1972). Expansion of the kinetic model of differentiation in *Dictyostelium discoideum*. *J. Biol. Chem.* **247**, 7875–7884.

Wu, L., and Franke, J. (1990). A developmentally regulated and cAMP-repressible gene of *Dictyostelium discoideum* – Cloning and expression of the gene encoding cyclic nucleotide phosphodiesterase inhibitor. *Gene* **91**, 51–56.

Wu, L. J., Gaskins, C., Zhou, K. M., Firtel, R. A., and Devreotes, P. N. (1994). Cloning and targeted mutations of Gα7 and Gα8, two developmentally regulated G protein α-subunit genes in *Dictyostelium*. *Mol. Biol. Cell* **5**, 691–702.

Wu, L., Franke, J., Blanton, R. L., Podgorski, G. J., and Kessin, R. H. (1995a). The phosphodiesterase secreted by prestalk cells is necessary for *Dictyostelium* morphogenesis. *Dev. Biol.* **167**, 1–8.

Wu, L., Hansen, D., Franke, J., Kessin, R. H., and Podgorski, G. J. (1995b). Regulation of *Dictyostelium* early development genes in signal transduction mutants. *Dev. Biol.* **171**, 149–158.

Wu, L. J., Valkema, R., van Haastert, P. J. M., and Devreotes, P. N. (1995c). The G protein β subunit is essential for multiple responses to chemoattractants in *Dictyostelium*. *J. Cell Biol.* **129**, 1667–1675.

Xiao, Z., and Devreotes, P. N. (1997). Identification of detergent-resistant plasma membrane microdomains in *Dictyostelium*: enrichment of signal transduction proteins. *Mol. Biol. Cell* **8**, 855–869.

Xiao, Z., Zhang, N., Murphy, D. B., and Devreotes, P. N. (1997). Dynamic distribution of chemoattractant receptors in living cells during chemotaxis and persistent stimulation. *J. Cell Biol.* **139**, 365–374.

Yagura, T., and Iwabuchi, M. (1976). DNA, RNA and protein synthesis during germination of spores in the cellular slime mold *Dictyostelium discoideum*. *Exp. Cell Res.* **100**, 79–87.

Yamamoto, A., and Takeuchi, I. (1983). Vital staining of autophagic vacuoles in differentiating cells of *Dictyostelium discoideum*. *Differentiation* **24**, 83–87.

Yamamoto, A., Maeda, Y., and Takeuchi, I. (1981). Development of an autophagic system in differentiating cells of the cellular slime mold *Dictyostelium discoideum*. *Protoplasma* **108**, 55–69.

Yang, C. Z., Brar, S. K., Desbarats, L., and Siu, C. H. (1997). Synthesis of the Ca^{2+}-dependent cell adhesion molecule DdCAD-1 is regulated by multiple factors during *Dictyostelium* development. *Differentiation* **61**, 275–284.

Yeh, R. P., Chan, F. K., and Coukell, M. B. (1978). Independent regulation of the extracellular cyclic AMP phosphodiesterase-inhibitor system and membrane differentiation by exogenous cyclic AMP. *Dev. Biol.* **66**, 361–374.

Yin, Y. H., Rogers, P. V., and Rutherford, C. L. (1994). Dual regulation of the glycogen phosphorylase 2 gene of *Dictyostelium discoideum*: the effects of DIF-1, cAMP, NH3 and adenosine. *Development* **120**, 1169–1178.

Yoder, B. K., Mao, J., Erdos, G. W., West, C. M., and Blumberg, D. D. (1994). Identification of a new spore coat protein gene in the cellular slime mold *Dictyostelium discoideum*. *Dev. Biol.* **163**, 49–65.

Yoder, J. A., Walsh, C. P., and Bestor, T. H. (1997). Cytosine methylation and the ecology of intragenomic parasites. *Trends Genet.* **13**, 335–340.

Yu, Y. M., and Saxe, C. L. (1996). Differential distribution of cAMP receptors cAR2 and cAR3 during *Dictyostelium* development. *Dev. Biol.* **173**, 353–356.

Yuen, I. S., Jain, R., Bishop, J. D., Lindsey, D. F., Deery, W. J., van Haastert, P. J. M, and Gomer, R. H. (1995). A density-sensing factor regulates signal transduction in *Dictyostelium*. *J. Cell Biol.* **129**, 1251–1262.

Yumura, S., Mori, H., and Fukui, Y. (1984). Localization of actin and myosin for the study of ameboid movement in *Dictyostelium* using improved immunofluorescence. *J. Cell Biol.* **99**, 894–899.

Zada-Hames, I. M. (1977). Analysis of karyotype and ploidy of *Dictyostelium discoideum* using colchicine induced metaphase arrest. *J. Gen. Microbiol.* **99**, 201–208.

Zamore, P. D., Bartel, D. P., Lehmann, R., and Williamson, J. R. (1999). The PUMILIO-RNA interaction: a single RNA-binding domain monomer recognizes a bipartite target sequence. *Biochemistry* **38**, 596–604.

Zeng, C., Anjard, C., Riemann, K., Konzok, A., and Nellen, W. (2000). gdt1, a new signal transduction component for negative regulation of the growth-differentiation transition in *Dictyostelium discoideum*. *Mol. Biol. Cell* **11**, 1631–1643.

Zhang, B., Gallegos, M., Puoti, A., Durkin, E., Fields, S., Kimble, J., and Wickens, M. P. (1997). A conserved RNA-binding protein that regulates sexual fates in the *C. elegans* hermaphrodite germ line. *Nature* **390**, 477–484.

Zhang, Y. Y., Brown, R. D., and West, C. M. (1998). Two proteins of the *Dictyostelium* spore coat bind to cellulose *in vitro*. *Biochemistry* **37**, 10766–10779.

Zhang, Y., Zhang, P., and West, C. M. (1999). A linking function for the cellulose-binding protein SP85 in the spore coat of *Dictyostelium discoideum*. *J. Cell Sci.* **112**, 4367–4377.

Zhu, Q. L., and Clarke, M. (1992). Association of calmodulin and an unconventional myosin with the contractile vacuole complex of *Dictyostelium discoideum*. *J. Cell Biol.* **118**, 347–358.

Zhu, Q. L., Hulen, D., Liu, T. Y., and Clarke, M. (1997). The $cluA^-$ mutant of *Dictyostelium* identifies a novel class of proteins required for dispersion of mitochondria. *Proc. Natl. Acad. Sci. USA* **94**, 7308–7313.

Zhukovskaya, N., Early, A., Kawata, T., Abe, T., and Williams, J. (1996). cAMP-dependent protein kinase is required for the expression of a gene specifically expressed in *Dictyostelium* prestalk cells. *Dev. Biol.* **179**, 27–40.

Zigmond, S. H., Joyce, M., Borleis, J., Bokoch, G. M., and Devreotes, P. N. (1997). Regulation of actin polymerization in cell-free systems by GTPγS and Cdc42. *J. Cell Biol.* **138**, 363–374.

Zimmermann, T., and Siegert, F. (1998). 4D confocal microscopy of *Dictyostelium discoideum* morphogenesis and its presentation on the Internet. *Dev. Genes Evol.* **208**, 411–420.

Zimmerman, W., and Weijer, C. J. (1993). Analysis of cell cycle progression during the development of *Dictyostelium* and its relationship to differentiation. *Dev. Biol.* **160**, 178–185.

Zinda, M. J., and Singleton, C. K. (1998). The hybrid histidine kinase *dhkB* regulates spore germination in *Dictyostelium discoideum*. *Dev. Biol.* **196**, 171–183.

Zuker, C., Cappello, J., Lodish, H. F., George, P., and Chung, S. (1984). *Dictyostelium* transposable element DIRS-1 has 350-base-pair inverted terminal repeats that contain a heat shock promoter. *Proc. Natl. Acad. Sci. USA* **81**, 2660–2664.

Index

AAC repeats, 40
aardvark, 194
ABP-120, 73–5, 82
ACA, *see* adenylyl cyclase of aggregation (ACA)
AcaA, 94, 96, 112
  null mutant, 112
Acanthamoeba, 37, 49, 67, 78
ACB, *see* adenylyl cyclase (ACB)
Acetobacter xylinum, 197
ACG, *see* adenylyl cyclase of germination (ACG)
acgA, 218–19
acrA, 206
  null mutant, 206
Acrasiales, 12, 16
Acrasidae, 7, 35
Acrasieae, 11–12
acrasin, 16–18, 98, 122
Acrasiomycetes, 35
Acrasis rosea, 7, 35
actin, 38, 70–87, 100, 162
  binding proteins, 62, 65, 71–8, 81, 131
  cytoskeleton, 60, 70–5, 78–80, 86, 101, 104, 118
  filaments, 71–85, 194, 196
  genes, 50, 71
  related protein, 69
  see also F-actin; G-actin
Acytosteliaceae, 7
Acytostelium leptosum, 33
adaptation
  evolutionary, 27
  receptor, 98–101, 106–9, 114, 123, 127–35, 180

adaptins, 67
adenosine, 106, 109–10, 169
adenylyl cyclase (ACB), 112, 158, 203, 206
adenylyl cyclase of aggregation (ACA), 60, 91, 95–6, 106, 108, 112–15, 117, 127, 129, 158, 168–9, 203–4
adenylyl cyclase of germination (ACG), 64, 112, 117, 158, 218–19
adhesion, 15, 18–19, 29, 140, 142, 158–65
  cell–cell, 60, 127, 158–62
  plaques, 78, 84
  to substrate, 60, 75, 78, 84, 162–3, 192
β-adrenergic receptor kinase, 113
Aerobacter (now *Klebsiella*) *aerogenes*, 13, 22
aggregate, 10, 13–14, 139–65
  adhesion in, 139–65
  and sheath formation, 4, 22, 140–5, 174, 180
  size, 2, 13, 125–7
  tip formation, 4, 139–41, 146, 156–7, 163–4, 166–71
  transition from loose to tight, 69, 141–3, 163–4, 185
aggregation, 2–7, 9, 16–18, 27–9, 91, 93–138
  centers, 2, 17–18, 127, 131, 135–8
  chemotaxis and, 2–4, 15–18, 27–9, 84–5, 93–138
  models of, 135–8
  number of genes required for, 56
  pattern, 2, 4, 126
  size of territories, 2, 13, 125–7, 135–8
  transition to loose mound, 141–3
aimless, 113–14, 117
ALCs, *see* anterior-like cells
altruism, 20, 29